Behavioral Teratogenesis and Behavioral Mutagenesis

A Primer in Abnormal Development

Behavioral Teratogenesis and Behavioral Mutagenesis

A Primer in Abnormal Development

Ernest L. Abel
Wayne State University
Detroit, Michigan

Plenum Press • New York and London

Library of Congress Cataloging in Publication Data

Abel, Ernest L., 1943–
Behavioral teratogenesis and behavioral mutagenesis: a primer in abnormal development / Ernest L. Abel.
p. cm.
Includes bibliographies and index.
ISBN 0-306-43053-3
1. Behavioral toxicology. 2. Developmental toxicology. I. Title.
[DNLM: 1. Chromosome Abnormalities. 2. Genetics, Behavioral. 3. Mutagens. 4. Mutation. 5. Teratogens. QH 457 A139b]
RA1224.A24 1989
616.89′071 – dc19
DNLM/DLC 88-38421
for Library of Congress CIP

A Division of Plenum Publishing Corporation
233 Spring Street, New York, N.Y. 10013

Printed in the United States of America

Preface

Most textbooks are cumbersome to carry, expensive to buy, difficult to read, and boring. They have no plot, no characterization, no suspense, no climax. What they have are facts. If *Dragnet's* Sgt. Friday were Scientist Friday, the script wouldn't be much different—"just the facts, ma'am."

Students can't escape textbooks. But like death and taxes, they are necessary evils. Death makes room for new people and the death of old ideas makes room for new ones. Taxes are the dues we pay to live in a country. Everybody gets stuck with paying some kind of dues and students are no exception. Students pay dues in the form of tuition to listen to professors lecture, and they also pay dues in what a former governor of California called "psychic bucks"—time, concentration, independent study, reading textbooks like this one—to come up with the correct answers to exam questions.

Textbooks on economics will tell you about where our tuition bucks come from. This book is about where our psychic bucks come from and the forces that can bankrupt our psychic nest eggs.

Animals and people do not pop up from nowhere. Instead, they are created by sexual unions that put into motion a series of progressive changes called development. These changes are controlled by the genetic code inside a single fertilized cell. Despite enormous progress in genetics, we know almost nothing about how this fertilized egg cell becomes a complex multimillion-celled organism. We know that genes carry codes for making proteins and we know how proteins are made in the cell, but we don't know how coding or protein building eventually results in different structures. This development continues until a fully formed creature appears and even then development never ceases.

Despite ignorance of the way development is orchestrated, we know there is a critical period during development when an organism is particularly vulnerable to chemicals or physical stimuli. If these agents cause damage to chromosomes they are called mutagens. If they interfere with the work of chromosomes during development but don't alter the chromosomes themselves, they are called teratogens. It is sometimes not possible to know if abnormal development is due to a mutagen or a teratogen. X-irradiation, certain chemicals, and a number of viruses are both teratogenic and mutagenic. If a mutation occurs in a sperm or egg cell, it has the potential for being passed on to another generation. If the mutation occurs in a nongerm cell it will only affect a single generation.

Until recently, teratologists (scientists interested in the causes of abnormal prenatal development) mainly concerned themselves with structural defects and the teratogens that caused them. Beginning in the 1960s, teratology expanded to include functional defects, especially behavioral anomalies. Just as genetics became divided into subspecialties such as cytogenetics and behavioral genetics, teratology developed its own subspecialty called behavioral teratology.

One of behavioral teratology's aims is to understand how prenatal exposure to damaging chemicals, drugs, or physical agents results in abnormal behavior. Since behavior is the endpoint, one could as easily speak of behavioral mutagenesis as behavioral teratology. The main difference between the two involves mechanism, not outcome, and in many cases the mechanism is simply not known. There is no point in creating artificially divisive barriers like the nature–nurture arguments with which developmental psychology was once and, to some extent, is still contending.

Behavioral teratology is now one of the fastest growing, most exciting, and most challenging fields in neuroscience. One reason for its phenomenal growth is our realization that the brain is the most sensitive organ in the body and that brain damage is forever. A missing arm or leg is a terrible misfortune, but artificial arms and legs are now available to allow people with these misfortunes to live full and satisfying lives. Disfigured bodies can be surgically corrected. A brain that is damaged is damaged for life.

Another reason for the excitement in behavioral teratology is the fact that it gives scientists in different fields the opportunity to work together. Although behavioral teratology has been characterized as a hybrid of teratology and experimental psychology, it is actually a hybrid drawn from many diverse fields. Identifying mutagens and teratogens and those areas of the brain and its workings that these agents damage, finding out how that damage occurs, and determining why it occurs in some people or animals and not in others are not one- or two-discipline jobs. Behavioral teratology is an area where psychobiologists, epidemiologists, neurophysiologists, neuroanatomists, neurochemists, pharmacologists, and other scientists can integrate their special disciplines and come to an understanding of how and why some people never reach their full potential.

Scope of the Book

In this book I have tried to create a perspective within which to examine prenatal determinants of behavior, rather than presenting an overwhelming panorama of details about specific factors that affect behavioral development. There is still no shortage of facts to master, but such facts are less important than principles. These principles are drawn from such diverse fields as molecular biology, genetics, embryology, pharmacology, biochemistry, toxicology, neuroanatomy, neurophysiology, psychology, and epidemiology. Focusing on principles rather than exploring specific agents allows us to at least understand why and how these forces do what they do.

Besides principles, another basic ingredient for appreciation of any science is a thorough understanding of the methodologies used and the problems and pitfalls associated with these methods. After all, the strength or weakness of any study depends on the procedures used. Treatments must be administered and outcomes measured in such a way that irrelevant factors are eliminated and relevant factors—other than those of immediate concern—are at the very least minimized, or at best controlled. These issues are discussed throughout the book.

In many cases, examples have been provided to illustrate various principles or procedures. That some studies are quoted and others are not does not imply that omitted studies were omitted because they were in any way inferior. There is a bias in my choice of examples that I readily acknowledge: most of my examples have been taken from the fetal alcohol literature. This is because I am most familiar with this literature and because alcohol's effects on development have been more thoroughly studied than the effects of any other chemical agent.

This book was originally conceived as a "how-to" text for students taking an introductory course in behavioral teratology. After reading it, many may feel that it should have stayed true to its purpose. But instead, my hidden muse took over and the book took off in a completely different direction, becoming broader in its scope and going farther away from its original purpose.

To my knowledge, no one else has ever taught a course in behavioral teratology before. Therefore, what I have emphasized may be completely different from what another of my colleagues would emphasize should he or she be writing a similar book. This is as it should be. If everyone thought the same, science would be dull and moribund.

Like all writers, I learned a lot in reading and rereading much of what I finally included in this book. I hope that I have been able to incorporate some of the interest and excitement I felt in reviewing these writings.

I have tried to make this text as readable as possible but some ideas just do not lend themselves to a breezy presentation. New concepts seem foreign and their mastery can make the going rough. More than a thousand years ago, Ptolemy, the king of Egypt, faced the same difficulty trying to master geometry. He complained to his teacher, Euclid, the genius who invented geometry, asking if there were any shortcuts that would make the subject easier. Euclid's reply was as true then as it is now: "There is no royal road to geometry." Or, he could have added, any other study.

During its long gestation, this book went through many drafts. Many colleagues were given various chapters and were kind enough to offer valuable suggestions. Those to whom special appreciation is due are F. Koppitch, E. Riley, M. Church, C. Zajac, Z. Djuric, and J. Jacobson.

Contents

CHAPTER 1

In the Beginning Was the Word and the Word Was . . . DNA

The ancient Egyptians believed their royal families were the descendants of a god. To keep that divine inheritance intact, the Egyptian royal family practiced incest between brother and sister and father and daughter. Four thousand years later, Dr. Svaante Paabo, a scientist at the University of Uppsala in Sweden, wondered if traces of the divine inheritance could still be detected in their mummified earthly remains. Using the latest techniques of genetic research Paabo was able to extract a tiny bit of intact immortality from a royal corpse. In this hodgepodge of 4000-year-old DNA, he discovered a gene sequence that we share with the kings and queens of ancient Egypt.

If the ancient Egyptians were right about their royal families, our common DNA makes us all godlike. The last ten years have seen such enormous advances in transplanting DNA from one organism to another that some people feel that scientists engaged in such recombinant DNA research are in fact playing God.

Even the scientists themselves think they have penetrated the origins of life. Commenting on *Double Helix,* the book in which he described how he discovered the structure of DNA, Nobel Prize winner James Watson told *Omni* (1984) magazine, "I'm in the process of rewriting the Bible—I'm going back to the origins and finding out what it's all about." Watson showed the way but it was Rebecca Caan, a geneticist at the University of Hawaii, who has followed the DNA trail back in time tracing our ancestry to Biblical Eve—a single African woman who lived about 200,000 years ago (Caan, 1987).

The first part of this chapter examines how Watson and Caan made this journey back in time. Paradoxically, it is a trip that begins in the 17th century, meanders into the present, and although the journey is still far from over, breaches the hitherto impenetrable curtain of our hoary past.

The Living Cell

The cell is the nub of life. To some scientists, it is also a microcosm of every facet of life. Several centuries of tedious, painstaking squinting through primitive microscopes had to pass before such recognition, but in its wake came a deep and abiding respect for the cell's mastery over our lives.

Since it is so basic, a cell cannot be meaningfully defined except in terms of itself. We have to describe it by way of what surrounds it, what's inside it, and what it does.

In its most elemental form, the surrounding is a lipid bilayer membrane the composition of which varies in different organisms. The interior is filled with a mixture of fluid and tiny membrane-bound structures collectively called cytoplasm. And what it does is provide us with the ability to grow, sense, think, act, reproduce, and die.

The fluid inside the cell is called *protoplasm* ("first form"); the tiny structures are called *organelles* ("tiny organs"), each of which has its own membrane, allowing them to conduct biochemical processes independently of other organelles and thereby making complex metabolic activities possible. The largest of the organelles is the *nucleus,* possession of which distinguishes primitive prokaryotic cells like bacteria from the eukaryotes that evolved from them. *Prokaryotes* are single-celled organisms with no organelles and no internal cellular organization. *Eukaryotes* have cell walls, an internal organization, organelles that perform vital tasks, and an internal source of energy (adenosine triphosphate).

Our current appreciation of the cell's uniqueness and importance is due, as are so many momentous discoveries, to technological developments that extended our abilities to see and do what previously had not been possible. The technological development that gave us our first glimmerings that there was such a thing as a cell at all was the microscope, invented around 1590.

The first microscopes were crude devices that were constantly being improved. One of the earliest microscopists was Robert Hooke, an English scientist, who described many of the objects he saw with this new instrument in his book *Micrographia,* published in 1665. One of these objects was an ordinary piece of cork. To the naked eye there was nothing remarkable about cork. But to an eye peering through a microscope, cork became a miniature honeycomb of tiny compartments. Hooke called these compartments cells (*cellulae,* Latin for "small rooms"). When other microscopists like Anton van Leeuwenhoek developed even better microscopes and began describing the structure of living cells in detail, cell biology was born.

The implications of these early observations did not sink in, however, until more than a century later. In 1838, Matthias Schleiden and a year later Theodor Schwann (whose name now surrounds nerve axons) both came to the same conclusion that the cell was indivisible and all organisms are collections of cells working together. In 1855, the German pathologist Rudolf Virchow added the final refinement to this new cell theory by showing that "all cells arise from preexisting cells" *(omnia cellula e cellula).*

Cell theory was accepted by the mid-1800s, but the role of the nucleus in cell division was yet to be clarified. As early as 1833, the Scottish botanist Robert Brown had called attention to a dense area inside cells that he referred to as the *nucleus* (from the Latin word for "kernel"). Although possession of a nucleus was recognized as a fundamental characteristic of all cells, other biologists were more interested in the cell wall because it was a distinctive feature whereas the interior of the nucleus was impenetrable to their microscopes.

The shift in focus from cell wall to nucleus followed a number of technological innovations that took place after 1860, among them the invention of synthetic dyes. In 1879, the German biologist Walther Flemming added one of these dyes to cells and saw small granules in the nucleus that he called *chromatin* (from the Greek word for "color"). With this new tool, Flemming made detailed studies of cells and discovered that strands of chromatin split longitudinally and that each of the halves transfers to a daughter cell where they are reconstituted into a nucleus. Flemming called this process *mitosis* (from the Greek word for "thread"). Four years later the German anatomist Wilhelm von Waldeyer began calling these threads *chromosomes* (''colored bodies''). By the last decade of the 19th century, most cell biologists realized that the physical basis of heredity is contained in the nucleus.

Another characteristic of chromosomes was independently discovered without the aid of any technological device. Two years before publication of Darwin's epochal *Origin of Species* in 1868, Gregor Mendel published his breeding experiments with pea plants in which he correctly deduced the existence of the gene and described how it controls hereditary transmission of traits. However, Mendel's work was not appreciated at the time and it was not until 1900 that its significance was finally recognized.

Mendel was born in Austria in 1822 and in 1843 entered the Augustinian monastery at Brno which, at that time, was a research center. Mendel's studies included agriculture and botany and he distinguished himself enough in these subjects to warrant admission to Vienna University. However, he failed to earn a teaching certificate and returned to the monastery where he became an instructor. During his summer vacations in 1856 to 1864 he busied himself doing experiments with pea plants in the monastery garden.

Mendel didn't choose pea plants by accident. Several practical reasons entered into this decision. First was their rapid growth and maturation. This meant he would have something to look at in a relatively short time. Second, pea plants have distinctive physical traits like the shape of their pods, the shape of their seeds, their height, and so on. This meant he could categorize them. Third was the way they reproduced. Normally, pea plants self-fertilize themselves, but Mendel knew he could cross-fertilize them for experimental purposes if he removed the stamens before they matured. The pea plant was therefore a good choice for the kinds of experiments Mendel intended to do.

Mendel wasn't the first to work with pea plants or the first to perform cross-fertilizing experiments. What he was the first to do was focus his attention on only one trait at a time and count how often these traits appeared. By quantifying his observations, he was able to proceed from systematic simple studies to more complex examinations and he developed a model of genetic inheritance that still forms the cornerstone of modern genetics.

Mendel's Model

Before he could start experimenting with his pea plants, Mendel had to make sure he was dealing with purebred specimens. The first step was to develop genetically pure sets of plants for a given characteristic like wrinkled or smooth seeds, by inbreeding them for several generations. Then he cross-fertilized them.

Mendel discovered that when a purebred pea plant with wrinkled seeds was bred with one pure bred for smooth seeds, all the progeny (later called F1, for first filial generation) had smooth seeds. The trait for wrinkling had disappeared! Mendel didn't know how or why it disappeared but he didn't stop experimenting. Instead, he interbred the F1 generation and found that three quarters of the F2 generation of plants had smooth seeds and one quarter had wrinkled. The trait had not disappeared. It had only been masked.

To account for these observations, Mended first postulated the existence of physical hereditary factors, which we now call *genes* (from the Greek for ''giving birth to''), for every trait. Mendel's second postulate was that each pollen and egg cell contains two of these genes. A third postulate was that each of the genes in a pair could be identical and therefore coded for exactly the same trait or could be slightly different and therefore each coded for an alternative form of the trait. If these genes are identical, the plant is *homozygous* for that trait; if not, it is *heterozygous* for that trait. We now call variations in genes *alleles,* and though there may be several different alleles at a given locus on a chromosome in a large number of people, each individual can have only two.

A fourth postulate was that there is no blending of genes. Mendel's fifth and final postulate was that only one of the traits coded for by the alleles is expressed. An allele whose trait appears is dominant; one whose trait does not, is recessive. In heterozygous individuals, the recessive traits are never expressed. Only if an individual is homozygous for the same two recessive genes will the trait coded for by those two alleles be expressed. In other words, if X represents the dominant allele and x the recessive, the first generation of peas from parents with XX and xx alleles will be Xx. When these hybrids produce pollen and egg cells, the F2 generation can contain only the following possibilities: XX, Xx, xX and xx. The combinations of Xx and XX will be physically indistinguishable because of the dominance of X and therefore the trait will be visible in a 3:1 ratio.

During the course of his breeding studies, Mendel performed many different experiments to test his model. One of those he devised was a way to determine if an individual that exhibits a dominant physical trait is either homozygous or heterozygous for that trait. The procedure is now called the *test cross.* The rationale is this: A homozygous individual with a dominant trait, e.g., smooth seeds will be XX for that trait; a heterozygous individual with that trait will look the same but will have a genotype that is Xx. In both instances the physical trait will be the same, like smooth seeds. A specimen with a recessive physical trait, e.g., wrinkled seeds, will have a xx genotype for that trait. If the specimens are bred and the physical trait of the offspring are all the same (e.g., smooth seeds), the test specimen must be XX because the only combination will be XX and Xx. If, however, the test specimen is heterozygous, then half the offspring will have smooth seeds and the other half will have wrinkled seeds because the only possible combinations are XX and xx.

Mendel's experiments with pea plants allowed him to gain considerable insight into general hereditary mechanisms. He correctly deduced the existence of genes, the fact that they come in pairs, the idea of dominance, the random segregation of dominant and recessive genes to gametes uncontaminated through each generation, and although not a principle, the distinction between an individual's genetic constitution (its genotype) and its physical appearance (its phenotype).

An intellect that could draw such broad inferences from the simple disappearance and reappearance of wrinkles on the seed of a pea plant deserves recognition. But if that intellect is too great, lesser intellects can't appreciate it. This is what happened to Mendel. He sent his results to one of the recognized experts of his time whose ho-hum comments discouraged Mendel. So instead of trying to publish his experiments in a major biological journal, Mendel contented himself with publishing his work in a local journal, where it received an even greater ho-hum.

In 1900, Mendel's work was rediscovered and two years later, Walter Sutton, a biologist at Columbia University, suggested a connection between genes and chromosomes. However, there were obviously far more physical characteristics for a species than there were chromosomes, so although cell biologists recognized that chromosomes had something to do with heredity, that particular something was still elusive.

Chromosome Theory

Mendel had correctly anticipated the genetic theory of inheritance, but he had no idea of the nature of genes or where they were located. The location problem was solved by Thomas Hunt Morgan, a zoologist at Columbia University.

Instead of pea plants, Morgan chose the common fruit fly *(Drosophila melanogaster)* as his experimental model. As with Mendel, Morgan's choice was a practical one: flies take up little space in a small laboratory (Morgan kept them in bottles), they are cheap, easily bred, have four chromosomes (compared with seven pairs in pea plants), and they have distinctive physical features, like eye color and wing shape.

In one sense, Mendel had been lucky in his choice of pea plants because each of the traits he worked with, except two, are located on separate chromosomes. Mendel didn't know this, of course, but the upshot of this separation was that there is virtually no "linkage" between pea-plant genes. Therefore the various traits Mendel was looking for assorted freely. We can say that Mendel was dealing with a very simple system.

When Morgan began breeding fruit flies, he noticed that some traits appeared together more frequently in offspring than should have occurred if genes assort independently as they did in pea plants. Because these traits seemed to be linked, Morgan felt that they must be physically bound together. And because geneticists were now of the opinion that genes are located on chromosomes, Morgan concluded that linked genes have to be located on the same chromosome. To test this hypothesis, Morgan bred more and more flies and carefully recorded how often various traits appeared together. As he suspected, some traits were paired more often than others. From this Morgan deduced that genes are arranged linearly on a chromosome; the closer the linear arrangement, the more likely the linkage.

In the course of these studies, Morgan discovered sex linkage in chromosomes. Most fruit flies have red eyes but Morgan came across a male with white eyes. When he bred this white-eyed male to a red-eyed female, all the F1 progeny were red eyed. Red was obviously dominant over white. Next he interbred the F1 flies. This time he got mostly red-eyed flies but there were also some white-eyed flies as he anticipated. However, all of the white-eyed flies were male. To see if it was possible to get a white-eyed female he made a test cross of the F1 progeny back to the original white-eyed male. This time he got an equal number of males and females with white eyes. This meant that the gene for white eyes is not peculiar to males, but is normally expressed only in males. The explanation, said Morgan, was that the gene for white eyes is sex-linked to the X chromosome.

The argument is as follows: A female has two X chromsomes; a male has one X and one Y chromosome. The X chromosome is the female-determining chromosome; the Y is the male-determining chromosome. Because the male carries different sex chromosomes, he is the one who determines the sex of offspring. If a female X and a male X chromosome unite, a female will develop; if a female X and a male Y chromosome unite, a male will develop. If the trait for white eyes were located on the X chromosome and was recessive (as he had already shown), then a female would only have white eyes if she had two recessive genes—a possibility, but a rare one, whereas a male would only need to have one recessive gene because there was no allele in the Y chromosome to offset it.

The most famous example of such a sex-linked disorder is that of Queen Victoria, who was a carrier for the "bleeding disease," hemophilia, a disorder associated with a defect on the X chromosome. The queen passed the defect on to her daughter, who was also a carrier. She married Czar Nicolas II of Russia and her son inherited the defect. Concern over the health of the boy drove the Czar's wife to distraction and may have contributed to the chaotic court life that eventually led to the overthrow of the Russian monarchy.

The idea that genes are located on chromosomes and that genetic inheritance is due to the transmission of genes by means of chromosomes was now firmly established. By syn-

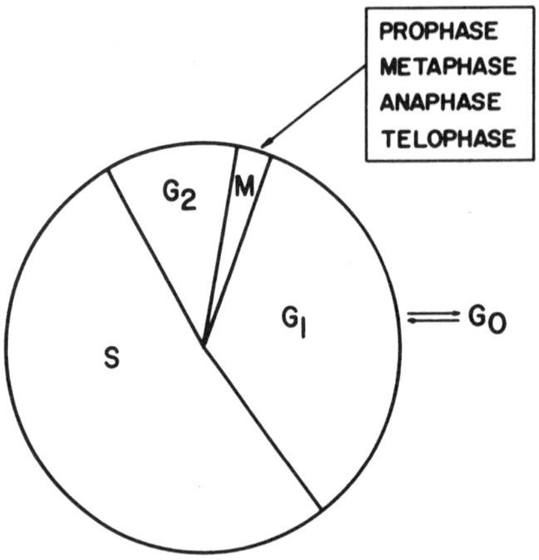

Figure 1.1. Life cycle of the cell. The cell cycle can be pictured as a recurrent series of events, most of which takes place during a period called *interphase*. Interphase consists of three subperiods, which have been arbitrarily labeled G_1, S, and G_2. G stands for gaps in the cycle when growth and cellular organization are taking place. S refers to synthesis of DNA. The shortest period in the cell cycle is mitosis, represented by M. In rapidly dividing cells, mitosis accounts for no more than 10 percent of the entire 12–24-hour cycle. G_0 represents a "time out" that sometimes occurs.

thesizing Flemming's, Mendel's, and Morgan's discoveries, cell biologists were now able to describe and explain how traits are passed on from one generation to the next.

The Cell Cycle

The moment-to-moment activities within cells are responsible for biological growth, development, well-being, repair of damage due to disease and accident, and death. To carry out those functions cells must grow and be continually replaced as needed and each new cell must be an exact duplication of the one it replaces. These growth and duplication activities take place during the cell's life cycle and have been divided into four phases (see Figure 1.1). The first, called the S-phase (synthesis phase), is when DNA replication occurs. During M-phase (mitosis phase), the chromosomes go through processes that result in the creation of two daughter cells. During G_1 (G = growth), immediately after mitosis and before DNA synthesis, cell growth occurs and cellular organelles double. During G_2, which occurs after synthesis and before mitosis, various enzymes and structural proteins required for mitosis are synthesized. Cells can also enter a nonproliferative phase known as G_0. During this phase, cell division processes come to a halt, which sometimes happens when the cell is exposed to adverse conditions.

The life cycle of a cell is a continuous process involving the interrelated activities described above. Most of the cell's life is taken up by growth (G_1-phase) and synthesis (S-phase). Because this chapter is concerned primarily with genes and chromosomes, events occurring during the S and M phases will be emphasized.

Mitosis

The various stages involved in cell duplication are illustrated in Figure 1.2. Most of this time is spent in a period called *interphase*. Individual chromosomes are not visible at this time, but this is when they replicate themselves.

By the end of interphase, the cell has synthesized two identical copies of each chromosome, called *sister chromatids*. However, at this stage the cell continues to hold them together by a protein structure called a *centromere*. The centromere is usually but not always located at the center of each chromosome. The splitting of the sister chromatid begins when the cell now delegates another of its specialized structures, called a *centrosome*, to split into two parts, called *centrioles*. These centrioles are microtubule-organizing centers from which a spindle will form. Each of them is dispatched to opposite sides of the cell.

The cell is now poised for its next stage, called *prophase*. Chromosomes now emerge as distinct structures composed of two chromatids. Simultaneously, the membrane around the nucleus etherealizes and the centrioles lock onto their positions at opposite sides of the cell. Two sets of microtubules made out of proteins called *tubulin* sprout from each centriole. One sprout is a short starlike array called an *aster*, which is directed toward the cell wall and which serves as a brace. The other is an intricate outgrowth of microtubules that migrates toward the aster, forming a spindle. Once the spindle has formed, the chromosomes move toward it. Mysteriously, some innate force makes them take up the same fixed position on the spindle each time the cell divides. This is not some sixth sense but a subtle technique that the cell has evolved over millions of years for getting its chromosomes to the right place at the right time: Instead of being passively drawn to the spindle, the protein centromere initiates its movement by sprouting its own set of microtubules, which are drawn, possibly by some chemical attraction, toward the centrioles of the spindle. When the microtubes make contact, they are pulled tight and the chromosome is drawn into position on the spindle equidistant from the two centrioles. Any mistake in the orientation of the centromere when it extends its microtubules, however, could be disastrous, for if the two sets of microtubules both attached to the same centriole, a nondisjunction would occur and the two sister chromatids would each be drawn to the same new cell.

Once the chromosomes are in position on the spindle, they are ready for their next experience. Now the cell passes into *metaphase*. Its nuclear membrane disintegrates and is absorbed by the cell's interior net of membranes, the *endopasmic reticulum*. In preparation for their impending schism, the chromatids enter their most visible state.

With the chromosomes firmly in place, the cell proceeds through *anaphase*. The microtubules attached to centromeres pull in opposite directions, splitting the centromere and the two sister chromatids separate. Using a sliding mechanism exactly like the one it uses to make muscle cells contract, the cell moves its chromatids to opposite sides of the cell. The splitting of the centromere and the separation of the sister chromatids from each other is a rite of passage earning each chromatid the status of a chromosome.

There is one more phase, called *telophase*. The cell has no more need for the spindle. It is now cellular rubbish and must be broken down. But instead of destroying it, the cell recycles the protein microtubules into protein for the cytoskeletons of the new cells. Meanwhile, the chromosomes elongate, entangle themselves, and lose their identity. From nowhere a nuclear membrane encircles each new set. Two sets of chromosomes are one set too many for any cell and the cell invaginates itself at its center. This invagination is called a *cleavage furrow* and it gets bigger and bigger until it finally pinches the cell in two.

Mitosis is the basic mechanism for growth and replacement of damaged cells. But not all cells are equal. Different types of cells differ in their capacity for mitosis. Skin cells constantly undergo mitosis; liver cells will do so only if damaged; brain cells will never experience mitosis. Because cells rely on mitosis to replace exhausted or damaged cells, cells like those in the brain, which do not experience mitosis, can never be replaced. Fortunately, we are born with about 15 billion brain cells so that there are usually more than enough "spares" to take the place of any that break down or die.

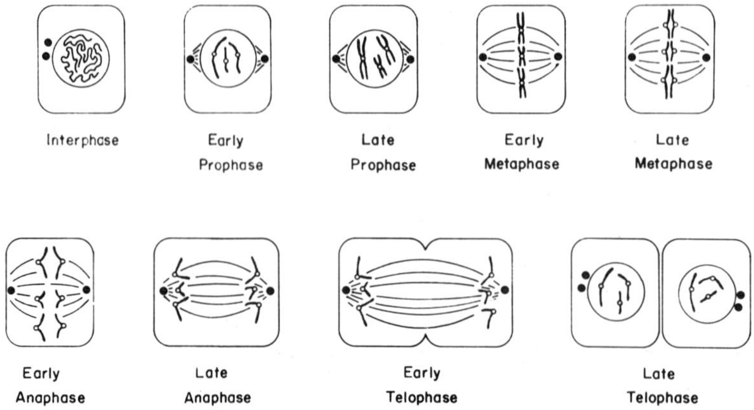

Figure 1.2. Various stages in mitosis. Duplication of chromosomes occurs during interphase. During prophase, the chromosomes begin to become visible under the microscope as threadlike filaments. By the end of prophase, the two duplicates joined by their centromere are visible. The two duplicates are called *sister chromatids* while they are joined. The beginning of metaphase is signaled by loss of the nuclear membrane and appearance of a spindle on which the chromosomes align themselves.

In anaphase, the centromere splits and two chromatids are now known as *chromosomes*. Each chromosome then moves toward opposite poles of the spindle. During telophase, the two sets of identical chromosomes congregate at their respective poles of the spindle and begin to revert back to their undifferentiated shapes. A nuclear membrane forms around each set and cell division is complete.

Meiosis

Early cell biologists knew that all individuals belonging to a species have the same number of chromosomes per cell, and with few exceptions each of these cells has the same number of chromosomes as every other. But these facts posed an enigma. If each individual is produced by the fusion of sperm and egg cells from its parents, why doesn't each new generation have twice as many chromosomes as the previous generation? And if genetic inheritance relies on a process like mitosis, how do individual differences arise?

The secret was pried from the cell a year after Flemming described mitosis, by a Belgian cell biologist, Pierre van Beneden. While peering at the chromosomes of the roundworm parasite, von Benden discovered that its sperm and egg cells each contained two chromosomes whereas the other cells in the parasite's body and in its newly formed embryos all contained four. This meant there must be two different kinds of cell division, one for body growth, involving duplication of the same material, and one for reproduction, involving formation of gametes. To prepare for reproduction, each sperm and egg cell has to lose half its genetic complement. When these cells unite, the normal number of chromosomes is restored, not doubled. This halving process was first called *reduction division* and then *meiosis* ("to make less").

Meiosis is a double division. In mitosis, chromosomes act as if they are oblivious of one another. In meiosis, chromosomes go through the same changes as in mitotic interphase—the cell synthesizes a new sister chromatid and the two sister chromatids are held together by a centromere. Then comes the difference. During the next stage, called *prophase I,* instead of homologous pairs separating and heading independently toward the spindle, the cell directs the two chromosomes, each with its two chromatids, to attach to a

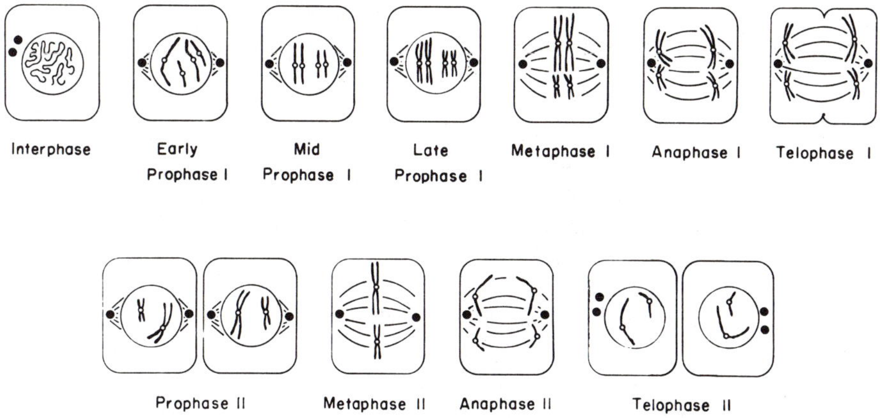

Figure 1.3. Meiosis is the process of nuclear division resulting in formation of gametes. During meiosis, cells divide twice but chromosomes divide only once. As a result, gametes wind up with half the number of chromosomes as the original cell.

Interphase in meiosis is identical to interphase in mitosis. Duplication of chromosomes occurs during the S phase.

Prophase I can be divided into five different substages, only three of which are shown in the figure. During this time, the chromosomes become visible and homologous chromosomes pair up next to each other like a zipper. This sideways pairing of homologous chromosomes is called *synapsis* and sets the stage for a possible exchange of homologous chromosome parts called *crossing over*. The two homologous chromosomes then move away from each other but the sister chromatids remain attached to each other at their centromere.

During metaphase I, the chromosomes become attached to the spindle fibers at their centromeres with each chromosome pair opposite each other. *The pairing of homologous chromosomes during this time has no counterpart in mitosis.*

In anaphase I, the centromeres of each pair of homologous chromosomes migrate toward each pole of the spindle, carrying both chromatids. *There is no splitting of the centromere.*

Telophase I represents the final formation of two daughter cells, each of which contains only one set of chromosomes rather than the two sets as in mitosis. The cell then moves into a brief interphase but no new synthesis of DNA occurs during this time. Prophase II sees the chromosomes becoming visible again as they prepare for metaphase. In metaphase II, the chromosomes attach to spindle fibers and align themselves in a vertical plane. During anaphase II, each centromere divides and the sister chromatids, now called *chromosomes,* move to opposite poles. The nuclear membrane forms around each haploid set of chromosomes during telophase II.

particular site on the nuclear membrane and then wraps them in a web of protein so that they are perfectly matched, gene to gene. Sometimes the chromatids on homologous pairs of chromosomes become so attracted to each other that parts of them exchange places with one another (see below).

The two homologous chromosomes, still joined together, are now ready for the transition to the next phase, *metaphase II*. Because they are still embraced, the inner part of their centromeres is not free to sprout microtubules and only the outer surfaces generate them. When the outer portions of the centromeres sprout their microtubules, the orientation of the chromosomes is random. Either a maternally or paternally derived chromosome will have its outer surface facing a particular centriole. This means each maternally and paternally derived chromosome will be randomly pulled to opposite ends of the cell when it divides. Each united pair of homologous chromosomes is then pulled into the spindle.

As in mitotic anaphase, during meiotic *anaphase I,* the microtubules attached to the centromere pull in opposite directions. However, instead of the centromeres splitting and causing the sister chromatids to separate, the two centromeres remain intact but separate. Each pair of homologous chromosomes with its two sister chromatids is tugged to opposite ends of the cell and the cell divides, completing *telophase I.* This separation results in two cells, each with the normal chromosomal complement, but each cell now contains a unique combination of maternally and paternally derived chromosomes. After this first meiotic division, the cell orchestrates a mitotic division in which the two sister chromatids split and a single set of chromosomes is packaged into gametes (see Figure 1.3). The end result is four haploid (single) sets of chromosomes, some of which originated in the father and some in the mother. If the cells are from a male, each of these sets will develop into a sperm, if from a female, three will degenerate and only one will develop into an ovum.

Gametogenesis

Mitotic cell division takes only hours. By contrast, it takes 70 days for a single human male gamete to achieve its manhood. In the process, it must pass through three phases, called *spermatogenesis.* The female counterpart, called *oogenesis,* can take over 13 years!

The Sperm Cell

Inside the human testes are yards of threadlike tubules called *seminiferous tubules.* The tubules make sperm in their outer germinal epithelium. In the first phase of sperm production, the germinal epithelium uses mitotis to produce two primitive spermatogonium cells, each with a complete set of chromosomes. One of these spermatogonium matures into a *primary spermatocyte;* the other remains in the germinal epithelium.

Next, the primary spermatocyte begins to move closer to the inner part of the tubule. As it moves along this journey, the primary spermatocyte undergoes its first meiotic separation and divides into two secondary spermatocytes, each of which contains 23 chromosomes, including one X or one Y chromosome. Next comes the second meiotic division. The chromatids previously joined together in the secondary spermatocyctes now separate and each goes into a spermatid. These spermatids continue to move toward the lumen of the tubule while developing into active motile sperm cells. The mature sperm cells then emerge from the tubules, enter the lumen, and swim to the epididymis, where they are stored while they continue to mature under the influence of androgens. Each primary spermatocyte thus gives rise to four sperm, two of which contain an X chromosome and two a Y chromosome. (see Figure 1.4).

The entire process of spermatogenesis is orchestrated by two pituitary hormones. Follicle stimulating hormone (FSH) has a stimulating effect on the seminiferous tubules; luteinizing hormone (LH) causes the Leydig cells in the testis to secrete androgens that assist in maturation of the sperm. Although spermatogenesis does not begin until puberty, Leydig cells are formed very early in embryological development and begin to secrete androgens prior to birth, the purpose of which is to masculinize the brain by affecting nerve cell development in an area of the hypothalamus (see Chapter 5).

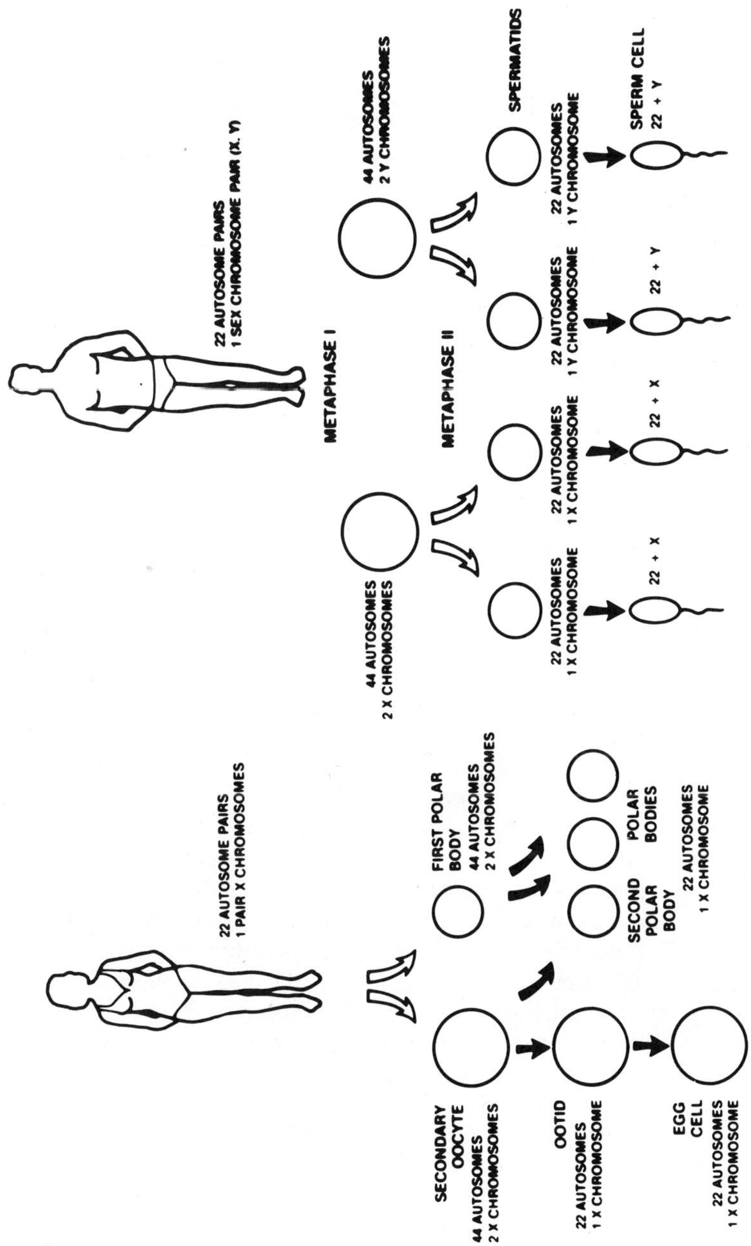

Figure 1.4. Formation of male and female gametes.

Sperm Imprinting

For a long time, geneticists believed the 22 paternally derived autosomal chromosomes acted in concert with their 22 maternally derived counterparts in a federation in which each gene was prepared to direct its particular role in the work of the cell. If the two sets of genes were homozygous, the output would be the same no matter which was expressed. If different, the dominant gene would prevail.

Then came a new discovery that explained a peculiar abnormality in some human pregnancies. For many years, obstetricians were puzzled by a rare anomaly in which a pregnancy occurred but only a placenta developed. This anomaly, called an *hydatidiform mole,* was eventually traced to the fertilization of an ovum that had no female nucleus. Instead of the sperm's chromosomes uniting with the ovum's, the sperm's chromosomes simply duplicated and continued to duplicate so that the entire genome came from the father. To study this phenomenon further, Surani and co-workers (1984) removed the male or female pronucleus from fertilized eggs and inserted one of them into an egg containing only a female pronucleus. When a male pronucleus was introduced, 38% of the zygotes reached term but none of the females did so, although most of the latter implanted, indicating development to at least the blastocyst stage.

The fact that a placenta needs paternally derived genes to develop means that these genes must become earmarked to play a special role in development should fertilization occur. As a result of this earmarking, paternal genes acquire a much more important role than maternal genes in the development of an embryo's placenta and other extraembryonic tissues, whereas a mother's genes are destined to exert much more of an influence on development of the embryo itself.

The mechanism for this earmarking or *imprinting* is rather simple (although the steps leading to its discovery were not). Using the new technology of transgenic mice and insertional mutagenesis, geneticists have discovered that imprinting involves the adding or elimination of methyl chemical groups from chromosomes when gametes are being made.

The transgenic mouse procedure involves removing genes from the genome of a host animal such as a mouse and then placing these into a fertilized oocyte. Such genes will often insert themselves into the genome of the host animal and will perform in the same way as endogenous genes. By mutating these insertional genes prior to placing them in the host, geneticists have been able to trace the effects of the foreign gene through subsequent generations. Using such techniques, geneticists have inserted "good" genes into mice with specific defects like thalassemia, hypogonadism, and myelination disorders and have corrected these disorders (Camper, 1987). Using the same kind of procedures, and enzymes that cut DNA one way if methylated and another if not methylated, they found that when a certain DNA patch was inherited from the mother, it was more methylated than if inherited from the father. When the same DNA patch was passed on to a son and inherited from him, it was less methylated, indicating reversal of methylation during each germ line transmission. For other foreign DNA patches, however, paternally derived DNA was more methylated than maternally derived DNA, whereas for still other patches, methylation did not change regardless of maternal or paternal origin.

Imprinting occurs in the germ cells and gives certain genes the place of *primus inter pares* in the gene federation. Although there are exceptions, higher status is conferred by a gene being undermethylated. Such genes have their messages transcribed by mRNA, whereas more methylated genes do not (Swain *et al.,* 1987). Methylation is now known to influence expression of genes involved in placenta formation, and it is possible that it may affect expression of genes involved in other functions as well.

The Egg Cell

The female counterpart to spermatogenesis is *oogenesis,* but there are several differences between sperm and egg formation. The first major difference is that around the sixth month of fetal life, mitosis of ovarian cells stops. All the germ cells that a female will ever have are now present in the ovaries. These female germ cells undergo the first stages of meiosis up to prophase I during this fetal period, resulting in formation of two cells, one of which is called a *polar body.* The polar body is very small and remains attached to the larger cell. The process then comes to a halt and does not begin again until puberty. After puberty, a single egg is released from the ovary and its first meiotic division, which has been hanging in limbo for many years, undergoes its second meiotic division, if fertilized. The fertilized cell then sets in motion millions of mitotic divisions, culminating in the creation of a new individual. If not fertilized, the immature egg cell disintegrates and is shed during menstruation.

Genetic Variability

Because each chromosome set pairs off independently of every other set during meiosis, considerable variations can occur within a species as a result of chromosome assortment. Variation can occur in other ways as well. As the homologous pairs of chromosomes assemble for their assortment during prophase I of meiosis, they line up in zipperlike fashion so that the four chromatids are positioned close to one another. This sidewise pairing is called *synapsis.* While in synapsis the chromatids make contact with one another at various points called *chiasmata,* and instead of their remaining as four independent chromatids, breakages and rejoinings occur between chromatids and mutual exchanges of chromatids take place at these chiasmata. Such an exchange is called a *crossover* (see Figure 1.5). As a result of crossovers, parts of the chromatid and their associated genes trade places. Crossovers can occur between the two chromatids of the same chromosome, in which case these crossovers are called *sister chromatid exchanges* (SCEs). If chromatids are transferred between homologous chromosomes, the crossover is called a *nonsister chromatid exchange.*

Yet another way genetic variation occurs is by genes hopping around. In the 1940s, when Barbara McClintock first began describing gene shifts in corn, few scientists believed her. But subsequently her observations were corroborated (she was awarded the Nobel Prize in 1983) and another mechanism for genetic variation was recognized.

Genetic recombination and crossing over means that a germ cell may express the genes

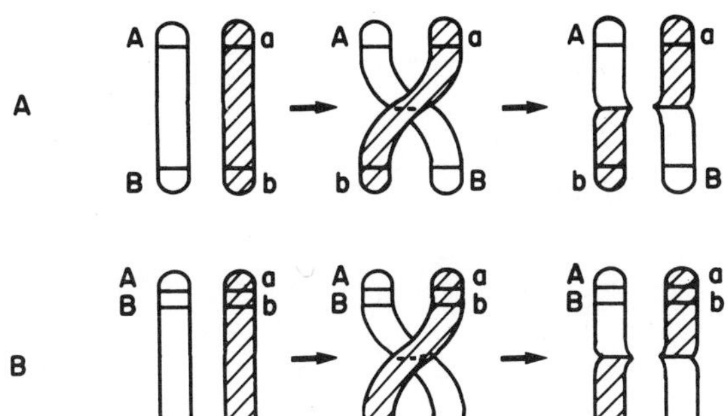

Figure 1.5. Sister chromatid exchange.
A. Crossing over between sister chromatids represents a recombination of maternal or paternal traits.
B. Crossing over can also occur at different places on chromosomes. This results in different possible recombinations of maternal and paternal characteristics in gametes.

for traits that were not expressed in either parent. For instance, a paternal or maternal chromosome might carry a mutation or several mutations. Because of the vast combinations that can arise during meiosis, genetic recombination can result in almost limitless possibilities and these mutations may or may not be expressed.

Gene Theory

Once the basic concepts of the chromosomal theory of heredity were determined, the next challenge was to identify the chemical composition of genes and to find out how they worked. These studies had begun in the 1870s when the Swiss biochemist Friedrich Miescher isolated a substance from the nucleus that he called *nuclein.* Twenty years later, nuclein was renamed *nucleic acid* after biochemical studies showed that nuclein was acidic in nature. Further analysis of nucleic acid by the German biochemist Albrecht Kossel showed that nucleic acid contained four carbon-nitrogen ring molecules, which Kossel called *adenine, guanine, cytosine,* and *thymine* (see Figure 1.6).

Kossel's student, Phoebus Levine, probed further into nucleic acid and found it consisted of two different acids, one of which contained a five-carbon sugar called *ribose* and another that resembled ribose except for having one less oxygen, which he called *deoxyribose.*

By 1934, Levine determined that the basic elements of nucleic acids were a phosphate group, a sugar made of ribose or deoxyribose, and either the two-ringed purines adenine and guanine or the single-ringed pyrimidines cytosine and thymine. These three basic elements were called a *nucleotide* (see Figure 1.7) and Levine concluded that nucleic acids were basically collections of these nucleotides.

Meanwhile, two scientists at the California Institute of Technology were probing into the secrets of how genes control specific traits. Using the fruit fly as an experimental model, geneticist George Beadle had been frustrated in his attempt to find out how genes determined the color of the fly's eye. Beadle's previous studies had taught him that genes produce some chemical that eventually produces a trait, but that was as far as he could go. The fruit fly's eye color seemed too complicated. (In fact eye color in the fruit fly is controlled by about 20 genes.) Stymied, Beadle tried a different strategy.

Instead of using a complicated model like the fruit fly, he switched to common bread

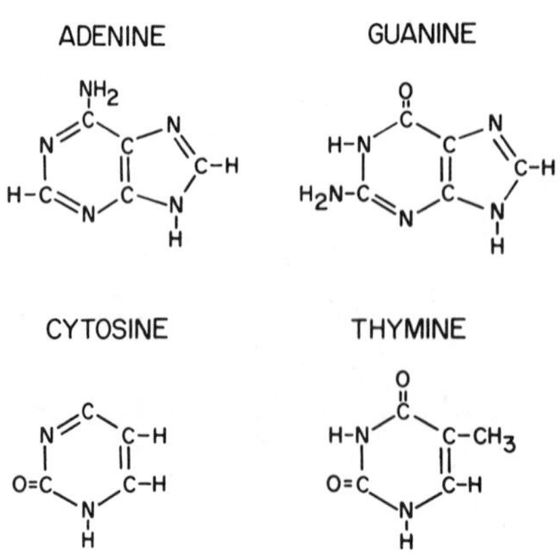

Figure 1.6. Structure of carbon-nitrogen organic bases in nucleic acids. Adenine and guanine have double-ringed structures and are called *purines;* cytosine and thymine have a single-ring structure and are called *pyrimidines.*

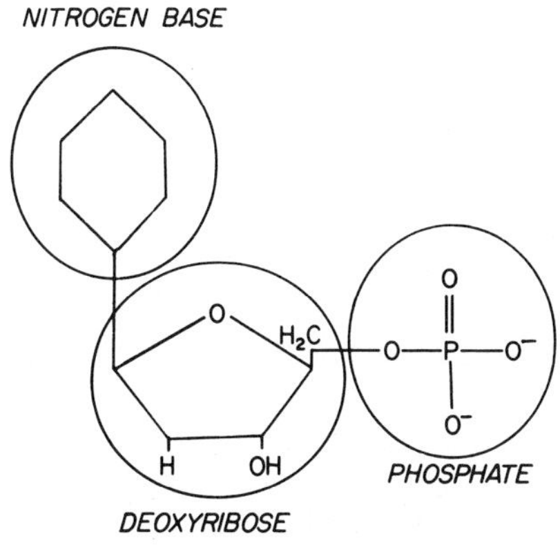

Figure 1.7. Components of a nucleotide.

mold. The fungus *(Neurospora)* is a primitive organism that needs nothing more than a little sugar, a little salt, and a little bit of a vitamin to exist. Yet the genes in such a simple organism allow it to synthesize all the chemicals it needs to exist. Joining forces with biochemist Edward Tatum, Beadle irradiated bread molds with x-rays to induce mutations that would render them unable to manufacture all their necessary chemical constituents. Then the two put these mutants into a broth.

Normal molds were able to use the ingredients in the broth to produce all their enzymes; mutants weren't. So Beadle and Tatum added different ingredients to the broth one at a time until they would grow. This told them which chemical the mold was no longer able to make for itself. For example, some mutants would grow only if they added arginine. Normal molds could synthesize arginine from sugar and ammonium salts. Because these mutants needed arginine, the gene that ordinarily controlled its synthesis must have been damaged by the x-rays. On the basis of this and observations of other mutant molds, they arrived at the one-gene one-enzyme hypothesis: a single gene directs the production of only one enzyme. Characteristics produced as a result of many enzymes are the product of many genes.

A footnote: About 30 years earlier, an English physician, Sir Archibald Garrod, had discovered the same one-gene, one-enzyme relationship in the course of his clinical work. Garrod had found that many peculiar hereditary diseases his patients suffered from were due to specific defects in their enzymes and concluded that these enzyme defects were due to mutated genes. Garrod described his observations in 1909 but Beadle and Tatum didn't know of them until after they published their own hypothesis. Mendel's ghost certainly could appreciate the oversight.

Whereas we now know that genes are made of DNA, cell biologists felt DNA was too simple to contain the hereditary principle. They knew that DNA was part of the chromosome but most of a chromosome contains protein. We now know that a large proportion of this protein, known as *histones,* is purely structural and allows DNA to be tightly compacted. But in the 1940s, most scientists believed it was the protein part of the chromosome, not DNA, that was involved in hereditary transmission because protein is so much more complex than DNA.

In 1944, three biochemists at Rockefeller University led by Oswald Avery finally proved that the hereditary material was DNA. The inspiration for Avery's studies began several years earlier when the English bacteriologist Frederick Griffith had reported an extraordinary finding involving the bacteria that cause lobar pneumonia ("double pneumonia").

There are two forms of these bacteria. One has a jellylike coat, the other has no coat at all. Those with coats cause pneumonia; those without are powerless. In one of his studies, Griffith killed some of the coated bacteria with heat and injected mice with a mixture of these dead bacteria along with live but harmless uncoated bacteria. Since the virulent strain of bacteria in the mixture was dead, nothing should have happened. But something did happen—the mice developed pneumonia and died. And in their blood were swarms of smooth-coated bacteria!

Perhaps he hadn't really killed the virulent bacteria? To make sure they were dead Griffith injected another group of mice only with the dead bacteria from the first batch. None of the mice died. Somehow, the dead virulent bacteria had passed on their virulence to the hitherto harmless bacteria, transforming them into virulent bacteria. The genius of hindsight makes it obvious that this conversion was due to a gene transference, but the idea evaded Griffith. He just couldn't conceive of bacteria having genes. Since Griffith was a very respected scientist, other scientists tried to repeat his experiments and did. The finding was real; the explanation, however, was elusive.

When Avery and his group decided to repeat Griffith's experiment, they were as much in the dark as everyone else. But they went about it differently. First they extracted a substance from the dead virulent strain of bacteria. Then they put this substance in a solution with the harmless noncoated bacteria and found that some of these grew coats and became virulent. And all the offspring of these transformed bacteria also had coats and were virulent. When the transforming substance was examined, it seemed to have the properties of DNA rather than protein. To make sure, they destroyed the protein component and repeated their studies. The experiment still worked. Then they destroyed the DNA. This time it no longer worked. This meant that the transforming principle was undoubtedly DNA.

At first, many scientists refused to accept Avery's work and argued that his DNA had probably been contaminated with protein, their own candidate for hereditary transmission. It was not until an elegant experiment at Cold Spring Habor in Long Island by Alfred Hershey and Martha Chase in 1952 that cell biologists finally admitted that DNA and not protein was the basic material of heredity.

Hershey and Chase were pioneers in what has come to be known as the phage group of molecular geneticists. *Phage* is short for *bacteriophage,* the term for a class of viruses that attack bacteria. A virus is the most primitive known form of life. It has just two parts—an outer coat of protein and an inner core of DNA. To reproduce, a virus has to find what is euphemistically called a host. But a phage is no ordinary guest. When it encounters a host bacterium, it needles its way through the bacterium's cell wall and then fires its DNA into the cell. The outer coat of protein stays outside; inside, a micro *coup d'etat* is staged. The virus's DNA destroys the bacterium's DNA and then commandeers all the bacterium's food and energy to create clones of itself, about a hundred of them, each with a new protein coat. These new viruses are as hungry for bacteria as their progenitor but they are trapped inside the bacterium. To live they have to escape; the bacterium, its DNA destroyed, obliges them by synthesizing enzymes that disintegrate its own cell wall. Freed, the marauding phages now search for new bacteria cells to destroy.

Because viruses are composed of only two parts, an outer coat of protein and an inner core of DNA, Hershey and Chase hit upon an ingenious way of deciding which part was the transmitting principle. They cultured one batch of viruses in a medium with a radioactive material that attached only to protein and another with radioactive material that attached only to DNA. Then they put the two different radioactively labeled viruses in with bacteria. Only radiolabeled DNA was found in the bacteria. This elegant study showed that it was DNA and not protein that entered bacteria.

Now that cell biologists were convinced that genes were made of DNA, the next challenge was to describe the structure of this DNA. They already knew that the three basic ingredients were a sugar, a phosphate, and a nitrogen base consisting of either a large double-ringed purine (adenine or guanine) or a smaller single-ringed pyrimidine (thymine or cytosine). In the late 1940s, biochemist Erwin Chargaff had also been able to determine that within each cell the total amount of purines equaled the total amount of pyrimidines and specifically the total amount of adenine was equal to the amount of thymine, whereas guanine equaled cytosine.

In 1951, Maurice Wilkins used x-rays instead of light to produce a picture of DNA that suggested a possible spiral-staircase structure. Two years later, in 1953, the English physicist Francis Crick and the American biochemist James Watson used Wilkins' photographs and Chargaff's rules to formulate a model that showed how DNA could carry genetic information.

The Double Helix

The model devised by Watson and Crick contained the three submolecular building blocks. Since Chargaff's rules dictated that the total amount of purines equals that of pyrimidines, Watson and Crick proposed that DNA is arranged in the form of a ladder. The vertical parts of the ladder contain the sugar and the phosphate molecules, whereas the rungs are made of bridges between purines and pyrimidines. Based on their differences in size, it is only possible for the larger purines to match up with the smaller pyrimidines and vice versa if the width of the ladder were to remain constant (see Figure 1.8).

Because of structural characteristics, adenine, which has two hydrogen binding sites, also has to match up with thymine, which has two hydrogen binding sites, whereas guanine and cytosine, each of which have three hydrogen binding sites, always have to have one another as a pair. Wilkin's x-ray photographs suggested that the two strands also twist around each other. When Watson and Crick incorporated all these facts, they came up with an arrangement like that shown in Figure 1.9, which has come to be known as the *double helix.*

One of the major contributions of this model was that it accounted for species differences through characteristic base sequences. We now know that a spider's DNA contains the same four bases as an elephant's or a college student's. The source of overall variability between species stems from the sequences of these bases on the double helix. You are what you are because of differences in these DNA base sequences, the proteins they make, and the way these proteins interact with other cytoplasmic components. There is a sequence for a spider, and elephant, and you.

A second major contribution of the Watson-Crick model was its ability to account for replication. DNA has to duplicate itself faithfully during cell division or the genetic information it contains will be lost. Because it does so flawlessly, different species have been able to retain their individuality over eons of time. The key is *restrictive base pairing.*

During interphase, the two strands of the double helix unravel and the hydrogen bonds

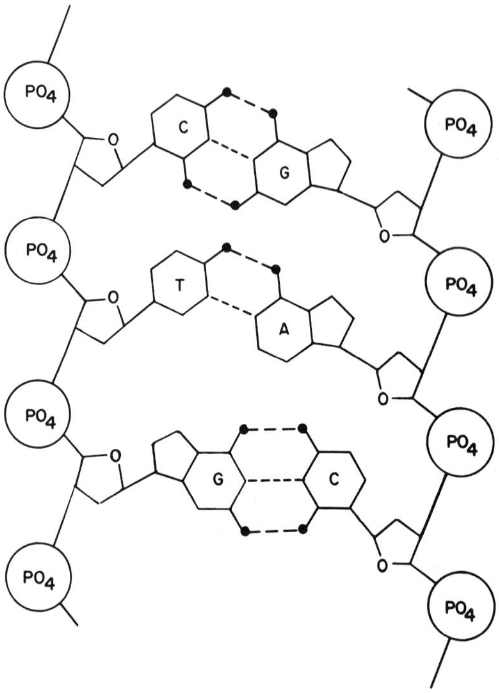

Figure 1.8. The basic unit in the DNA molecule is a pair of nucleotides bonded to each other by hydrogen. This is known as a *base-pair* (bp). The nucleotide is made up of a sugar (s), a phosphate (p), and a base. The backbone of each strand contains the sugar and the phosphate and the horizontal lines represent the bases made of adenine (A), cytosine (C), guanine (G), and thymine (T). Each phosphate submolecule of one nucleotide is joined to a sugar submolecule above it and below. The nitrogen bases project inward. A bonds with T, and G with C. As a result, each double-stranded DNA contains equal proportions of A and T and of G and C. Two hydrogen bonds keep A and T together, whereas G and C are kept together by three hydrogen bonds.

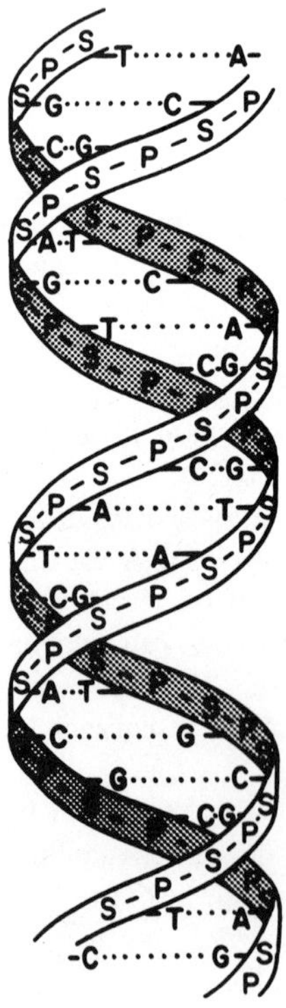

Figure 1.9. Organization of nucleotides in DNA. The nucleotides are arranged in the form of two double strands that are twisted around each other in the form of a spiral.

between bases break in orderly succession causing the helix to separate into two strands. Each of the original strands straightens out and acts as a template for the synthesis of a complementary strand. The process is called *seimconservative replication.*

Free nucleotides take up positions in accord with the restrictions on base pairing, A with T, C with G. The sequence of bases in each daughter DNA is exactly like that in the parent DNA, and each human chromosome contains about 100 million of these base pairs. Replication is completed when nucleotides in the cytoplasm have all been attracted to the unpaired bases on the DNA template (see Figure 1.10). By selecting material from within the cell to match each base, two new double helixes are created resulting in duplication of genetic information. The replication is called ''semi-conservative'' because only one strand of the spiral is preserved.

Every cell (except red blood cells) in the human body contains DNA. Most of this DNA is arranged on chromosomes but some DNA is also located in mitochondria. In prokaryotes like bacteria, all the DNA in the cell is contained in a molecule packaged into a single closed circle. In most eukaryotes, the cell's DNA is partitioned between a number of chromosomes. Chromosomes in every cell in the body are all identical to those in every other cell (exceptions are germ cells and a genetic disorder called *mosaicism*)—liver cells have the same chromosomes as brain cells or lung cells but cells in different organs lose or use different parts of their DNA. How this happens is still a mystery.

Since every human cell is only about 20 microns in size and yet has about 13 feet of DNA coiled within it, an awesome amount of DNA has been packaged in a tiny space. This is a squeezing job to boggle the imagination. There is so much DNA in the trillions of cells that make up the human body that if the strands were joined end to end, it would form a single filament stretching from the earth to the sun—2,000 times! (McAleer, 1985).

The secret of a cell's capacity for growth and repair lies in the linear sequencing of this DNA on its chromosomes. The basic element of the code (called a *codon*) is three bases long. Each triplet codon contains the formula for making 1 of 20 amino acids or a stop-or-go signal for stringing these amino acids together into a protein. The fact that a codon is three bases long was known before any experiments actually proved it. Table 1.1 shows that if one base coded for one amino acid, only four amino acids would be possible. A duplicate code with two bases coding for one amino acid would result in only a possible 16 different amino acids. A triplicate code allows for 64 possible rearrangements of the four bases, which is more than enough to make 20 different kinds of amino acids.

A gene is thus a sequence of three nucleotides that contains information for the synthesis of proteins out of amino acids. The 20 different kinds of amino acids are listed in Table 1.2.

Proteins are long chains (called *polymers*) made up of these amino acids. In a way,

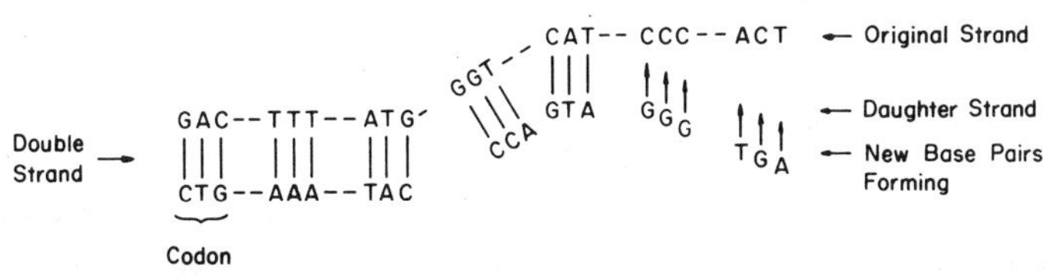

Figure 1.10. When a cell divides, each new cell contains a complement of DNA identical to the parent (except for germ cells). This is done by separating the two strands and matching of appropriate base pairs.

Table 1.1. Number of Minimal Bases Needed to Produce 20 Amino Acids

Number of bases in code	Number of possible amino acids
1	4
2	16
3	64

Table 1.2. The Twenty Amino Acids That Make Up Proteins

Amino acid	Abbreviation
Alanine	Ala
Arginine	Arg
Asparagine	AspN
Aspartic acid	Asp
Cysteine	Cys
Glutamic acid	Glu
Glutamine	GluN
Glycine	Gly
Histidine	His
Isoleucine	Ileu
Leucine	Leu
Lysine	Lys
Methionine	Met
Phenylalanine	Phe
Proline	Pro
Serine	Ser
Threonine	Thr
Tryptophan	Tryp
Tyrosine	Tyr
Valine	Val

DNA is not a hereditary code but a protein code. The kind of protein manufactured by DNA is what makes individuals and species different from one another. Each human cell contains about 10,000 different kinds of protein. Certain kinds of these proteins are used to make the structural materials that go into the cell's cytoskeleton. Others are used for transport like the hemoglobin molecule that carries oxygen. The majority of proteins, however, are enzymes whose job is to expedite chemical reactions in the body that otherwise would occur very slowly. During development, these biochemical reactions must occur quickly or the precise timetable coordinating cell assemblies would be disrupted and deviations in the formation of organs would occur. Different types of protein classified according to their function are listed in Table 1.3.

A major function of the genetic code is to make these proteins out of amino acids. The endless variations in nature are due to differences in the amino acid sequences in

Table 1.3. Functional Classification of Protein

Class of protein	Example
Enzyme	The largest and most varied of the different classes
Structural	Cytoskeleton of the cell; connective tissue (e.g., collagen)
Transport	Movement of gases through the blood (e.g., hemoglobin)
Contractile	Microtubules of spindle and muscle cells
Receptor	The "locks" into which chemical "keys" fit to initiate cell activities (e.g., receptor for insulin)
Hormones	Growth hormone
Storage	Reservoir for materials needed for embryonic growth (e.g., ovalbumin)
Regulatory	Repressor substance that controls gene expression
Protective	Antibodies

proteins. The unique characteristics of an individual or a species, including its behavior, ultimately is due to the sequence of amino acids that make up its protein and the jobs these proteins perform. Environment may shape behavior but the raw material it has to work with is dependent on these proteins.

How DNA Sends Messages

The code contained in DNA uses two steps to make a protein. The first step is called *transcription.* During transcription the sequences of bases making up a gene are copied or transcribed onto a strand of messenger ribonucleic acid (mRNA). RNA resembles DNA but differs in important characteristics (summarized in Table 1.4).

What happens during transcription is this:

RNA polymerase attaches to one strand of the double helix at a transcription site and a base-by-base copy of one of the structural genes is made by messenger RNA (mRNA). When the copy has been made, mRNA detaches itself from the DNA. However, before mRNA slips through the nuclear membrane into the cytoplasm, the introns (see *"Junk*

Table 1.4. Comparison of DNA and RNA

	DNA	RNA
Sugar	Deoxyribose	Ribose
Bases	Adenine	Adenine
	Guanine	Guanine
	Cytosine	Cytosine
	Thymine	Uracil
Number of strands	2	1
Helix	Yes	No
Function	Dictates protein synthesis	Implements instruction for protein synthesis

Genes,'' p. 24) are removed and the loose ends are joined. When the introns have been removed in this way, mRNA is said to be ''mature'' or ''processed.''

Now translation has to occur. The first step requires mRNA to attach itself to ribosomes. The ribosome then travels along the mRNA, moving three base pairs at a time. At the same time, free amino acids are ''activated'' by becoming attached to a different form of RNA called *transfer RNA* (tRNA). Each amino acid has its own tRNA, which contains three bases, called an anticodon, that pair complementarily to the three-based codon on mRNA. As the ribosome travels along the mRNA, a new codon is briefly exposed on its surface. The exposed codon attracts a tRNA with the complementary anticodon and they join. The tRNA then releases its amino acid, which is spliced to the incoming amino acid by a peptide bond. Stripped of its amino acids, the first tRNA vacates its ribosomal attachment. The ribosome then moves to the next codon and the process is repeated until all the codons have been read. When the reading is complete, the ribosome releases the amino acid chain (see Figure 1.11).

In 1961, Drs. Marshall Nirenberg and Heinrich Matthaei at the National Institutes of Health created a synthetic mRNA containing sequences of three uracil bases and used it to

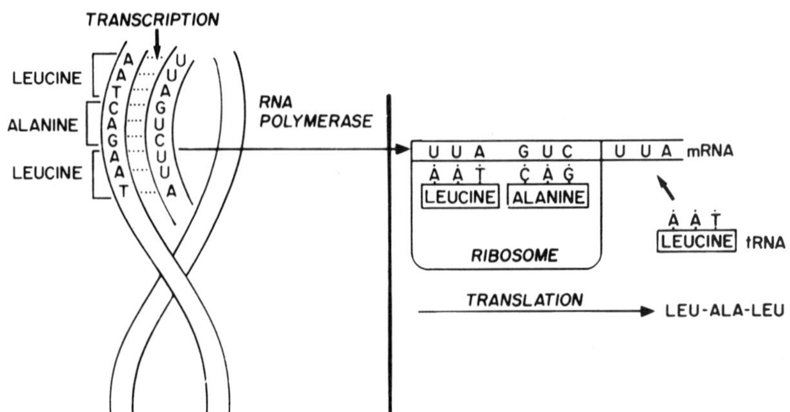

Figure 1.11. Transcription and translation of DNA. The helix is split and a copy of a gene is made by mRNA, which then moves through the nuclear membrane to the cytoplasm and binds to ribosomal RNA (rRNA) initiating protein synthesis. After mRNA has attached to a ribosome, the ribosome moves along the mRNA a codon at a time. Amino acids are activated by tRNA whose anticodons are attracted to the codons exposed by the ribosome as it travels along the mRNA. The tRNA ''transfers'' its amino acid to the incoming amino acid. The stripped tRNA then is released from its binding site and the ribosome moves to the next codon, where the next tRNA is received.

Table 1.5. The Genetic Code[a]

DNA	RNA	Amino acid	DNA	RNA	Amino acid	DNA	RNA	Amino acid	DNA	RNA	Amino acid
AAA	UUU	PHE	AGA	UCA	SER	ATA	UAU	TYR	ACA	UGU	CYS
AAG	UUC		AGG	UCC		ATG	UAC		ACG	UGC	
AAT	UUA	LEU	AGT	UCA		ATT	UAA	Stop	ACT	UGA	Stop
AAC	UUG		AGC	UCG		ATC	UAG		ACC	UGG	TRP
GAA	CUU	LEU	GGA	CCU	PRO	GTA	CAU	HIS	GCA	CGU	ARG
GAG	CUC		GGG	CCC		GTG	CAC		GCG	CGC	
GAT	CUA		GGT	CCA		GTT	CAA	GLN	GCT	CGA	
GAC	CUG		GGC	CCG		GTC	CAG		GCC	CGG	
TAA	AUU	ILE	TGA	ACU	THR	TTA	AAU	ASN	TCA	AGU	SET
TAG	AUC		TGG	ACC		TTG	AAC		TCG	AGC	
TAT	AUA		TGT	ACA		TTT	AAA	LYS	TCT	AGA	ARG
TAC	AUG	MET	TGC	ACG		TTC	AAG		TCC	AGG	
CAA	GUU	VAL	CGA	GCU	ALA	CTA	GAU	ASP	CCA	GGU	GLY
CAG	GUC		CGG	GCC		CTG	GAC		CCG	GGC	
CAT	GUA		CGT	GCA		CTT	GAA	GLU	CCT	GGA	
GAC	GUG		CGC	GCG		CTC	GAG		CCC	GGG	

[a] A = adenine; C = cytosine; G = guanine; T = thymine; U = uracil (in RNA U replaces the T in DNA).

crack the DNA code for the amino acid phenylalanine. When they put this "Poly U" into a test tube containing various amino acids, they produced a protein containing phenylalanine only. This meant that the code for this amino acid was UUU. Because UUU on mRNA was the code for phenylalanine and because the uracil on mRNA always pairs with adenine on DNA, the DNA codon for phenylalanine had to be AAA. Using similar procedures, the codons for all 20 amino acids were deciphered by 1967 (see Table 1.5).

You Are a Degenerate

Because there are only 20 amino acids and the four bases can be arranged into 64 possible codons, there must be a number of synonyms. The synonym for AAA on DNA, for example, is AAG. The amino acid glycine has four synonyms (GGU, GGA, GGC, and GGG). There are also three codons that do not code for any amino acids. These codons are UAA, UAG, and UGA and they function as "terminators." When they appear on mRNA, the joining of amino acids stops, presumably because the chain is complete.

Because more than one codon can produce the same amino acid, the genetic code is said to be *degenerate*. Degeneracy is sometimes not a bad thing. If you look at Table 1.2 carefully, you will see that the first two bases in a codon have the major role in specifying which amino acid will be incorporated into a protein. This means that a change in the third base will not necessarily cause any damage. For example, if the adenine in the codon AAA were changed to guanine to make the codon AAG, phenylalanine would still be made. When such changes (called *point mutations*) occur but have no biological consequences, they are sometimes called "silent" mutations. If you and I were not DNA degenerates, we would be at the mercy of many more potential catastrophes than we are.

Junk Genes?

Only about 2% of the existing 3 million genes in the human genome produce proteins (Omenn, 1976). Although some of the other 98% act as regulator genes (see below), most of our genes have no known function. Why do we have them? Apparently, most of our DNA is simply redundant "junk genes" called *introns*. Whereas gene variation is the rule in most species, much of this variation may be due to these introns. When RNA makes a copy of DNA, it also copies these introns along with the functional genes, but before RNA takes its copy into the cytoplasm, it somehow splices them out.

Conceivably, junk genes are safety valves and may be one reason extrapolation of mutagenic and teratogenic experiments from animals to humans is less than straightforward. For example, Neel's (1982) analysis of data from Hiroshima and Nagasaki indicated that 250 rem of atomic radiation exposure produces the same frequency of mutations as occurs spontaneously in every human generation. (Units and mechanisms for radiation-induced mutation are discussed in Chapter 3.) Earlier studies based on mutation rates in mice had estimated this dose at 40 rem in humans. The new calculation showed that humans are about 6 times less sensitive than the mouse to radiation. One reason we may be less sensitive to mutagens or teratogens may be because "junk genes" confer some protective effect to our "valuable" gene pool.

What Turns You On?

Two of the main questions still associated with genes are how they are turned on and off and how their activities are coordinated—essential processes in growth and development.

In the late 1950s, two biochemists at the Pasteur Institute in Paris, Francois Jacob and Jacques Monod, conducted a number of experiments with *Escherichia coli (E. coli)* bacteria that suggested a possible way for genes to be activated. *E. coli* uses lactose as an energy source. To break lactose down into a usable energy source, the bacterium produces a enzyme called *Beta-galactosidase.* If there is no lactose in the cell, the bacterium doesn't make the enzyme because it would have nothing to act on. Only when lactose is present is the enzyme beta-galactosidase induced.

Ordinarily, the genes that make beta-galactosidase are turned off when the cell has no need for lactose. (Similarly, during growth and development, certain enzymes are needed at certain times and not others. There is no point in their being active before or after they are needed; for maximum and optimal efficiency, they must be present only when needed.)

The group of genes that contain the code for an enzyme like beta-galactosidase is called a *structural gene.* The activity of the structural gene is controlled by two groups of genes in *E. coli;* one group is called a regulator, the other an operator. Bordering the edges of these genes are start and stop signals. This entire group of genes and signals controlling lactose utilization in *E. coli* is called the *lac operon.*

The general organization of the operon is like a business with two levels of management. The structural genes are the workers, the operator is middle management, and the regulator is upper management.

In this metaphor, upper management is always concerned with efficiency. In business it costs money to keep the lights on but you need them to work or else you couldn't see. After work hours when everyone has gone home, having the lights on is inefficient. It costs money. To keep costs down upper management (the regulators) tells middle management (the operators) to make sure the lights are turned off at night. Middle management then passes the message on to the employees (the structural gene). If, however, business is good and it's important to work late, upper management wouldn't send such a message to middle management and the lights would stay on. When business slows down, the message about lights would be reissued.

The regulator (upper management) controls gene expression by preventing mRNA transcription of the structural gene and it exerts this control through the operator (middle management). The main components of the regulator are a promoter and a regulatory gene. The *promoter* is a site at which an RNA polymerase attaches to the regulator so that mRNA can transcribe a *repressor* protein from the regulator genes. Adjacent to the regulator is the *operator,* which has three basic components, the CAP, the lac promoter, and an operator gene.

Ordinarily, the repressor produced by the regulator occupies the operator gene and in so doing, it covers part of the lac promoter, thereby preventing RNA polymerase from attaching to it, and this in turn keeps mRNA from transcribing the structural gene. However, the repressor is a protein whose structure can change and does so when lactose binds to it. When the structure of the repressor is altered, it is no longer able to interact with the operator and therefore it no longer keeps RNA polymerase from the promoter. The presence of lactose in the cell thus acts to induce its metabolism; lactose is said to be an *inducer.*

However, before RNA polymerase can gain access to the *lac* promoter at all, the DNA helix must be opened. This is accomplished by the other point of control in the operon, the CAP component of the operator.

The structure of the CAP is such that it is normally unresponsive to a protein called an *activator,* which has the ability to turn it on. However, when there is a need for the product produced by the structural gene, the shape of the activator is altered slightly so that it can interact with the CAP. The chemical that "activates" the activator is cyclic AMP (cAMP). Ordinarily, the amount of cAMP in the cell is low, but when energy such as that provided by lactose is needed, the level of cAMP increases and it binds to the activator, altering its shape and thereby enabling it to interact with the CAP and in essence, turning it on. When the CAP is turned on, the DNA helix unravels, making the *lac* promoter and operator susceptible to other substances in the cell.

The lac operon is thus under dual control. The regulator normally keeps the structural genes from making an enzyme (beta-galactosidase) via a repressor, but this repression can be overcome by an inducer (lactose). This overriding occurs when the cell has need for energy it can get from the inducer.

Although this model was based on cellular activities taking place in a bacterium, it represents an important conceptual basis for explaining how genes are turned on and off in all cells. The model also suggests how microenvironments—the chemical milieu of the cell—interact with gene activity. If an environmental influence like lactose is present, it will induce a sequence of events that will utilize it; if absent, relevant gene activity is repressed. Genes do not operate in a vacuum but in response to environmental factors. This interaction will again be emphasized in embryological development.

Pleiotrophy

Male fruit flies come in a number of colors, some of which are more sexy to females than others. Gray fruit flies seem to have the advantage in this area compared with yellow competitors. However, yellow flies don't lose out to their more successful gray rivals because of their color. A female fruit fly couldn't care less about the color of her suitor. No, it's not what the male looks like that counts, it's how fast he can vibrate his wings. To a female fruit fly, "it don't mean a thing if it ain't got that wing." Because yellow males can't vibrate as fast as gray males, they aren't as sexually attractive.

The gene for yellow codes not only for color but also for sexual behavior and sexual success in the world of the male fruit is related to the neural pathways that control how fast he beats his wings (Bastock, 1956). A vibrating wing is what turns female fruit flies on. Such *pleiotropic* (multiple) gene effects are the norm and because a single gene can affect several aspects of an organism's physical appearance and behavior, identifying the gene or genes responsible for a particular activity is not straightforward.

Epistasis

Most characteristics, especially behavioral characteristics, do not depend on single genes and behavior does not occur in an all-or-none fashion. Instead, differences between individuals are thought to be quantitative and the result of many genes operating in concert, each with a small influence. Collectively, these small differences work together to produce the potential for a given characteristic. Genes that work together in this way are called *polygenes* and their influence is called "polygenic." Polygenic factors can be examined at present only by statistical analysis of characteristics in a population.

When genes located at different loci interact, it is called *epistasis.* In epistasis, one homologous pair of genes may mask the effects of another, and it is said to be epistatic to it. A classic example of epistasis occurs in a strain of mouse that carries a gene (C or c) responsible for pigment being deposited in its fur. If a mouse is recessive (c) for this gene, no pigment will be deposited and it will be albino. If the C allele is present and a second dominant gene, B, is also present, the pigment that is deposited is black; if the recessive allele (b) is present, a brown pigment will be deposited. A third gene for color in the mouse controls depositing of a yellow band near the tip of its hairs. When the dominant gene (A) is present, the pigment is deposited; when only the recessive (a) gene is present it is not.

If mice recessive for color (cc) are crossed with pigmented mice containing B, b, or A alleles, the mouse will be albino and despite the fact that these albino mice offspring carry a gene for color, the phenotypic expression of the other allele is prevented. If the recessive albino gene (c) were replaced by the dominant gene for color (C), the genes for color would be expressed.

The Molecular Clock

Cell division is very much like making a photocopy of this page. You have already opened the book (unzipped the DNA) and now you lay the page on the transparent plate of the copier. Press the start button and seconds later you have a copy. The copy is always the same—it may be smudged or darker or lighter than the original, but it's the same. The photocopy never has words on it that weren't on the original, unless something goes wrong (e.g., a page wrinkled over as you pressed it down and now a new word appears in the copy that wasn't in the same place as on the original and other words have disappeared).

But sometimes an error does occur in the replication of DNA and the code is not passed on exactly to the next generation. An extensive change may involve an alteration in the structure of the chromosomes so that parts of them are lost. The linear order of genes along a chromosome may be rearranged or the number of chromosomes passed on to new cells may be altered. These kinds of changes are discussed more fully in Chapter 3 and can have very serious consequences for individuals.

Rarely is a mutation beneficial. However, if it is beneficial, it may have an influence on evolutionary process.

For instance, an error involving the replacement of one nucleotide base pair by another (called a *point mutation*) could result in the replacement of one amino acid by another. A substitution would result in subtle changes in the structure and function of a protein. The accumulation of a number of such relatively minor point mutations and their resulting progressive changes in protein function could result in a new species like Adam and Eve and ultimately you.

The fact that individuals are not all homozygous for the same allele at a given locus indicates that mutations have occurred. Genetic variation is the rule not the exception. This is why almost every physical and behavioral characteristic exhibited by an individual can be increased or decreased over succeeding generations by selective breedings.

Because dominant alleles will prevent expression of recessive, and possibly mutated alleles, it would be necessary to monitor families over several generations to determine if a mutation has occurred purely on the basis of physical characteristics. This is obviously impractical. One alternative is to examine children for specific proteins and to compare these protein profiles with parental proteins. The rationale is that proteins are made by amino acids, which in turn are made by genes. A difference in the composition of a particular protein would reflect a mutation in the genes that code for that protein. If most individuals in a population have the same proteins, the genes coding for those proteins must be the same in each individual, whereas variability in protein structures reflects variation in gene structures. Such comparisons were made possible by the development of a technique called *gel electrophoresis* developed in the late 1940s.

Electrophoresis depends on the movement of molecules through a strip of jelly that has an electric voltage at either end. Different protein molecules will take up different positions in the gel. By looking for bands in the gel of a child that are not present in his parents, detection of mutations is possible.

In 1949, Linus Pauling used this technique to show that the hemoglobin of people with sickle cell anemia is abnormal. When the abnormal hemoglobin in sickle cell anemia is placed on the gel and current is applied, it travels slower through the jelly than normal hemoglobin. The movement is only millimeters per hour, but after a number of hours, there is a clear separation, which appears as distinct bands when the gel is stained. In 1955, the English biologist Vernon Ingram pinpointed the anomaly. Instead of glutamic acid in position 6 of the 146 amino acid chain, sickle cell hemoglobin had valine (see Table 1.6). One single amino acid out of 146! Yet for want of a nail, a kingdom may be lost. Because valine does not have an electric charge and glutamic acid does, sickle cell hemoglobin moves at a different pace in an electrified gel. And because of this single amino acid substitution, not enough oxygen may be delivered to the cells of the body and someone may die.

Twenty-five years earlier, Beadle and Tatum had proposed their one-gene, one-enzyme hypothesis. Ingram's studies refined that hypothesis. A single gene specifies the synthesis of a single enzyme, by controlling the position of an amino acid in a polypeptide chain.

In 1962, Pauling proposed that amino acid substitutions in protein represent mutations that reflect the molecular basis of evolution and that these mutations occur at a constant rate. According to this "molecular evolution clock hypothesis" (Zuckerkandl & Pauling, 1962; Wilson *et al.*, 1977), differences in proteins between different species can be utilized to estimate rates of molecular evolution by using fossil evidence as a reference. If proteins

Table 1.6. Sequence of Amino Acids in Normal Hemoglobin and Hemoglobin Associated with Sickle-Cell Anemia

Normal sequence:	Valine .	Histidine .	Leucine .	Threonine .	Proline .	Glutamic acid .	Glutamic acid. . . .
Sequence in sickle-cell anemia:	Valine .	Histidine .	Leucine .	Threonine .	Proline .	Valine .	Glutamic acid. . . .
Position in chain:	1	2	3	4	5	6	7

change at a constant rate over time, it should be possible to trace the evolutionary past of a species by comparing its amino acids with another species and using that difference to see when these two species diverged from a common ancestor. Conceptually, hemoglobin from two monkeys that shared a common ancestor 10 million years ago should be twice as different in their hemoglobin as a species that diverged only 5 million years before. On the basis of these kinds of studies with other proteins like albumin, molecular biologists estimate that humans diverged from chimpanzees and gorillas about 5 million years ago, a time much earlier than had previously been estimated.

In the 1970s, further advancements in "genetic engineering" and electrophoresis allowed molecular biologists to study DNA directly. This was a more sensitive procedure because of the degeneracy of the DNA code. Because introns are present in DNA but are removed from mRNA before translation, examining amino acids will not reveal mutations in introns, which make up the builk of DNA. And because introns are nonfunctional, changes in their structure will have been passed on more readily than changes in functional DNA. Therefore, introns are a better place to look for mutations.

Finding Eve

In 1979, Rebecca Cann, a geneticist at the University of Hawaii, used these new techniques to compare mitochondrial DNA from placentas of children born in different areas of the world. Mitochondrial DNA codes for enzymes that are different from those coded for by nuclear DNA and has only about 16,000 base pairs compared with nuclear DNA, which has many hundreds of millions of bases. Mitochondrial DNA is also inherited only from the mother because only the nucleus of a sperm fuses with an egg during fertilization. Because mitochondrial DNA is not subject to the random shuffling that characterizes nuclear DNA inherited from both parents, any changes in mitochondrial DNA have to be due to mutations. By comparing mitochondrial DNA from different primates and relating this information to fossil evidence, molecular biologists were able to construct a molecular DNA clock that suggested that we have all evolved from a single African female—Eve—tens of thousand years ago.

On the basis of molecular and fossil evidence, Cann calculated that mitochondrial DNA mutates at a rate of 2 to 4% every million years. She concluded that Eve was born 200,000 years ago. This is some 196,000 years older than Bishop James Ussher calculated on the basis of the internal evidence of the Bible. (Ussher put the date of creation at precisely 4004 B.C.)

The Origin of Species

We now know that gene variation is very common and these variations have resulted from mutations. But how did these variations occur?

There are two main hypotheses. The Darwinian hypothesis argues that evolution is due to random heritable modifications in individuals. When these modifications were advantageous, those possessing them survived and passed them on to their offspring. Individuals with disadvantageous modifications disappeared through natural selection. Because populations are closely adapted to their environments, most modifications must have been disadvantageous. Advantageous modifications are rare but when they occur, those that have them are superior and will pass them on to offspring and these will eventually characterize the population. For example, black skin in the tropics is an advantage because it protects the underlying tissues from sunburn. In the northern latitudes sunburn is not a danger. Instead, lack of the ultraviolet component in sunlight that must penetrate tissues to help them manufacture vitamin D is a problem. Without vitamin D there is a danger of rickets. So in the northern latitudes, a pale, translucent skin is an advantage (Patterson, 1978).

Since genetic variation in a natural population is enormous, Darwin's idea that evolution is solely due to transmission of advantageous traits had to be reevaluated. The neutral mutation-random drift hypothesis of "neutral theory" proposed by Kimura (1968) and King and Jukes (1969) contends that not all gene mutations are adaptive. Some are advantageous whereas others are deleterious or of no consequence. Large numbers of allelles are present in populations but may not be optimal for a given epoch. These allelles are maintained at a low heterozygous frequency. If the environment changes, creating conditions in which these gene products become optimal, their frequency will increase as a result of natural selection. The more important the gene, however, the more constraints against its mutation because natural selection would result in elimination of deleterious changes. This is why mutations are more likely to be found in introns that functional genes.

Genetic Engineering

One of the most dramatic advances in molecular biology has been the development of recombinant DNA technology, often called *genetic engineering*. This technology became available following the discovery of enzymes called *restriction endonucleases*. These enzymes can recognize specific sequences of DNA and allow for removal of a specific area of DNA that can then be characterized in terms of its nucleotide sequences. Once the sequences have been identified, molecular biologists can make that gene in the laboratory by forming an identical sequence of nucleotides. The sequence can also be deliberately altered to see what such a mutated gene might do once it is spliced back into a living cell like *E. coli*.

One application of this technology is known as the *DNA probe*. This is basically a fragment of test DNA that can recognize and hybridize (i.e., attach to) a complementary DNA sequence in a biological sample. The test DNA may be derived from someone with a genetic disease that is "labeled" with radioactive isotopes. When the radioactive DNA probe attaches to a complementary site in tissue samples, it means that the person from whom the tissue samples were taken has a similar genetic disorder.

Other nonradioactive labeling techniques have also been devised that rely on color changes or formation of specific antibodies. Probes can be formed for either DNA or RNA. Because DNA is a double strand it has to be denatured for it to interact with the probe. If an RNA probe is used instead, no denaturing (i.e., separation) is required because RNA is already single stranded and therefore ready to hybridize with the probe.

DNA probes can be used to detect multiple base changes, rearrangements of bases, additions or deletions of bases, and even changes in gene expression. Using this kind of advanced technology, three research laboratories within weeks of one another independently discovered the gene possibly responsible for Alzheimer disease (see Chapter 3).

Although recognizing that evolutionary changes are the result of a gradual accumulation of point mutations, few scientists have extended this notion to differences between individuals. It is ironic (even extreme hubris) to admit that accumulated point mutations can explain the divergence between mice and men yet refuse to consider that these same kinds of changes might be responsible not only for Alzheimer disease, but also for proneness to alcoholism, variations in intelligence, and other differences. If Eve was born 200,000 years ago, is it not possible that her sons and daughters and their progeny experienced slight mutations in their gene pools since then?

For many people the whole idea that genes influence behavior means that we are robots controlled by computerlike gene programs. But genes are not destiny. A genetic influence does not imply fatalism. Even in films like *The Boys from Brazil,* in which Dr. Mengele creates genetic carbon copies of Hitler by cloning the Führer's DNA, the clones have to grow up in environments like those that nurtured the maniacal German leader. However, while genes and the amino acids or proteins they make do not directly cause any behavior to occur, as the next chapter shows, they may influence the expression of behavior in a certain direction.

References

Alberts, B., Bray, D., Lewis, J., Raff, M., Roberts, K., and Watson, J. *Molecular Biology of the Cell.* New York: Garland Publishing, 1983.

Bostock, M. A gene mutation which changes a behavior pattern. *Evolution,* 1956, *10,* 421–439.

Caan, R. In search of Eve. *The Sciences,* 1987, *27,* 30–37.

Camper, S. A. Research applications of transgenic mice. *Biotechniques,* 1987, *5,* 638–650.

Casperson, T., Zech, L., Johansson, C., and Modest, E. J. Identification of human chromosomes by DNA-binding fluorescent agents. *Chrosoma,* 1970, 30, 215–227.

Comings, D. E. Mechanisms of chromosome banding and implications for chromosome structure. *Annual Review of Genetics,* 1980, *12,* 25–46.

Embury, S. H., Scharf, S. J., Saiki, R. K., Gholson, M. A., Golbus, M., Arnheim, N., and Erlich, H. A. Rapid prenatal diagnosis of sickle cell anemia by a new method of DNA analysis. *New England Journal of Medicine,* 1987, *316,* 656–661.

Goltz, S., Todd, J., Kline, S., Pollice, M., Watson, H., Jou, L., Thalenfeld, B., and Yang, H. L. DNA probes for diagnosis of sexually transmitted diseases. *American Clinical Products Review,* 1986 *5,* 30–35.

Goodman, M., Romero-Herrera, A. E., Dene, H., Czelusniak, J., and Tashian, R. E. Amino acid sequence evidence on the phylogeny of primates and other eutherians. In M. Goodman (Ed.), *Macromolecular Sequences in Systematic and Evolutionary Biology.* New York: Plenum Press, 1982, pp. 115–119.

Hawkins, J. D. *Gene Structure and Expression.* Cambridge: Cambridge University Press, 1985.

Jacob, F., and Monod, J. Genetic regulatory mechanisms in the synthesis of proteins. *Journal of Molecular Biology,* 1961, *3,* 318–356.

Kimura, M. Evolutionary rate at the molecular level. *Nature,* 1968, *217,* 624–626.

King, J. L, and Jukes, T. H. Non-Darwinian evolution. *Science,* 1969, *164,* 788–798.

Kohne, D. E. Application of DNA probe tests to the diagnosis of infectious disease. *American Clinical Products Review,* 1986, *5,* 20–29.

Li, W. H., Luo, C., and Wu, C. Evolution of DNA sequences. In R. J. MacIntyre (Ed.), *Molecular Evolutionary Genetics.* New York: Plenum Press, 1985, pp. 1–94.

McAleer, N. *The Body Almanac.* New York: Doubleday, 1985.

Neel, J. V. The wonder of arm presence here: A commentary on the evolution and maintenance of human diversity. *Perspectives in Biology and Medicine,* 1982, *25,* 518–558.

Nirenberg, M. W., and Matthaei, J. H. The dependence of cell-free protein synthesis in *E. coli* upon naturally occurring or synthetic polyribonucleotides. *Proceedings of the National Academy of Sciences,* 1961, *47,* 1588–1601.

Omenn, G. S. Inborn errors of metabolism: Clues to understanding human behavioral disorders. *Behavior Genetics,* 1976, *6,* 263–284.

Omni,. Interview with James Watson. *Omni,* 1984, p. 77.

Paabo, S. Molecular cloning of ancient egyptian mummy DNA. *Nature,* 1985, *314,* 644–645.

Patterson, C. *Evolution.* London: Butler & Tanner, Ltd, 1978.

Swain, J. L., Stewart, T. A., and Leder, P. Parental legacy determines methylation and expression of an autosomal transgene: A molecular mechanism for parental imprinting. *Cell,* 1987, *50,* 719–727.

Wilson, A. C., Carlson, S. S., and White, T. Biochemical evolution. *Annual Review of Biochemistry,* 1977, *46,* 573–639.

Zuckerkandl, E., and Pauling, L. Molecular disease, evolution, and genetic heterogenecity. In M. Kasha and N. Pullman (Eds.), *Horizons in Biochemistry.* New York: Plenum Press, 1962, pp. 189–225.

CHAPTER 2

Behavior Genetics

The search for the genetic determinants of human behavior has become a proverbial "red flag." At best there is a basic skepticism about such influences; at worst, there is a deep-rooted philosophical antipathy toward any premise that seems to limit personal achievement. For some critics, acknowledgment of a genetic basis for behavior, regardless of its extent, implies that heredity is destiny; for others, there are sinister motives behind the collection of such data. Dr. Leon J. Kamin, for example, claims research looking into and finding a genetic basis for intelligence as reflected in I.Q. scores "has been fostered by men committed to a particular social view" (1974, p. 2).

Behavior geneticists have for the most part steered away from polemics about the implications of their work. For instance, Plomin and his colleagues (1980) urge their readers only to "consider the evidence on its own merits and avoid being unduly influenced by your own particular social view. We have no ax to grind on this or any other controversial issue in behavior genetics and will strive to remain apolitical" (p. 2).

Behavior genetics research has also met with considerable resistance because of a basic premise in psychology that learning and environment are responsible for all individual differences. The beginnings of this viewpoint go back about 50 years ago to John B. Watson, the founder of the behavioristic school of psychology, who claimed that environment and environment alone was the only factor influencing behavior:

> Give me a dozen healthy infants, well-formed, and my own specified world to bring them up in and I'll guarantee to take any one at random and train him to become any type of specialist I might select—doctor, lawyer, artist, merchant-chief and, yes, even beggar-man and thief, regardless of his talents, penchants, tendencies, abilities, vocations, and race of his ancestors (1930, p. 104).

Some scientists still adhere to this rigid environmentalistic view—a view that leaves almost no place for genetic influences.

Another reason such research has not been widely accepted is that behavior cannot be interpreted in terms of monogenic Mendelian principles (see Chapter 1). Because behavioral characteristics are polygenic rather than monogenic there is no simplistic genetic paradigm within which to explain such influences.

Despite these criticisms, data has been accumulating at a steady pace supporting the influence of genetic factors on behavior, especially behavioral abnormalities. The sources of these data range from epidemiological studies using rigorous and sophisticated statistical analyses to isolation and identification of genes involved in behavioral anomalies previously thought to be solely due to environmental factors.

Such studies have eroded the philosophical antagonism against recognition of a genetic role in behavior. But even if we had the complete genetic code for any individual and knew all the steps from code to assembly of cells in an individual's brain, we would still be unable to predict that individual's behavior. Genetic and environmental influences do not act alone; each interacts with the other. The nature–nurture distinction was always arbitrary at best; at the molecular level it is meaningless. Take fingerprints for example. Not only does every person have a different fingerprint pattern from every other person, but every finger on your own hand has a different fingerprint pattern from every other finger. The genes on each of your fingers are identical. If genes were the only influence directing formation of fingerprints, your fingerprints should all be identical. Obviously, some subtle microchemical influence during development must act in concert with the genes for the whorls, loops, and ridges on the surface of your fingers in order for them to be different from each other. In this situation, what is nature and what is nurture?

If genetic and environmental influences are inseparable, why then a chapter on behavior genetics? The answer is that behavior is the most sensitive measure of individual differences. If we assume that behavior is a reflection of the type and organization of cells in the brain, behavior represents a way of understanding that cellular influence. Behavior genetics is simply an extension of traditional studies in genetics to behavior. It depends on genes to the extent that behavior is the end result of interactions between a myriad of gene messages that begin with the making of an amino acid.

An example of this influence is language. Spoken language is an exclusively human behavioral ability. Nearly every human learns to speak a language. Although there are still arguments about how this happens, it is fundamental. Language is a genetically endowed human characteristic represented by anatomical structures in the human brain common to everyone and known to be concerned with language (Chomsky, 1965). The language we speak, however, is completely determined by our environment. Typically, this means our schools and parents. French, German, English, or any other language is not genetically endowed.

For each language there are those who speak and write it better than others. Among the many thousands of genetic disorders, there are several hundred in which speech and language are abnormal. This has led to the identification of several major genes influencing human speech and development (McKusick, 1986). One such gene carries the code for an enzyme that metabolizes the amino acid histidine. When this enzyme is absent or nonfunctional, a condition called histidinemia occurs. Almost all people with histidinemia have speech and language problems; e.g., they say "less" when they mean "yes," and they can't repeat words added one at a time, although they have normal hearing. This leads to the question of how, *not* how much, this facility is related to genetically endowed language capabilities. Was Shakespeare's eloquence simply the by-product of an amino acid? Or were his words merely the slings and arrows of an outrageous genetic fortune?

The Heritability of Behavior

Suppose you tested a number of people for a certain behavioral characteristic and found some had a lot of it whereas others only had a little of it. If you suspected the characteristic was due to genetic factors, you might select those people exhibiting the extremes of the particular characteristic and mate them, each to their own extreme. For each generation, you would continue to select out and breed the extremes only. After about 20 generations, you would probably find that the two final groups differed a lot more than the two original groups. This separation occurred because you allowed the genes for the characteristic to be passed on in one group and eliminated in the other. Basically, this is what behavioral geneticists do when they create strains of animals that exhibit high or low frequencies of a particular behavior.

Obviously, no one could do this kind of study with people. For one thing, grant money supporting such research would run out long before the 20th generation even if there were someone to inherit the project. People simply take too long to breed. For another, outside of science fiction, selective human breeding is impossible. Finally, there is no way of imposing experimental control over factors apart from genetic influences that could influence the behavior you were interested in studying.

One viable alternative is to focus on the correlation among relatives for certain behavioral characteristics to determine how much of a characteristic is due to genetic factors. If genetic factors are important, pairs of relatives who are most genetically alike, such as monozygotic twins, ought to resemble one another more closely on a particular characteristic than other relatives. If heredity does not contribute to the characteristic, differences in genetic background should not affect the resemblance of relatives of that characteristic. Heritability (H) refers to the amount of variation in a characteristic of a population due to genetic variance divided by this same variance plus environmental variance. This is represented by the formula

$$H = \frac{\text{genetic variance}}{\text{genetic variance} + \text{environmental variance}}$$

For heritability to be 100%, all the variance would be genetic. In such a situation, every genotype would be phenotypically different but those with the same genotype would be alike. If heritability were 50%, genetic and environmental influences would each contribute a similar amount to the characteristic. A zero heritability would mean individuals with the same genotype were phenotypically different.

Behavioral geneticists exploring the genetic basis of human behavior look for heritabilities. If characteristics are heritable, relatives will resemble each other; that is, the correlation for a particular characteristic will be higher than the correlation between pairs of unrelated individuals because relatives share more genes. Monozygotic twins ought to resemble one another most because they share identical genes. However, relatives can resemble one another not only because they share the same genes but also because they share the same experiences. To tease apart these two contributions behavioral geneticists have devised three basic designs for examining heritability in humans. These are the family study, the twin study, and the adoption study. These designs provide ways of estimating the genetic influences on measured characteristics; environmental influences are considered to account for the remaining influence.

Paradigms Used in Human Behavioral Genetics Research

Family Study

Because members of a family are genetically more similar to one another than to strangers, a particular behavioral disorder or characteristic with a genetic basis should occur more often among family members than among the general public. For instance, if someone in a family is schizophrenic and schizophrenia is due to some genetic factor, other family members should have a greater than chance probability of also being schizophrenic. The same is true for alcoholism or any other disorder. In addition, the frequency of the disorder should be higher among first-degree relatives compared with second- or third-degree relatives. However, the obverse is not necessarily true. Just because one's parents are not schizophrenic, does not mean that one or both do not carry a recessive gene or set of genes for schizophrenia (provided such a gene or gene set exists). To be fairly confident of a negative family history researchers often include in their analyses not only the parents of the "index" or control case, but first-degree relatives as well.

However, members of a family generally share a similar family life. These "within-family" characteristics include everything that goes with a particular socioeconomic class, race, religion, child-rearing practices, and so on. Although each family member is not treated exactly like every other, life within a family is generally more similar than "between families." Because people within a family share both a similar genetic and environmental background, the covariance among family members on a particular trait should be greater than the covariance between a particular family member and someone picked at random from outside the family. The latter share neither the heredity nor environment of the "index" person.

Families are an intangible environmental unit. When behavioral geneticists refer to environment, they usually mean "between-family" differences. They characterize such differences in terms of social class, religion, and the like, and assume that they are partially controlling for such influences by matching families on these variables. However, this is a superficial kind of control at best (Wachs, 1983). For one thing, the characteristics chosen may have little or no relevance to the questions being asked nor is there any way of knowing to what extent the covariance between family members is due to genetic or environmental influences or their interaction. For another, it is extremely naive to assume that people are like one another simply because they are Catholics or Protestants. Studies showing an increased frequency of some disorder among family members are only a starting point; they are not conclusive evidence. Such studies can only suggest possible genetically determined behaviors.

Twin Study

Twin studies offer a way of teasing apart heredity and environmental influences, but such studies also have their shortcomings.

In every 90 births, a set of twins will be born. One third of these twins will be genetically identical—the result of a single fertilized ovum dividing in two, each giving rise to two separate individuals with exactly the same genes. The other two thirds are fraternal twins. These are children created as a result of two separate ova being fertilized by two separate sperm. These separate sperm can even come from different fathers, as in a case a few years ago, of a German woman who gave birth to two boys, one white and one black. Fraternal twins are no more genetically alike than any other brothers or sisters. If environment were the only influence on their behavior, identical twins should be as alike in their behavior as fraternal twins. If, however, identical twins resemble one another in their behavior more than fraternal twins, this might imply that genetic factors contribute more to the behavior being evaluated than environmental factors.

What may start off as genetic identity, however, may be subverted. Being monozygotic is no guarantee of an identical genome. This is known from case reports in which one monozygotic twin has a genetic disorder like Turner syndrome (in which one of the X chromosomes is missing) and the other does not (Pedersen *et al.*, 1980).

Twin studies compare the frequency of some factor between identical and fraternal twins. If genetic factors influence some behavior, identical twins will have a much higher concordance (the behavioral geneticist's term for percentage of twins with the same outcome) for that behavior than fraternal twins. If concordance between identical and fraternal twins were similar, it would mean genetic factors were minimally responsible for that behavior.

However, neither identical nor fraternal twins grow up in a vacuum. Identical twins may be more similar than fraternal twins in their behavior because they are treated more similarly than fraternal twins; the similarity of their experiences may contribute to their greater concordance. Another possibility is that identical twins become more alike because they share the same placenta and therefore receive the exact same nutrients, dosage of drugs, and other substances taken by the mother, whereas fraternal twins each have their own placenta. (The placenta originates from the fetus not the mother.) Because there are two placentas, each with its own genetically related capabilities of nutrient transport, drug metabolism, and other functions, one twin may receive different levels of some blood-borne element whose presence affects development.

Adoption Study

Adoption studies are the most sensitive of all behavioral genetic designs because they compare twins raised from birth in different homes and genetically unrelated children raised together. Only in such cases is the role of genetic factors unambiguous because monozygotic twins will have a genetic correlation of 1 compared with a genetic correlation near 0 for unrelated siblings. Fraternal twins will have a genetic correlation somewhere in between. A high correlation between identical twins raised apart suggests a genetic contribution; a low correlation implies an environmental contribution. Unfortunately (for the behavioral geneticist), there are very few situations where identical twins or other siblings are raised by families that are so different from those of their biological parents that environment can be totally excluded. This is because adoption agencies try their best to match adoptive parents as closely as possible to biological parents.

Using these three basic kinds of research paradigms, behavior geneticists have been searching for genetic influences in many different human behaviors. Several examples, such as schizophrenia, Alzheimer's disease, alcoholism, and intelligence, will now be examined as examples of such research. The reasons for focusing on only a few examples are space limitations that preclude a more comprehensive survey and the greater attention devoted to these behaviors compared with others that have been examined.

Schizophrenia

Mark J. is a 22-year-old accountant. Usually cheerful and good natured, he has become suspicious of those he works with. He believes they are secretly altering his figures to make him look bad. Some nights he broods over their treachery and worries about what new tricks they will play on him. To protect himself he no longer socializes with people at work. He makes excuses not to eat with his former friends because he thinks they are trying to poison him. The more he broods, the more overwhelmed he becomes at the rush of thoughts that pour through his mind.

When Mark started voicing his thoughts out loud he was fired. But the thoughts wouldn't go away. Instead he began to believe his former colleagues were secretly watching and following him. Then he began hearing voices and feeling detached from his body.

Mark J. is a schizophrenic. About 75% of all schizophrenics begin developing symptoms between 16 and 25 years of age. Few begin to do so after age 30, almost none after age 40. Male schizophrenics usually develop symptoms at an earlier age than female schizophrenics.

Schizophrenia is one of the most common forms of psychosis in the United States, but its incidence has been hard to determine because the diagnosis cannot be made objectively and because schizophrenia varies considerably in form and intensity. Worldwide estimates presently range from a low of 0.30 per 1,000 among a group of Amish farmers in the United States to a high of 17 per 1,000 for a group of people in a remote part of Sweden (Torrey, 1987).

The "rule of thirds" held by most American psychiatrists is that one third of people with schizophrenia who are hospitalized will recover completely, one third will improve and will need only occasional hospitalization, and one third will not improve.

Schizophrenia definitely runs in families. If someone in your family has schizophrenia, you are four times more likely to become schizophrenic than someone at random (Rosenthal, 1970). Children of schizophrenics are at the highest risk for becoming schizophrenic, followed by brothers and sisters of schizophrenics and first cousins (Rosenthal, 1970).

Schizophrenia is not a new disorder. During the 1800s, psychiatrists were bewildered by the many different and unrelated kinds of mental disorders they observed. To create some order out of this chaos, Emil Kraepelin, a German psychiatrist, organized the various psychiatric syndromes into categories. By categorizing these disorders he hoped to identify their causes and eventually their cures. One criterion Kraepelin used in his diagnostic scheme was the prognosis (cause and outcome) of a disorder. In distinguishing between schizophrenia (which was then called *dementia praecox*) and affective psychosis like manic-depression, Kraeplin claimed that the former was degenerative and ended in deterioration of the personality, whereas the latter was episodic and personality structure was preserved. To Kraepelin, schizophrenia's inexorable outcome meant the disorder probably had a single biological cause.

Kraepelin's classification scheme brought organization into psychiatry, but it also generated arguments still unsettled today. For one thing, dementia praecox did not always result in deterioration; some patients recovered and later exhibited no signs of the disorder. Such recoveries prompted a Swiss psychiatrist, Eugen Bleuler, to diagnose psychotic illnesses such as dementia praecox in terms of symptoms rather than outcome. Because he regarded the splitting of emotion and intellect as the major clinical characteristic of dementia praecox he renamed the disorder *schizophrenia*, literally "split mind."

Another of Kraepelin's premises that now seems incorrect is that schizophrenia is a discrete mental disorder. Instead, it appears to include a spectrum of mental disorders. Some patients with schizophrenia also suffer from depression, especially in the early years of their illness. Patients with affective disorder also have higher rates of schizophrenia in their families than patients with pure schizophrenia (Andreasen, 1987).

After lithium began to be used successfully to treat mania, some clinicians began treating their schizophrenic patients who had depression with lithium and found they too improved. As a result, schizoaffective disorder is now considered to be intermediate between schizophrenia and affective disorders. Overall, schizoaffective patients don't respond to combinations of lithium and antipsychotics as well as people with pure affective disorders but do better than people with pure schizophrenia (Andreasen, 1987).

Family and Twin Studies

The search for a genetic basis for schizophrenia began about 60 years ago when the German psychiatrist Franz Kallmann began his initial studies using records from a Berlin mental hospital. From these records, he identified about 1,000 patients with schizophrenia. He then tried to locate or obtain information about the relatives of these patients.

When Kallmann came to the United States, he studied patients from psychiatric hospitals in New York State. His first study of the patients appeared in 1946, the second in 1953. Although he had not been able to include twins in his earlier study in Germany, such information was available to him from the New York data. These data presented strong evidence for a genetic basis for schizophrenia, especially the 86% concordance rate among monozygotic twins. Subsequent studies have also found a very high concordance for schizophrenia among identical twins (see Table 2.1). These data have been "age-corrected" to take into account the possibility that younger relatives may not have developed the disorder when examined but might do so at a later age. (For example, some genetic disorders like Huntington's disease do not make themselves evident until middle age. Looking for the disorder among teenagers would result in a gross underestimation of its hereditary basis.)

The general concordance rate among monozygotic twins has usually been in the range of 50%, considerably below Kallmann's results, but recent studies have found equally high rates. For example, Farmer and her co-workers (1987) reported a heritability of 85% for schizophrenia when they compared monozygotic and dizygotic twins. Ratings of schizophrenia were done blindly and the researchers used specific criteria from the *American Psychiatric Association's Diagnostic and Statistical Manual* (DSM-III). They also combined various DSM-III categories with that for schizophrenia to see if they could increase the concordance ratio and found that the greatest MZ/DZ concordance ratio was produced when certain additional diagnostic categories such as schizoaffective disorder were combined with schizophrenia.

Table 2.1. Pooled European Family and Twin Studies of Schizophrenia, 1920–1978[a]

Family relation to affected case	Affected	Concordance	No. studies
Monozygotic twins	44	85	3
Dizygotic twins	12	50	3
Full siblings	7	38	10
Half siblings	3	20	3
First cousins	2	9	3
Spouses	1	3	4

[a]From Gottesman *et al.*, 1987.

Adoption Studies

Many studies of schizophrenia among adoptees have been conducted in Denmark because the population is relatively homogeneous and because the Danes are meticulous record keepers. They maintain a population register that lists name, date of birth, date of death, and addresses for everyone who has ever lived in the country for even a few months. The Danes also keep detailed lists of every legal adoption in the country including the names of biological and adoptive parents. And finally, they also keep records of everyone who has received professional psychiatric treatment. Using this record system, researchers can study biological and environmental contributions to certain disorders by comparing adopted and nonadopted individuals with their biological and adoptive parents and siblings.

One of the first researchers to take advantage of this natural experiment was a group interested in possible links between schizophrenia and heredity headed by Seymour Kety (Kety *et al.*, 1968). To examine the possible role of genetic factors in schizophrenia, Kety's group used a design called the "adoptees' family method" (see Table 2.2).

The study involved wading through thousands of institutional records to find adoptees in the greater Copenhagen area who had been admitted to psychiatric hospitals and who could be characterized as schizophrenic on the basis of these records. The criteria Kety used for identifying schizophrenia were based on characteristics described in the DSM-III.

On the basis of these criteria, 33 "index" cases were identified. For each index case, Kety located a control adoptee who had no psychiatric history and matched each to the index case as closely as possible in terms of age, sex, social class of adoptive family, and age at time of adoption. Next he searched through the psychiatric records for all of Denmark for relatives of the index and control adoptees. The researchers identified 150 biological relatives of the index cases and 156 biological relatives of the controls. Because there was only one clearly identified case of schizophrenia described in the records, Kety relied on a schizophrenic spectrum of disorders. Kety then determined the incidence of schizophrenia among the biological and adoptive relatives of the index and control cases.

Kety's two hypotheses in this study were (1) if schizophrenia has a genetic component, there should be a higher percentage of schizophrenics among biological relatives (SB) than biological relatives of controls (CB), and (2) if schizophrenia is primarily due to environmental factors, the percentage of schizophrenics among adoptive relatives (SA) should be higher than among adoptive relatives of controls (CA).

The results supported the first hypothesis (see Table 2.3). There was a higher percentage of schizophrenia among biological relatives of the index cases (8.7%) than among

Table 2.2. Design for Adoptee's Family Study Comparing Relatives of Schizophrenics and Controls[a]

	Relatives		
	Biological(B)		Adoptive(A)
Probands:			
Schizophrenics(S)	SB	⟷	SA
Controls	CB	⟷	CA

[a] ⟷, Groups compared.

Table 2.3. Number and Percent of Schizophrenics in Biological Families of Schizophrenic and Control Adoptees[a]

	Relatives	
Probands	Biological	Adoptive
Schizophrenics	13/150 (8.7%)	2/74 (2.7%)
Controls	3/156 (1.9%)	3/83 (3.6%)

[a]Data from Kety *et al.*, 1968.

adoptive relatives (1.9%). By contrast, the incidence of schizophrenia among biological and adoptive relatives of controls was very low and the incidence was much more similar.

A word of caution about this and related studies. Because only hospital records were used in the Kety study many relatives with schizophrenia may have escaped detection. Such studies also have an innate bias because prospective adoptive parents are screened before they are permitted to adopt children and people with behavioral problems are less likely to receive adoptees. Similarly, children born to known schizophrenics may not be put up for adoption. Instead, they may be placed in foster homes or institutions. As a result of such screening, there could be a bias that underestimates the concordance among biological and adoptive relatives.

Family, twin, and adoptee studies now leave little doubt that there is a genetic contribution to schizophrenia. In fact, a heritability as high as 85% leaves almost no room for environmental contribution. Adoption away from a monozygotic twin with schizophrenia does not decrease an individual's risk for the disorder.

The concordance rate between monozygotic twins is increased even further if cerebral laterality is taken into account. For example, when Boklage (1977) compared monozygotic twins with schizophrenia on the basis of handedness, he found the concordance rate for schizophrenia among monozygotic twins who were right handed was 92% compared with 25% among left-handed monozygotic twins. This association has frequently been noted in other studies as well (e.g., Largen *et al.*, 1983; Reveley & Reveley, 1987) and implies an association of the disorder with lateralization of brain function.

There is one other condition frequently associated with schizophrenia worth noting—a higher than expected incidence of the disorder among individuals born during the winter months (Hare & Walter, 1978; Bradbury & Miller, 1985; Torrey, 1987). This circa-annual cycle, combined with its heritability and lateralization, has led some researchers to propose that schizophrenia results from a mutation caused by a virus. In this regard, Mednick and his co-workers (1988) have detected a link between second-trimester exposure to a particular influence virus and schizophrenia in Finland.

Like other Scandinavian countries, Finland keeps meticulous records of its schizophrenic patients. During a month in the fall of 1957, the city of Helsinki was victimized by a Type A2 influenza virus epidemic. About two thirds of the inhabitants were infected. Curious as to the impact of the viral epidemic, Mednick identified 216 cases hospitalized for psychotic care before they had reached 26 years of age who had also been born nine months after the epidemic. He then compared these 216 patients with 1,565 controls hospitalized for psychiatric disorder who were also 26 years or younger at the time of admission and who had been born in the same county during the same month during the previous six years. About 36% of the index patients exposed during their second trimester of fetal development were schizophrenic compared with 22% of index or control patients exposed during the first or third trimesters of gestation.

The virus hypothesis is also congruent with Weinberger's (1987) hypothesis that schizophrenia is due to a neurodevelopmental disorder caused by a brain lesion occurring in early life. Because many schizophrenics have neuropathological damage but no evidence of recent damage, Weinberger has argued that the damage is most likely indicative of a congenital lesion.

One final characteristic association with schizophrenia worth noting involves the neurotransmitter dopamine. Drugs like amphetamine that cause the release of dopamine from nerve endings can precipitate a reaction closely resembling paranoid schizophrenia, and all of the effective antipsychotic drugs, such as haloperidol, block dopamine receptors. Dopamine agonists (compounds that mimic dopamine) also bind to dopamine receptors in subcortical areas of the brain more readily in schizophrenics, and dopamine levels on the left side of the limbic system are increased in schizophrenics (see review by Meltzer, 1987). These findings suggest schizophrenia is associated with increased activity in the dopaminergic neurons located in the left side of the brain.

Brain asymmetry represents a late evolutionary development, possibly related to the singularly human capacity for speech. Crow (1987) argues that genes responsible for brain asymmetry (and dopamine function) are a "hot spot" for viral mutation. Crow also raises the interesting question as to why a disorder such as schizophrenia should be so persistent and ubiquitous. The answer, he suggests, is pleiotropy (i.e., different phenotypic expressions of the same gene). Crow cites an Icelandic study showing that relatives of patients with psychosis tend to be overrepresented in Iceland's *Who's Who.* In the United States, a significant percentage of people in the *Who's Who* are born during the winter months (Huntington, 1938), the same time of year the majority of schizophrenics are born. In the absence of infection (or comparable damage), this gene locus may confer intellectual advantage and is transmitted to subsequent generations. Nonschizophrenics with the same "hot spot" pass it on to subsequent generations. If this "hot spot" becomes activated, schizophrenia will develop, hence the persistence and ubiquity of the disorder.

Alzheimer Disease

Senile dementia is a progressive deterioration of intellectual function starting with increased forgetfulness, followed by loss of other cognitive functions, motor disabilities and rigidity, and eventual dementia. When Dr. Alois Alzheimer found an association between senile dementia and specific brain lesions, senile dementia was renamed Alzheimer disease and a search was begun for the causes of these lesions. These studies showed that Alzheimer disease is not an inexorable fate associated with aging but a distinct behavioral disorder that expresses itself in old age.

Alzheimer disease and its associated medical complications is now the fourth leading cause of death in the Western world with age-dependent prevalence rates as high as 5.8 per 100 people per year (St. George-Hyslop *et al.*, 1987). In the United States, about $25 billion is spent annually for institutional care for Alzheimer patients (St. George-Hyslop *et al.*, 1987).

The brains of people with Alzheimer have characteristic lesions that result from plaques and neurofibrillar tangles that form in and around neurons. The senile plaque is made up of an insoluble protein called amyloid surrounded by abnormal neuronal and glial cells. The protein can actually be visualized because of its peculiar straining property when treated with Congo Red dye. Amyloid plaques generally form extracellularly, whereas the neurofibrillar tangles form inside the nerve cell cytoplasm. Both emerge as a result of either an increased synthesis or a decreased degradation of protein, a change in susceptibility to enzymes that break it down, or because of a structural change that affects it adherence to cell surfaces (Breakefield & Cambi, 1987). Symptoms appear when the amyloid deposits become large enough to destroy neurons.

Alzheimer disease is now known to be a hereditary disorder. Mohs and his co-workers (1987) examined the incidence of the disease among first-degree relatives of 39 men and 11 women with Alzheimer. A control group was chosen from the spouses and in-laws of these patients and from other unrelated adults with no evidence of psychiatric, neurological, or dementia problems. The older the first-degree relatives of the index case, the more likely they were to have Alzheimer. At 60 years of age, less than 1% of the relatives had Alzheimer; at 70 years, the incidence rose to 7%; by 75, it was 15%, by 80, it was 24%, and by 86 (the age of the oldest case), the incidence was 46%. In contrast, the incidence among controls was 12% at 85 years of age.

Not only does Alzheimer disease run in families but the familial incidence increases rapidly by the eighth and ninth decades. By 90 years of age, people who have a relative with Alzheimer have a 50% chance of having the disorder. Since spouses of affected individuals are no more likely than control relatives to have the disorder it is unlikely that it is due to environmental influences.

Because 50% of relatives of Alzheimer victims are also affected it appeared that this disorder was due to an autosomal dominant gene, an inference subsequently borne out by the discovery of such a gene by Kang and his associates (1987). The gene was isolated from brain tissue taken from senile plaques and codes for a protein containing 695 amino acids including the 43 amino acids that make up the amyloid protein in Alzheimer. Surprisingly, it is located on chromosome 21 (Robakis *et al.*, 1987; St. George-Hyslop *et al.*, 1987), the same chromosome responsible for Down syndrome. Because nearly all individuals with Down syndrome also have the same characteristic plaques and neurofibrillar tangles (Masters *et al.*, 1985), the two disorders may be due to a comparable genetic influence. One such common influence could stem from the extra gene in chromosome 21 in Down syndrome and its presence in Alzheimer. In both cases there is too much amyloid protein produced resulting in destruction of brain tissue and the associated cognitive impairments.

Alcoholism

Suspicion of a hereditary basis for alcoholism goes back to ancient times (Abel, 1984). Plutarch, for example, said that "one drunk begets another," but it is only recently that solid evidence supporting a genetic influence on alcoholism has been reported.

Although alcoholism affects millions of individuals, there is still no widely accepted definition for the disorder. Most definitions emphasize habitual drinking of large amounts of alcohol and adverse physical and social effects; the emphasis is on the drinking behavior rather than the physical or social consequences.

Several family and twin studies suggest a genetic influence on alcoholism. Schuckit and his co-workers (1972) found that having an alcoholic biological father increased eight-fold the probability of a son becoming an alcoholic compared with a two-fold increase for being raised by an alcoholic adoptive father. Twin studies have also shown a high concordance for alcoholism. In Sweden, Kaij (1966) identified 174 male twin pairs where one of the twins was convicted of drunkenness. The concordance for drunkenness among monozygotic twins was 70% compared with 32% for dizygotic twins. Partanen *et al.* (1966) found monozygotic twins showed greater resemblance than dizygotic twins in the amount of their drinking but not in their ability to stop drinking. Loehlin (1972) likewise found that monozygotic twins were more similar than dizygotic twins in heavy drinking, having had a hangover, and similar symptoms.

Adoption studies reported by Don Goodwin and his colleagues in 1973 produced some of the best evidence for a genetic influence on alcoholism. Like Kety and his co-workers, Goodwin's team made valuable use of Danish records. Using specified criteria for alcoholism, the researchers found that among the 5,483 adoptees they studied, adopted-out sons of biological alcoholics were four times more likely to become alcoholic than adopted sons of nonalcoholics.

Although Goodwin and his co-workers were able to demonstrate a genetic influence on alcoholism, most of those who had alcoholic fathers did not become alcoholics. To find out some of the possible ways genetic influences interact with environment, a Swedish research team headed by Michael Bohman began a series of studies to clarify the interaction (Bohman *et al.*, 1981; Bohman *et al.*, 1984; Cloninger *et al.*, 1981; Cloninger *et al.*, 1983).

Like Denmark, Sweden keeps detailed social and medical records of adoptees and their biological and adoptive parents. By moiling through these records Bowman's team tried to find (1) how genetic and environmental factors interacted to contribute to alcohol abuse, (2) the characteristics of the biological parents associated with risk for alcohol abuse in offspring, and (3) the characteristics of adoptive parents that contribute to alcohol abuse in offspring. To do so, the researchers divided adoptees into three categories of alcohol abusers—mild, moderate, and severe. They were able to do this because in Sweden, there are Temperance Boards in each community that are legally responsible for maintaining sobriety. These Boards impose fines and supervise treatment for alcoholics and keep records in such actions. Mild abusers were defined as those with one registration for abuse and no record of treatment for alcoholism; moderate abusers had two or three registrations and no treatment; severe abusers had four or more registrations, had been treated for alcoholism, or had been hospitalized with a diagnosis of alcoholism. To examine the role of environment, the researchers divided the families into severe, mild, or nonprovocative (defined in terms of parental alcohol abuse, criminality, and occupational status). The results of their analysis are shown in Table 2.4.

For sons with mild alcohol abuse, both genetic and adoptive family background contributed significantly to their drinking—adoptees with both genetic and environmental contributions were four times more likely to have a mild abuse problem than those with neither. For moderate abusers, genetic factors were associated with a ninefold increase regardless of environmental background. For severe abusers, neither genetic nor environmental factors alone contributed significantly to their drinking.

On the basis of these findings, the researchers concluded that there are two types of alcohol abuse, each with different genetic and environmental causes. They labeled the more common type *milieu-limited* because its occurrence and severity is related to genetic pre-

Table 2.4. Percentage of Mild, Moderate, and Severe Alcohol Abuse in Adoptees[a,b]

	Mild alcohol abuse		Severe alcohol abuse		Moderate alcohol abuse	
Adoptive parents	Alcoholic biological parent	Non-alcoholic biological parent	Alcoholic biological parent	Non-alcoholic biological parent	Alcoholic biological parent	Non-alcoholic biological parent
Alcoholic	26.7	7.1	10.3	6.7	17.9	4.1
Nonalcoholic	10.4	6.1	7.1	4.5	16.9	1.9

[a]Parental alcoholism includes parental alcoholism, criminality, and occupation status.
[b]From Cloninger *et al.*, 1981.

disposition and environmental provocation. Milieu-limited alcoholism characterizes most alcohol abusers and is associated with extensive treatment for alcohol abuse but not criminality in the biological parents. The other type they called *male-limited.* Male-limited alcoholism is very heritable and is associated with extensive alcohol abuse and criminality in biological parents. The differences between the two are summarized in Table 2.5.

Recognizing that some degree of alcoholism is inherited leads immediately to the question of what is inherited? One line of research aimed at this question has focused on genetically based differences in the enzymes involved in alcohol metabolism.

Table 2.5. Prominent Characteristics Distinguishing Two Types of Genetic Predisposition to Alcoholism[a]

Distinguishing characteristics	Type of genetic predisposition: Milieu-limited	Male-limited
Prevalence in adopted men	13%	4%
Biological father	Mild alcohol abuse Minimal criminality No treatment	Severe alcohol abuse Considerable criminality Treatment for alcohol abuse
Biological mother	Mild alcohol abuse Minimal criminality	No alcohol abuse No criminality
Family environment	Affects frequency and severity of alcoholism in susceptible sons	No affect on frequency Possible effect on severity
Severity of alcoholism	Generally isolated or mild, but could be severe	Recurrent or moderate, but could be severe
Relative risk in genetically predisposed sons to others (risk of 1 means no difference)	2, with postnatal provocation; 1, with no postnatal provocation	9, regardless of family environment

[a]Adapted from Cloninger *et al.*, 1981

Caucasians and Orientals have different forms of the enzymes alcohol dehydrogenase or aldehyde dehydrogenase. These differences are believed to be responsible for the unpleasant flushing reaction on the part of Orientals, as their enzyme form results in high levels of acetaldehyde when they drink (Stamatoyannopulos *et al.*, 1975). As a result, heavy drinking is more aversive for Orientals than Caucasians and may be responsible for the much lower rate of alcoholism in countries like Japan compared with the United States. If there are individual differences in these enzymes within different racial groups as well, then such differences might be associated with different susceptibilities for alcoholism. For example, alcoholic Caucasians achieve about 70% higher blood acetaldehyde levels than nonalcoholics (Korsten *et al.*, 1975), and blood acetaldehyde levels among relatives of alcoholics are about 60% higher after a challenge dose of alcohol compared with nonalcoholics (Schuckit & Rayses, 1979).

It is also possible that some individuals have a higher innate tolerance for alcohol than others and therefore can drink more. Since the potential alcoholic must be able to drink a lot, this innate tolerance could be another genetic basis for developing alcoholism. Support for this hypothesis comes from behavioral genetic studies of inbred strains of mice like those developed at the University of Colorado Institute for Behavioral Genetics (Erwin *et al.*, 1976). These strains differ considerably in their response to alcohol but originated from a heterogenous group of mice.

To develop these strains, researchers injected the ancestors of these mice with alcohol and recorded the "sleep time"—the time it took them to right themselves after being placed on their sides. The researchers then selected animals who slept for only a short time, designated "short sleep" (SS), and bred these to one another and bred animals who slept for a long time, designated "long sleep" (LS), to other LS mice. For each succeeding generation, the lowest of the low and the highest of the high were bred to one another. A large number of animals in each group were bred to one another to avoid interbreeding so that characteristics correlated with alcohol sensitivity would be randomly distributed among animals. By the 14th generation, SS mice were only sleeping for 6 minutes compared to 156 minutes for LS mice after injection with the same amount of alcohol (Erwin *et al.*, 1976). Additional studies showed that this difference was not due to differences in metabolism but to differences in sensitivity of the brain to alcohol.

These breeding studies indicate that sensitivity to alcohol is a genetically determined phenomenon. Whether such a difference in sensitivity contributes to male-limited alcoholism is still very speculative, but it represents a "working" hypothesis.

Phenylketonuria

One of the best-known genetically transmitted causes of mental retardation is the *inborn error of metabolism,* phenylketonuria (PKU), which affects 1 in every 15,000 newborns. The disorder is due to an autosomal recessive gene on chromosome 12 (Lidsky *et al.*, 1985). Thus, there is a 1-in-4 chance of acquiring the condition should both parents be carriers.

PKU results from a deficiency of phenylalanine hydroxylase, the enzyme that metabolizes phenylalanine to tyrosine. Phenylalanine is an essential amino acid we get from certain foods and is the main ingredient in NutraSweet. An inability to metabolize phenylalanine results in a buildup of phenylalanine in the brain. The defect does not express itself until shortly after birth when phenylalanine hydroxylase activity usually begins. This is also the time of rapid biochemical differentiation of the brain. When levels are about eight times higher than normal, myelination of nerves is interfered with and nerve cells are damaged.

If PKU is not treated it can lead to growth retardation, irritability, hyperactivity, and mental retardation. The infant with PKU looks normal at birth and behaves normally until about one to three months of age. He or she then becomes irritable, apathetic, and listless and developmental delay of abilities begins to become evident. By the first year, developmental delays are very noticeable.

Because PKU is due to an increase in phenylalanine and phenylalanine comes from food, one of the ways to prevent someone with PKU from becoming mentally retarded is by early detection and subsequent dietary management. (Products containing NutraSweet carry a warning to phenylketonurics that the product contains phenylalanine.) Although proper dietary management can prevent mental retardation and can lead to at least average intelligence, children with PKU still have slightly lower IQs than nonaffected siblings, possibly due to a residual effect of the disorder.

PKU screening is one of America's most successful public health measures such that PKU now accounts for less than 1% of mental retardates. The screening procedure is called the Guthrie test and it is now mandatory in 47 states and voluntary in two others. PKU testing involves pricking a newborn's heel to get a drop of blood. This drop is applied to a specially treated piece of paper that is sent to a special laboratory for analysis. The test costs only $1.50. If the test shows a baby has PKU, treatment can begin immediately. If treatment begins soon after birth, there will be no adverse effect of PKU.

Fragile X

A chromosomal anomaly involving breakage that is currently receiving considerable attention is *Fragile X syndrome.* The reason the syndrome is called Fragile X is that when cells taken from people with this syndrome are placed in a particular media that lacks folic acid, part of the X chromosome near the end breaks. People with this disorder do not always have distinct physical features, although a prominent jaw and forehead, large ears, and poorly developed teeth are not uncommon, especially in adults. Adult Fragile X males often also have large testes (called macroorchidism).

The most serious feature of this syndrome is mental retardation. The extent of retardation ranges from borderline to severe. Other behavioral anomalies include excessive shyness, mild self-mutilation (biting), autism, and attention deficit disorder. Stuttering and repetitive speech appear to be a common behavioral problem. IQ ranges from 20 to 80 but in most cases is in the 50–60 range. Verbal IQ is usually lower than performance IQ. Some of the more common physical and behavioral characteristics are shown in Table 2.6.

Fragile X women do not appear to be affected as seriously as men but they can also be mentally retarded. Although Fragile X is now recognized as a distinct syndrome, there is considerable variation among affected individuals. Veenema and co-workers (1987) assessed intelligence (using the Wechsler tests) among 52 members of a large family with Fragile X syndrome and compared these scores with chromosomal evaluation. The greater the percentage of Fragile X positive cells, the greater the mental retardation. Nonretarded Fragile X negative family members had an average IQ of 102. One nonretarded male with 6% Fragile X positive cells had an IQ of 98, whereas his retarded brothers had an average of 27% Fragile X positive cells. The highest IQ among the retarded males was 31. Non-

Table 2.6. Common Characteristics Associated with Fragile X Syndrome in Males

Parameter	Features
Birthweight	Normal, sometimes larger than normal
Head circumference	Slightly increased in childhood
Forehead	Prominent
Jaws	Prominent
Ears	Prominent
Genitals	Large
Cognitive effects	Low IQ, greater verbal than performance impairment
Speech	Stuttering, repetitive, echolalia
Behavior	Stereotyped, hyperactive, shyness, autistic, self-mutilation

retarded carrier females had an average IQ of 87; the highest score among retarded females was 41. Because this anomaly is associated with the X chromosome, all males positive for Fragile X should be affected, but as this and other studies (e.g., Dakar *et al.*, 1981) indicate, Fragile X can occur in nonretarded men and the impact of the chromosomal disorder depends on some as yet unknown factor affecting how many cells express this disorder.

Fragile X does not occur as a new mutation. All Fragile X males have mothers who are Fragile X carriers. Fragile X females also receive their Fragile X from their mothers, not their fathers. Because the proportion of affected males is 0.4 rather than the expected 0.5 for an X-linked recessive gene, it is possible the mutation is "switched on" in the egg and once activated the affected individual becomes a Fragile X. One of the agents that may be responsible for this switching is folic acid (Sutherland & Hecht, 1985). When cells with Fragile X are cultured in media free of folic acid or in media that inhibit folate metabolism or thymidylate synthetase, the enzyme that converts folate to thymidine, Fragile X occurs.

This suggests that Fragile X occurs because of a deficiency of folate or thymidylate synthetase. However, mothers of children with Fragile X syndrome have seldom been found to have folate levels below normal.

Additional studies have shown that either too little or too much thymidine may be responsible for inducing Fragile X. A similar paradox between folate and other anomalies has also been noted. For instance, maternal folate deficiency causes birth defects in rats (Hurley, 1980) and folate antagonists, e.g., aminopterin cause neural tube defects in humans (Hurley, 1980) but maternal folate deficiency has not been associated with neural tube defects in humans.

Intelligence

There is no more controversial topic in behavioral genetics than that relating to intelligence. In large part, this is because some behavioral geneticists have argued that most of the differences in IQ scores between black and white children is due to genetic factors.

There is no unequivocal evidence suggesting that any race is superior or inferior to any other. Many behavior geneticists attribute black–white differences in IQ scores to cultural rather than genetic differences. Culture represents a learned adaptation to environment. Since IQ tests are developed primarily for and are standardized in a white-dominated culture, it is not surprising that black children do not do as well as white children on such tests. However, some black children score higher on such tests than other black children and some white children score higher than other white children. It is these differences among children from similar cultural backgrounds that appear to support a genetic basis for intelligence.

There is no argument that genes determine intelligence. People with untreated PKU or with Fragile X syndrome are mentally retarded and these conditions are passed on through generations. (Down syndrome, by contrast, arises as a mutation only in the affected individual.) Although there are still arguments about the nature of intelligence, there is no doubt that people with these disorders are less capable of taking care of themselves in society than people who do not have these or other related genetic disorders. But do these genetic anomalies have any implications for normal intelligence?

Unquestionably, Fragile X chromosome makes for a less than capable individual, but why is it that people with the same Fragile X chromosome don't all have the same IQ? The same is true for PKU. Although people with PKU lack the same enzyme, the variation in their IQs can be as much as 80 points (Scarr & Carter-Saltzman, 1983). The same variation occurs among unaffected populations.

Modern behavioral genetics studies of intelligence have a long history dating back to the mid-1800s when Francis Galton published his "Hereditary Genius. An Inquiry into Its Laws and Consequences" (1869/1962). Using bibliographical information, published works, and personal inquiries, Galton identified 1,000 of the most eminent men of his era and determined their family backgrounds and the closeness of their family relationship and found eminence was a family trait.

In the United States, a similar study was conducted by R. L. Dugdale (1877) for feeble-mindedness. While inspecting country jails in New York, Dugdale found six prisoners who were related. Through a meticulous survey of their family backgrounds, he traced the family back to six sisters and gave them the pseudonymous surname "Jukes." These sisters, he discovered, had a legacy of mental retardation, criminality, immorality, and destitution that Dugdale concluded were hereditary traits.

In 1912, H. H. Goddard published a similar study about a family that he named the "Kallikaks" (from the Greek *kalos,* "good," and *kakos,* "bad") whose roots went back to the American Revolution. At that time, a Martin Kallikak took a brief time off from his soldiering duties to woo a feeble-minded girl he met in a tavern. The girl gave birth to a son that the girl named Martin Kallikak, Jr. After the war, Martin Sr. started another family in his home town and two branches of his family emerged. Goddard located about 480 descendants of the first wife and found nearly all of them were feeble minded or engaged in criminal activities, whereas the descendants of Martin Sr.'s second marriage were nearly all normal.

Although both family studies were regarded as evidence of a hereditary basis for intelligence, critics argued that their intelligence ratings could just as likely have been the result of the environment they grew up in as heredity. Goddard was also accused of deliberately falsifying his data, a charge subsequently disproved (Fancher, 1987).

These early studies generated considerable controversy about the hereditary basis of intelligence, which is still heated. In a recent review, Bouchard and McGue (1981) exam-

Table 2.7. Familial Correlations for IQ[a]

Biological	Median correlation
Parent offspring	.475
Biological mother and offspring	.38
Biological father and offspring	.43
Adopting mother and offspring	.195
Adopting father and offspring	.155
Monozygotic twins reared together	.85
Monozygotic twins reared apart	.67
Dizygotic twins reared together	.58
Siblings reared together	.45
Siblings reared apart	.24
Nonbiological siblings	.29

[a]Data from Bouchard and McGue, 1982.

ined 111 such studies published since 1963. This summary included 526 familial correlations based on 113,942 pairings. The results, shown in Table 2.7, indicate that the greater the proportion of genes shared by family members, the greater the correlation between their IQs. The most interesting comparison concerns monozygotic twins, whose average correlation in over 30 studies is about 0.85 whereas that for dizygotic twins is about 0.58.

Many studies have compared IQ scores of siblings, but relatively few have compared parental and offspring scores for adoptees. In one of the few to do so, Skodak and Skeels (1949) tested the IQs of biological mothers at the time of their giving birth and later tested the IQs of their adopted-out children. Adopted-out children had IQ scores above 105 at every age they were tested whereas average maternal score was about 85, indicating that the adoptive family environment had had a positive influence on IQ. However, the influence of environment depended on the child's genetic endowment because there was a higher correlation between biological mothers and their children than between adoptive parents and children despite the advantages of the adoptive home setting.

In a more recent study, Scarr and Weinberg (1976) tested IQ scores of black children adopted into white homes. Although black children raised in black homes generally score below average on such tests, black children raised in white homes scored above average. This study showed that black children raised in white homes do as well as white children and therefore genetic factors are not responsible for the major differences between whites and blacks on IQ tests. Instead, they are related to the culture in which the tests are standardized and aimed.

However, as in the Skodak and Skeels (1949) study, the correlation between black adoptees' IQ and their natural parents' education level (used as a proxy for IQ) was higher than that between adoptive parents' educational level and adoptee IQ, indicative of a genetic contribution to intelligence. When adoptees were retested at a later age, the correlations between biological parent and adoptees remained generally unchanged, whereas the correlation between adoptive parent and adoptee fell to about 0.03, suggesting that family environment has no lasting effect on IQ after adolescence. These adoption studies indicate that there are both genetic and environmental contributions to IQ test performance.

A Flutist Needs a Flute to Be a Flutist

Genes set the upper limit for a particular characteristic like intelligence; environment determines how close or how far from these limits one may come. Suppose you inherit a gene pool that is optimal for playing the flute. If you were given a flute when you were young and were given flute lessons by the best teachers and practiced diligently, you would have the potential for being the best flute player in the world. If you did achieve this distinction, you would realize your maximal genetically endowed potential. However, suppose you inherit the same gene pool but grow up in a home where your parents struggle for existence. There is no way they can afford to buy you a flute or pay a teacher to train you. And in your neighborhood, the closest thing to a flute is a peashooter. This is a flute-playing potential that will never be realized or recognized.

Consider a more likely possibility: You inherit a gene pool with a maximal potential less than optimal for playing the flute. You simply lack the timing, rhythm, pitch, finger dexterity, breath control, embouchure (shape of your mouth), or any of the countless other attributes of a great flutist. Even if you were given the same training and teachers as our world-class flutist, you would never be in the same class. You might become very good, but a world-class flutist—never (see Figure 2.1). In the same way, intelligence is genetically endowed, but environment determines whether maximal intelligence will be realized. This indicates nothing about the superiority or inferiority of different races.

Studies in animals allow researchers to do many of the things they can't do with humans. For one thing, they can carry out the kind of selection study mentioned earlier in which individuals with more or less of a particular characteristic are bred over many generations. By selective breeding, genes controlling a particular characteristic can be retained in one group and eliminated in another. The goal is to generate a group of animals with some specific characteristic.

In a now-classic selective breeding study, Dr. R. C. Tryon, a psychologist at the University of California at Berkeley, tested rats in mazes and identified the "bright" and the "dull" maze learners. Then he selectively bred the bright to the bright and the dull to the dull. When the progeny of these rats were tested, the maze-bright animals were much better than the maze-dull animals in learning the maze.

Selection studies in animals have also permitted behavioral geneticists to examine the interaction between heredity and environment because in animal studies both factors can be manipulated. In one such study (Cooper & Zubek, 1958), maze-bright and maze-dull rats were reared in an "enriched" environment in which cages were colored and contained toys

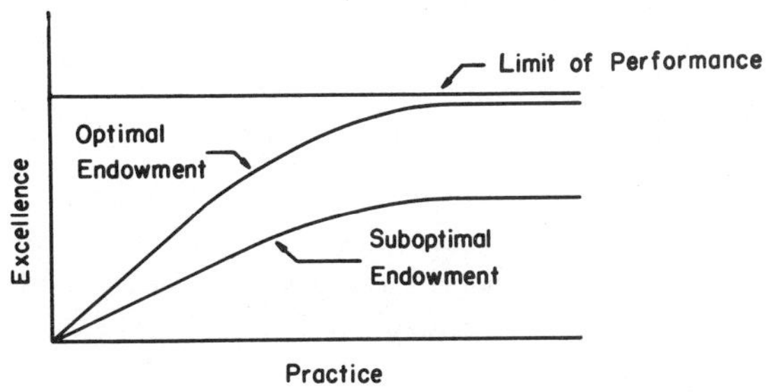

Figure 2.1. Hypothetical relationship between optimal genetic potential for flute playing and practice. If gene pool carries the potential for world-class ability (e.g., timing, rhythm, embouchure) and is nurtured (practice, good teachers, encouragement, etc.), this optimal endowment will realize its limit of performance. If not nurtured, suboptimal performance will be achieved.

to interact with, or in an "impoverished" environment in which cages were gray and there were no movable objects. After living under these conditions from 25 to 65 days of age, the animals were retested for maze-learning behavior.

Maze-dull animals raised in the enriched environment improved significantly and didn't differ from maze-bright animals. Maze-bright animals didn't benefit from living in the enriched environment—they had already achieved their optimal performance through selective breeding. However, if maze-bright animals lived in the impoverished environment, their genetic advantage was lost and they did not differ from maze-dull animals. Maze-dull animals also did not get any worse because they were already at their worst.

Studies such as this underscore the dual contribution of genetic and environmental factors to performance. It is not at all certain that the "bright" rats were actually smarter than the "dull" rats. When these animals were retested on a variety of other maze-learning problems, the "bright" animals outperformed the "dulls" on only two out of five tests (Searle, 1949). As in maze learning, a good score on an IQ test does not necessarily reflect anything more than an ability to correctly answer IQ test questions.

Prenatal Learning in the Classwomb

In behavioral genetics, phenotypic variance that can't be accounted for by genes is attributed to environmental influences. Such influences are usually viewed in a broad way and include all an individual's experiences after he or she is born. However, an even broader view includes prenatal events as well, such as being exposed to and remembering certain sounds and voices while still in the womb. Very often this premise seems to be supported by human-interest stories appearing in newspapers, magazines, and popular books. One such example is the story of two identical twin boys, adopted at 4 weeks of age by different families, who seemed nevertheless to have lived identical lives (e.g., Cassill, 1982). One twin grew up in Lima, Ohio, the other Dayton, Ohio. For over 40 years, each twin lived completely ignorant of the other's existence. When reunited, the two twins not only looked alike, they

- Both married women named Linda
- Both divorced these Lindas
- Both remarried women named Betty
- Both named their sons James Alan and James Allen
- Both owned Chevrolets
- Both had driven their Chevrolets to St. Petersburg, Florida, to spend their vacation there
- Both named their dogs Toy
- Both had been poor spellers in school
- Both liked mathematics at school
- Both used the same slang words

Is it possible that these similarities are nothing more than coincidence?

If these are not coincidences, does it follow that these similarities are due to inheritance of genes that predestined these men to marry women named Linda or Betty or to name their sons Alan (Allen)? Is there a gene for naming dogs Toy? This, of course, is absurd.

Consider the following study: 33 women, each about 7½ months pregnant, read aloud one of three stories twice a day in a quiet room. The three stories were "The King, The Mice and The Cheese," the first 28 paragraphs of "The Cat in the Hat," and a variant of this story with salient nouns changed called "The Dog in the Fog." Shortly after they were born, the babies of these mothers listened to two of these stories, each recorded by the infant's own mother. The question being asked was whether *in utero* exposure to a particular story makes it more appealing to an infant after birth.

To see if the babies had a preference for any story they were given a nipple to suck on. Even when they get no nourishment, babies make sucking movements. Often such nonnutritive sucking movements occur in bursts. The longer the infant sucked while he or she listened to a particular story, the longer the infant was allowed to hear it. And if he or she made more sucking movements, that would suggest that the story was acting as a reinforcer—just as food acts as a reinforcer for a hungry baby or animal. The results indeed showed that when a newborn baby hears two stories, each read by its own mother, the infant prefers the one his or her mother read aloud while he or she was still in the womb (DeCasper & Spence, 1986). This means that fetuses learn and remember sounds and recognize their mothers' voices. More important, the sounds they hear in the womb can influence subsequent preference for sound patterns after birth.

However, the sounds fetuses hear while in the womb are not exactly like those they hear once they are born. A lot of sound energy is lost as sound passes through the mother's body wall and the lower the frequency the greater the attenuation (Horner *et al.*, 1987). Noises within the amniotic sac such as the mother's heart, blood flowing through the placenta and uterus, and the fetus's own sloshing in the amniotic fluid all mask externally generated sounds to some degree. Despite these distortions, the fetus hears and remembers what he or she hears even as far as recognizing the source of those sounds.

Is it possible that the names we give our children or our pets, or the names of our spouses, are in some way related to the sounds we hear in our mothers' wombs? If so, the remarkable coincidences between the twins described earlier, e.g., their preference for wives named Linda and Betty, for sons named James Allen, for naming their pet Toy, are neither coincidences nor a result of genetic programming but a reflection of a common preference for certain sounds stemming from intrauterine experiences.

References

Abel, E. L. *Fetal Alcohol Syndrome and Fetal Alcohol Effects.* New York: Plenum Press, 1984.

American Psychiatric Association. *Diagnostic and Statistical Manual of Mental Disorders* (DSM-III). Washington, D.C.: APA, 1980.

Andreasen, N. C. The diagnosis of schizophrenia. *Schizophrenia bulletin,* 1987, *13,* 9–21.

Bohman, M. Some genetic aspects of alcoholism and criminality: A population of adoptees. *Archives of General Psychiatry,* 1978, *35,* 269–276.

Bohman, M., Cloninger, C. R., von Knorring, A. L., and Sigvardsson, S. An adoption study of somatoform disorders. *Archives of General Psychiatry,* 1984, *41,* 872–878.

Bohman, M., Sigvardsson, S., and Cloninger, C. R. Maternal inheritance of alcohol abuse: Cross-fostering analysis of adopted women. *Archives of General Psychiatry,* 1981, *38,* 965–969.

Boklage, C. E. Schizophrenia, brain asymmetry development, and twinning: Cellular relationship with etiological and possibly prognostic implications. *Biological Psychiatry,* 1977, *12,* 19–35.

Bouchard, T. J. and McGue, M. Familial studies of intelligence: A review. *Science,* 1987, *212,* 1055–1059.

Bradbury, T. N., and Miller, G. A. Season of birth in schizophrenia: A review of evidence, methodology and etiology. *Psychological Bulletin,* 1985, *98,* 569–594.

Breakefield, X. O., and Cambi, F. Molecular genetic insights into neurologic diseases. *Annual Review of Neurosciences,* 1987, *10,* 535–594.

Cassill, K. *Twins.* New York: Atheneum, 1982.

Chomsky, N. *Aspects of the Theory of Syntax.* Cambridge, Mass.: M.I.T. Press, 1965.

Cloninger, C. R. Genetic and environmental factors in the development of alcoholism. *Journal of Psychiatric Treatment and Evaluation,* 1983, *5,* 487–496.

Cloninger, C. R., Bohman, M., and Sigvardsson, S. Inheritance of alcohol abuse. *Archives of General Psychiatry,* 1981, *38,* 861–868.

Cooper, R. M., and Zubek, J. P. Effects of enriched early environments on the learning ability of bright and dull rats. *Canadian Journal of Psychology,* 1958, *12,* 159–164.

Crow, T. J. Psychosis as a continuum and the virogene concept. *British Medical Bulletin,* 1987, *43,* 754–767.

Dakar, M. G., Clidiac, P., Fear, C. N., and Berry, A. C. Fragile X in a normal male: A cautionary tale. *Lancet,* 1981, *1,* 780.

DeCasper, A. J., and Spence, M. J. Prenatal maternal speech influences newborns' perception of speech sounds. *Infant Behavioral Development,* 1986, *9,* 133–150.

Dugdale, R. L. *The Jukes.* New York: G. P. Putnam's Sons, 1877.

Erwin, V. G., Heston, W. D. W., McClearn, G. E., and Deitrich, R. A. Effects of hypnotics on mice genetically selected for sensitivity to ethanol. *Pharmacology, Biochemistry and Behavior,* 1976, *4,* 679–683.

Francher, R. E. Henry Goddard and the Kallikak family photographs. *American Psychologist,* 1987, *42,* 585–590.

Farmer, A. E., McGuffin, P., and Gottesman, I. I. Twin concordance for DSM-III schizophrenia. *Archives of General Psychiatry,* 1987, *44,* 634–641.

Galton, F. *Hereditary Genius: An Inquiry Into Its Laws and Consequences.* Cleveland, World Publishing, 1962. (Original work published 1869)

Goddard, H. H. *The Kallikak Family.* New York: Macmillan, 1912.

Goldgaber, D., Lerman, M. I., McBride, O. W., Saffiotti, U., and Gajdusek, D. C. Characterization and chromosomal localization of a cDNA encoding brain amyloid of Alzheimer's disease. *Science,* 1987, *235,* 877–880.

Goodwin, D. W., Schulsinger, F., Hermansen, L., Guze, S. B., and Winokur, G. Alcohol problems in adoptees raised apart from alcoholic biological parents. *Archives of General Psychiatry,* 1973, *28,* 236–243.

Goodwin, D. W., Schulsinger, F., Muller, N., Hermansen, L., Winokur, G., and Guze, S. Drinking problems in adopted and nonadopted sons of alcoholics. *Archives of General Psychiatry,* 1974, *31,* 164–169.

Gottesman, I. I., McGuffin, P., and Farmer, A. E. Clinical genetics as clues to the "real" genetics of schizophrenia. *Schizophrenia Bulletin,* 1987, *13,* 23–47.

Hare, E. H., and Walter, S. D. Seasonal variations in admission of psychiatric patients and its relation to seasonal variation in their birth. *Journal of Epidemiology and Community Health,* 1978, *32,* 47–52.

Honzik, M. P. Developmental studies of parent–child resemblance in intelligence. *Child Development,* 1957, *28,* 215–228.

Horner, K. C., Serviere, J., and Granier-Deferre, C. Deoxyglucose demonstration of in-utero hearing in the guinea pig foetus. *Hearing Research,* 1987, *26,* 327–333.

Huntington, E. *Season of Birth: Its Relation to Human Abilities.* New York: Wiley, 1938.

Hurley, L. C. Developmental Nutrition. Englewood Cliffs, New Jersey: Prentice-Hall, 1980.

Kaij, L. *Alcoholism in Twins.* Stockholm: Almquest & Wiksell, 1966.

Kallmann, F. J. *The Genetics of Schizophrenia.* Locust Valley, N.Y.: J. J. Augustin, 1938.

Kallmann, F. J. The genetic theory of schizophrenia: An analysis of 691 schizophrenic twin index families. *American Journal of Psychiatry,* 1946, *103,* 309–322.

Kallmann, F. J. *Heredity in Health and Mental Disorder.* New York: Norton, 1953.

Kamin, L. J. *The Science and Politics of I.Q.* Potomac, Md.: Erlbaum, 1974.

Kang, J., Lemaire, H. G., Unterbeck, A., Salbaum, J. M., Masters, C. L., Grzeschik, K., Multhaup, G., Beyreuther, K., and Muller-Hill, B. The precursor of Alzheimer's disease amyloid A_4 protein resembles a cell-surface receptor. *Nature,* 1987, *325,* 733–736.

Kety, S. S., Rosenthal, D., Wender, P. H., and Schulsinger, F. The types and prevalence of mental illness in the biological and adoptive families of adopted schizophrenics. *Journal of Psychiatric Research,* 1968, *6,* 345–362.

Kety, S. S., Rosenthal, D., Wender, P. H., and Schulsinger, F. Studies based on a total sample of adopted individuals and their relatives: Why they were necessary, what they demonstrated and failed to demonstrate. *Schizophrenia Bulletin,* 1976, *2,* 413–428.

Korsten, M. A., Matsuzaki, S., Feinman, L., and Lieber, C. S. High blood acetaldehyde levels after ethanol administration. *New England Journal of Medicine,* 1975, *292,* 386–389.

Largen, J. W., Calderon, M., and Smith, R. C. Asymmetries in the densities of white and gray matter in the brains of schizophrenic patients. *American Journal of Psychiatry,* 1983, *140,* 1060–1062.

Leohlin, J. An analysis of alcohol-related questionnaire items from the National Merit Twin Study. *Annals of the New York Academy of Sciences,* 1972, *197,* 117–120.

Lidsky, A. S., Law, M. L., Morse, H. G., Kao, F. T., Rabin, M. Ruddle, F. H., and Woo, S. Regional mapping of the phenylalanine hydroxylase gene and the phenylketonuria locus in the human genome. *Proceedings of the National Academy of Sciences,* 1985, *82,* 6221–6225.

Masters, C. L., Simms, G., Weinman, N. A., Multhaup, G., McDonald, B. L., and Beyreuther, K. Amyloid disease and Down syndrome. *Proceedings of the National Academy of Sciences,* 1985, *82,* 4245–4249.

McClearn, G. E. The inheritance of behavior. In L. Postman (Ed.), *Psychology In The Making.* New York: Knopf, 1963. pp. 144–252.

McKusick, V. A. *Mendelian Inheritance in Man.* Baltimore: Johns Hopkins University Press, 1986.

Mednick, S. A., Machon, R. A., Huttunen, M. O., and Bonnet, D. Adult schizophrenia following prenatal exposure to an influenza epidemic. *Archives of General Psychiatry,* 1988, *45,* 189–192.

Meltzer, H. Y. Biological studies in schizophrenia. *Schizophrenia Bulletin,* 1987, *13,* 77–111.

Mohs, R. C., Breitner, J. C. S., Silverman, J. M., and Kavis, K. L. Alzheimer's disease: Morbid risk among first-degree relatives approximates 50% by 90 years of age. *Archives of General Psychiatry,* 1987, *44,* 405–408.

Partanen, J., Bruun, K., and Markkanen, T. *Inheritance of Drinking Behavior: A Study on Intelligence, Personality, and Use of Alcohol in Adult Twins.* Helsinski: Finnish Foundation for Alcohol Studies, 1966.

Pedersen, I. K., Philip, J., Sele, V., and Starup, J. Monozygotic twins with dissimilar phenotypes and chromosome complements. *Acta Obstetrica Gynecologica Scandinavica,* 1980, *59,* 459–462.

Plomin, R., DeFries, J. C., and McClearn, G. E. *Behavioral Genetics.* San Francisco: Freeman, 1980.

Reveley, A. M., Reveley, M. A., and Murray, R. M. Cerebral ventricular enlargement in nongenetic schizophrenia: A controlled twin study. *British Journal of Psychiatry,* 1984, *144,* 89–93.

Reveley, M. A., and Reveley, A. M. Left cerebral hemisphere hypodensity in discordant schizophrenic twins: A controlled study. *Archives of General Psychiatry, 1987, 44,* 625–632.

Robakis, N. K., Wisniewski, H. M., Jenkins, E. C., Devine-Gage, E., Houck, G. E., Yao, X., Ramakrishna, N., Wolfe, G., Silverman, W. P., and Brown, W. T. Chromosome 21q21 sublocalization of gene encoding beta-amyloid peptide in cerebral vessels and neurite (senile) plaques of people with Alzheimer disease and Down syndrome. *Lancet,* 1987, *1,* 384–385.

Rosenthal, D. *Genetic Theory and Abnormal Behavior.* New York: McGraw-Hill, 1970.

Rosenthal, D., Wender, P. H., Kety, S. S., Schulsinger, F., Welner, J., and Ostergaard, L. Schizophrenics' offspring reared in adoptive homes. *Journal of Psychiatric Research,* 1968, *6,* 377–391.

Scarr, S., and Carter-Saltzman, L. Genetics and intelligence. In J. L. Fuller and E. Simmel (Eds.), *Behavior Genetics.* Hillsdale, N.J.: Erlbaum, 1983, pp. 217–335.

Scarr, S., and Weinberg, R. I.Q. test performance of black children adopted by white families. *American Psychologist,* 1976, *31,* 726–739.

Scarr, S., and Weinberg, R. Attitudes, interests, and IQ. *Human Nature,* 1978, *5,* 29–36.

Schuckit, M. A., Goodwin, D. A., and Winokur, G. A study of alcoholism in half siblings. *American Journal of Psychiatry,* 1972, *128,* 1132–1136.

Schuckit, M. A., and Rayses, V. Ethanol ingestion: Differences in blood acetaldehyde concentrations in relatives of alcoholics and controls. *Science,* 1979, *203,* 54–55.

Searle, L. V. The organization of hereditary maze-brightness and maze dullness. *Genetic Psychology Monographs,* 1949, *39,* 279–325.

Skodak, M., and Skeels, H. M. A final follow-up of one hundred adopted children. *Journal of Genetic Psychology,* 1949, *75,* 85–125.

St. George-Hyslop, P. H., Tanzi, R. E., Polinsky, R. J., Haines, J. L., Nee, L., Watkins, P. C., Myers, R. H., and Feldman, R. G. The genetic defect causing familial Alzheimer's disease maps on chromosome 21. *Science,* 1987, *235,* 885–890.

Sutherland, G. R., and Hecht, F. *Fragile Sites on Human Chromosomes.* New York: Oxford University Press, 1985.

Stamatoyannopulos, G., Chen, S. H., and Fukui, M. Liver alcohol dehydrogenase in Japanese: High population frequency of atypical form and its possible role in alcohol sensitivity. *American Journal of Human Genetics,* 1975, *27,* 789–796.

Swinson, R. P. Genetic polymorphism and alcoholism. *Annals of the New York Academy of Sciences,* 1972, *197,* 129–133.

Torrey, E. F. Prevalence studies in schizophrenia. *British Journal of Psychiatry,* 1987, *150,* 598–608.

Veenema, H., Veenema, T., and Geraedts, J. P. M. The fragile X syndrome in a large family. II. Psychological investigations. *Journal of Medical Genetics,* 1987, *24,* 32–38.

Wachs, T. D. The use and abuse of environment in behavior-genetic research. *Child Development,* 1983, *54,* 396–407.

Watson, J. B. *Behaviorism.* New York: Norton, 1930.

Weinberger, D. R. Implications of normal brain development for the pathogenesis of schizophrenia. *Archives of General Psychiatry,* 1987, *44,* 660–669.

Wender, P. H., Rosenthal, D., Kety, S. S., Schulsinger, F., and Welner, J. Crossfostering: A research strategy for clarifying the role of genetic and experimental factors. *Archives of General Psychiatry,* 1974, *30,* 121–128.

CHAPTER 3

Some DNA Is Born Great, Some Never Achieves Its Greatness, and Some Has Its Greatness Poisoned before It Can Be Recognized

The common male fruit fly, an insect with the formidable scientific name of *Drosophila melanogaster,* is a sexual stud.

When a male fruit fly meets a female, he starts courting by tapping her abdomen with his forelegs. If she ignores this advance, he will fly away, reconnoiter, and try again. Eventually, this amorous tapping will succeed in getting her attention. Certain, as only a fruit fly can be certain, that his next advances will not be rejected, he approaches his lady love from front, side, or back and starts vibrating his wings rapidly. He then stops for a few thousandths of a second and then vibrates again. In the world of the fruit fly, this off-again on-again pattern of wing vibrations is a courting song with its own rhythms and beat.

Her sexual appetite whetted by this serenading foreplay, the female fruit fly allows her suitor to approach. Eventually, he edges behind her, always keeping the wing closest to her head vibrating as fast as he can. This is because in the eye of a female fruit fly, beauty is a rapidly vibrating wing.

Once in position behind her, the male licks her genitals with his proboscis, mounts her, grasps her abdomen with his forelegs, and copulates in that position for a full 20 minutes. This ordinary, common fruit fly, as insignificant a creature on God's green earth as you can imagine, has a marathoner's sexual stamina. This fruit fly couldn't care less about powdered rhinoceros horn or ground-up Spanish fly. He is a satyr in his own right and has no need for aphrodisiacs.

The basis for this incredible feat of fruit fly passion is a submicroscopic fleck of DNA buried deep within the cells of its tiny body on a hairlike chromosome. Because such sexual prowess is genetic, such studs are born not made. But they can be unmade.

The first chapter of this book focused on DNA, how it is coded, how its codes are expressed, and how these codes are passed on to the next generation. Some people are fortunate enough to inherit what we might imaginatively call ''great'' DNA—DNA that will give them the neurobehavioral potential to excel far beyond any of their less-endowed rivals. The second chapter focused on some of these neurobehavioral possibilities. Even though someone inherits ''great'' DNA, he or she may never achieve that greatness because opportunities, what we call environment, do not allow for the expression of this greatness. The present chapter follows up on this theme by focusing on genetic poisons.

These are agents that directly alter DNA so that what may once have been ''great'' DNA loses that ''greatness.'' We can also think of ''great'' DNA in a less grandiose way as simply the genetic blueprint for average physical and intellectual capabilities. If this ''great'' DNA is poisoned, the loss of this ''greatness'' could be death, malformation, subnormal intelligence, and behavioral anomalies. Such changes are called *mutations;* the agents that cause them are called *mutagens* or genetic poisons.

Mutations in DNA are nearly always harmful, although what is detrimental to some may be seen as an improvement to others. One such problematical mutation turns a so-so fruit fly into a super stud. As prodigious as is the 20-minute copulation for the average male fruit fly, poison his DNA in a very special way and he remains clutched to a female for much longer than 20 minutes. Such males have been unimaginatively dubbed ''stuck'' by the geneticists who induce such mutations. Poison a fruit fly's DNA one way and you get a ''stuck,'' poison it another and you get what geneticists' call a ''coitus interruptus''—a fruit fly who leaves his love after only a paltry 10-minute copulation (Benzer, 1973).

Other kinds of behavior are also subject to mutation. A group of fruit fly aficionados at Princeton University (Quinn *et al.*, 1979) have trained male fruit flies to avoid a specific odor by associating it with electric shock. The flies are then allowed to feast on sugar water to which the behavioral geneticists have added ethyl methane sulfonate (EMS). This is a chemical that induces a large number of point mutations in the chromosomes of fruit fly sperm cells, one of which causes the memory of their offspring to become impaired. These offspring are still able to learn the avoidance problem. But when tested 45 minutes after learning, they seem to have completely forgotten what they had learned and go right back to the tubes where they got shocked. Normal flies stay away for hours. Because the mutated fruit flies are able to learn the task in the first place, their sensory and motor skills and their short-term memory are obviously not damaged by the mutation. The fact that they can't remember what they have learned means that the mutation interfered either with the storage of the learned material in long-term memory or its retrieval.

Sources of Mutations: Spontaneous versus Induced

It's one thing to induce mutations deliberately in the laboratory, it's another to have them induced accidently by environmental agents. When mutations are induced this way, the result is rarely beneficial.

Although many mutations result in similar outcomes, the mechanisms producing them are often different. Mutations may be spontaneous or induced. Spontaneous mutations are those that occur in nature by chance. They are essentially random changes in nucleotide sequences. Induced mutations arise from three main sources: ionizing radiation, ultraviolet radiation, and chemicals. Some sources like x-rays act at many different sites; others are very specific. The kind of mutation that occurs also depends on the complexity of the organism. In bacteria, only point mutations involving additions, deletions, or substitutions of single bases are possible. If larger effects involving several bases occur, these organisms cannot survive them because they don't have the capability for repair (see below).

Table 3.1. Types of Mutations

1. Point mutations
   a. Silent
   b. Missense
   c. Neutral
   d. Nonsense
2. Sister chromatid exchange
3. Chromosomal mutations
   a. Breaks
   b. Deletions
   c. Translocations
   d. Inversions
4. Genome mutations

Types of Mutations

Mutations are changes in DNA sequences and can be classified in various ways. These classifications are somewhat arbitrary and are not mutually exclusive but they are useful for making general distinctions.

A mutation involving a change in a single base pair is called a *point mutation.* Such changes occur when a base in DNA is replaced by another, a base is altered causing it to mispair with another, or a base is so damaged it is no longer able to pair with any base no matter what the conditions. Although a point mutation may be transmitted to offspring, it won't necessarily be expressed unless the mutation is dominant or the offspring inherits the same recessive gene from each parent.

An interchange between chromatids on the same chromosome without the loss of any genetic material is called a *sister chromatid exchange;* repositioning of a whole segment of DNA from its original site on a chromosome to another is a *translocation,* whereas a change in number of chromosomes is a *genome mutation* (see Table 3.1).

Somatic versus Gametic Mutations

If a mutation occurs within a somatic cell, such as in skin, lung, or bowel, it may be passed on mitotically to all the cells that arise from it but it may have little or no effect because there are millions of other cells performing the same function. If many cells undergo a mutation, the cumulative effect of this mutation could be an uncontrolled growth of cells (a cancer) in the skin, lung, bowel, etc.

A mutation in somatic cells will only affect the individual. It will not be transmitted to future generations. If a mutation occurs in the germ cells and these mutant cells are involved in fertilization, the mutation may become an integral component of an individual's hereditary endowment to his or her progeny, thereby increasing its frequency in the population. Dominant autosomal mutations will result in some phenotypic change in the first generation. An autosomal recessive mutation may not be phenotypically expressed for many

Table 3.2. Percentage Increase of Males Relative to Females with Nonspecific Mental Retardation in British Columbia, Canada[a]

Degree of mental retardation, males	Excess percent
Mild	39
Moderate	50
Severe	38
Profound	9
Unspecified	28
Total	34

[a]From Herbst and Miller, 1980.

generations and may become widely distributed in the population. It would become noticed only when two individuals with the same recessive mutation mated and the two copies of the mutation were combined homozygously.

Sex-linked recessive mutations (which are all X-linked) will be expressed phenotypically in the first generation if the offspring is a male because there is no other gene to offset their actions. If a gene mutation occurs on an X chromosome, it is less likely to be expressed in a female because the other X chromosome may keep it from being expressed. In such cases, the female will not be affected but will be a ''carrier.'' Only if a female inherits both recessive genes will the trait be expressed.

Nearly all sex-linked birth defects are thus due to recessive X gene inheritance and are seen much more often in males (2.6 per 1,000) than in females (1.5 per 1,000). The higher incidence of birth defects in males seems to be especially true for mental retardation (see Table 3.2).

Categories of Mutations

Mutations are generally detected as a result of some associated phenotypic effect. This may be a subtle change in the structure of a protein that is undetectable except by special biochemical procedures or may be so severe as to cause death. *Biochemical mutations* refer to a genetic change preventing production of an amino acid, e.g., sickle-cell anemia. The bread molds used by Beadle and Tatum (see Chapter 1) are another example. Beadle and Tatum took special advantage of the fact that the growth of organisms with biochemical mutations can be restored by supplementing their growth medium with specific nutrients. *Lethal mutations* are detected by the death of offspring, usually prior to birth. Biochemical mutations might also be included in this category; for example, sickle-cell anemia is often a lethal mutation. In between, are *morphological mutations,* which affect visible characteristics of an organism, and *behavioral mutations,* which are very subtle and the most difficult to detect. Studies of the latter are still in their infancy. Thus far they have mainly been confined to organisms with primitive nervous systems and relatively short life spans like *Drosophila*. Information gained from these studies will, however, undoubtedly shed light on how mutagens adversely affect development of the nervous systems of mammals including the human brain.

Conditional Mutations

Some mutations are *conditional mutations,* so-called because the mutant allele expresses its mutant phenotype only under *restrictive* conditions, such as an increase in temperature. Under *permissive* conditions, such as normal temperature, the typical phenotype is expressed. An example of a conditional mutation is the dominant heat-sensitive lethal mutation in *Drosophila.* Heterozygotes with this condition die when the temperature is raised to 30°C (restrictive condition) but are normal at 20°C (permissive condition).

Point Mutation

When James Watson and Francis Crick first described their DNA model in 1953, they recognized the possibility of base substitution. If this happened, the DNA code might be altered and an incorrect message would be transcribed by mRNA resulting in an abnormal structural or functional protein being produced. The name given to this kind of substitution is *point mutation* (also called *single gene mutation*). Such mutations occur in about 12 out of every 1,000 live births (Hook, 1984).

A point mutation is a change in a nucleotide base sequence wherein one base is either replaced, inserted, or deleted by another base. A substitution between similar bases (e.g., a pyrimidine base replacing another pyrimidine or a purine replacing a purine) is called a *transition;* an interchange between different kinds of bases (e.g., purine for a pyrimidine) is called a *transversion.*

Point mutations frequently result in easily detected behavioral abnormalities due to inborn errors of metabolism. When metabolic pathways are blocked or misdirected, one or more of the products of the metabolic reaction prior to the defect will begin to accumulate and these accumulations may be toxic to the cell.

A *silent mutation* is a base change that has no consequences. For example, a change in the codon AGG to CGG has no effect because both codons code for arginine. A *missense mutation* is a base change that can have serious consequences, e.g., in sickle-cell anemia the *C* in the sequence *GAC* is replaced by a *G* giving *GAG* (see Chapter 1). A *neutral mutation* is a base change in which a different but functionally equivalent amino acid is produced. A *nonsense mutation* is one in which a base in a stop codon is altered.

Replacement of a base will affect only one codon and therefore only a single amino acid. Addition or deletion of a base is usually more serious than a replacement because all the codons after the change are altered resulting in a *frameshift* mutation, that is, an alteration in the codon frame. The result is a drastic alteration of amino acid sequences in proteins.

In contrast to point and frameshift mutations, polygenic disorders are caused by two or more genes that each have small effects but whose effects are additive. Identifying mutagens that cause polygenic disorders and the genes involved has not been possible as yet but polygenic disorders are believed to be responsible for many birth anomalies such as neural tube defects and cleft lip.

Death by Beans

An interesting anecdote concerning a lethal type of point mutation goes back to the time of Pythagoras, the great Greek mathematician who lived in the 5th century B.C. Mixed in with the theorems Pythagoras passed on to his pupils was the cryptic warning "Don't eat the fava bean; don't even walk through a field of fava beans."

Pythagoras didn't explain why he had this grudge against fava beans, he just insisted his pupils keep away. Pythagoras was also not the kind of teacher whose motto was "do as I say, not as I do." Not at all. He scrupulously avoided fava beans and fava bean fields himself, so much so that when he was being chased by a Sardinian mob that had become infuriated by his religious teachings, he stopped at the edge of a bean field and refused to cross it. The mob caught up with him there and killed him.

Centuries later the reason for Pythagoras' injunction was discovered. It happens that historically, Sardinians have generally kept to themselves. Despite frequent invasion, they still generally intermarry only with other Sardinians. Even today, Sardinians who have left the island for work return to choose a spouse. This has made for a lot of inbreeding.

One of the curiosities about Sardinia is that every February through May, the people become very lethargic and some start urinating blood and die. This disorder is due to hemolytic anemia, one of several forms of blood disorders that also include thalassemia (from the Greek word for "sea") and sickle-cell anemia. Hemolytic anemia occurs when red blood cells in the bloodstream burst. These ruptured cells are then removed by the kidney and are excreted in urine. If relatively few cells burst, the effect is lethargy; if a lot of cells burst, death can occur.

In 1956, researchers at the University of Chicago found that in many cases individuals suffering from hemolytic anemia had a point mutation that resulted in absence of the enzyme glucose-6-phosphate dehydrogenase (G-6-PD), which keeps red blood cells from bursting. Subsequent studies traced the location of this point mutation to the X chromosome.

But why does the outbreak of the disease occur only in springtime? The reason is that beginning in late February the fava bean begins to flower in Sardinia. Eating fava beans or inhaling pollen from its flowers introduces a chemical into the body that doesn't cause any discomfort to most people but can be damaging to people with the missing enzyme. The puzzle of Pythagoras' enigmatic warning was now solved. Apparently, he was one of those who lacked G-6-PD and became lethargic when exposed to fava beans or their pollen.

Since the discovery of this enzyme deficiency, researchers have found that it is not only Sardinians who are affected. On a per capita basis, the incidence is very high in Sardinia because of inbreeding, but millions of other people including millions of Americans with a Mediterranean background also lack the enzyme. And not only does the fava bean trigger an attack, many other substances including aspirin and vitamin K can also do so. That's the bad news. The good news is that just as people with sickle cell anemia are resistant to malaria, so too are people with G-6-PD deficiency. The same gene that makes their blood cells sensitive to destruction by an ingredient in the fava bean also makes them resistant to the parasites that carry malaria.

The abnormal point mutation (A) in hemolytic anemia may be inherited from one or both parents resulting in a gene pair NA or AA. If the person has the double AA gene, he or she will have hemolytic anemia and will have a lesser chance of growing up and having children. This should have eliminated the A gene from the human gene pool. But in some areas of the world, like the Mediterranean, the frequency of A hemoglobin is very high. In these regions, about as many people who didn't have the abnormal hemoglobin died from malaria as people who did. Those who had one normal N gene and one abnormal A gene had the best chance of survival. As a result, a condition called a *balanced polymorphism* evolved and people with NA were the most likely to live and have children. In America, hemolytic anemia has no beneficial effect but in the Mediterranean it had survival value because of the resistance to malaria.

Sister Chromatid Exchange

Sister chromatid exchanges (SCEs) represent a type of crossing over (see Chapter 1) in which symmetrical interchanges of DNA occur between chromatids. Presumably, such exchanges involve DNA breakage and reunion but such lesions do not usually result in aberrant development. Neither the mechanism responsible for SCE nor its biological significance is known. Some SCEs appear to occur spontaneously but in some cases agents capable of inducing SCEs have been identified and these do so at doses below those that produce translocations (e.g., Kato & Shimada, 1975) or teratogenic effects (Reimer & Singh, 1983).

Despite uncertainties about cause and significance, SCE analysis is among the most sensitive of all procedures for detecting mutagens with few false positives (Wolff, 1982). The U.S. Environmental Protection Agency has also concluded that positive results for SCE indicate a compound is a mutagen (Latt *et al.*, 1981). However, agents like x-rays that produce a 100-fold increase in translocations at a given dose produce only a twofold increase in SCEs (Perry and Evans, 1974). This suggests that either the mechanisms by which mutagens cause SCE are different from those causing translocations or there are mechanisms in the cell that can repair minor damage but are unable to cope as well with major damage. Nevertheless, the method is very sensitive for many chemicals like alkylating agents and agents that distort DNA base pairing (Latt *et al.*, 1981).

Chromosomal Mutations

In 1959, a major breakthrough occurred in genetic cell biology. Using new staining techniques for visualizing chromosomes, French geneticists discovered that cells from three people with Down syndrome contained 47 chromosomes instead of the normal 46 and the extra one was a number 21 chromosome. This discovery ushered in a new branch of genetics now known as *cytogenetics*. Before this discovery, geneticists and cell biologists worked relatively independently. After this discovery, the two separate fields merged and cytogeneticists started looking for other correlations between chromosome number, structure, and disease.

Chromosomes can't be visualized very well until the cell begins to divide. Until then, they appear like a ball of twine in the nucleus. To make chromosomes more distinct geneticists place cells in tissue culture, stimulate them to divide (mitosis), and then stop this division at metaphase with a mitosis-arresting agent like colchicine. A hypotonic salt solution is then added to make the arrested cells swell and to give the chromosomes room to separate. They are then fixed with a methanol/acetic acid mixture and a drop of the mixture is put on glass slides to dry. The chromosomes can then be stained with various dyes.

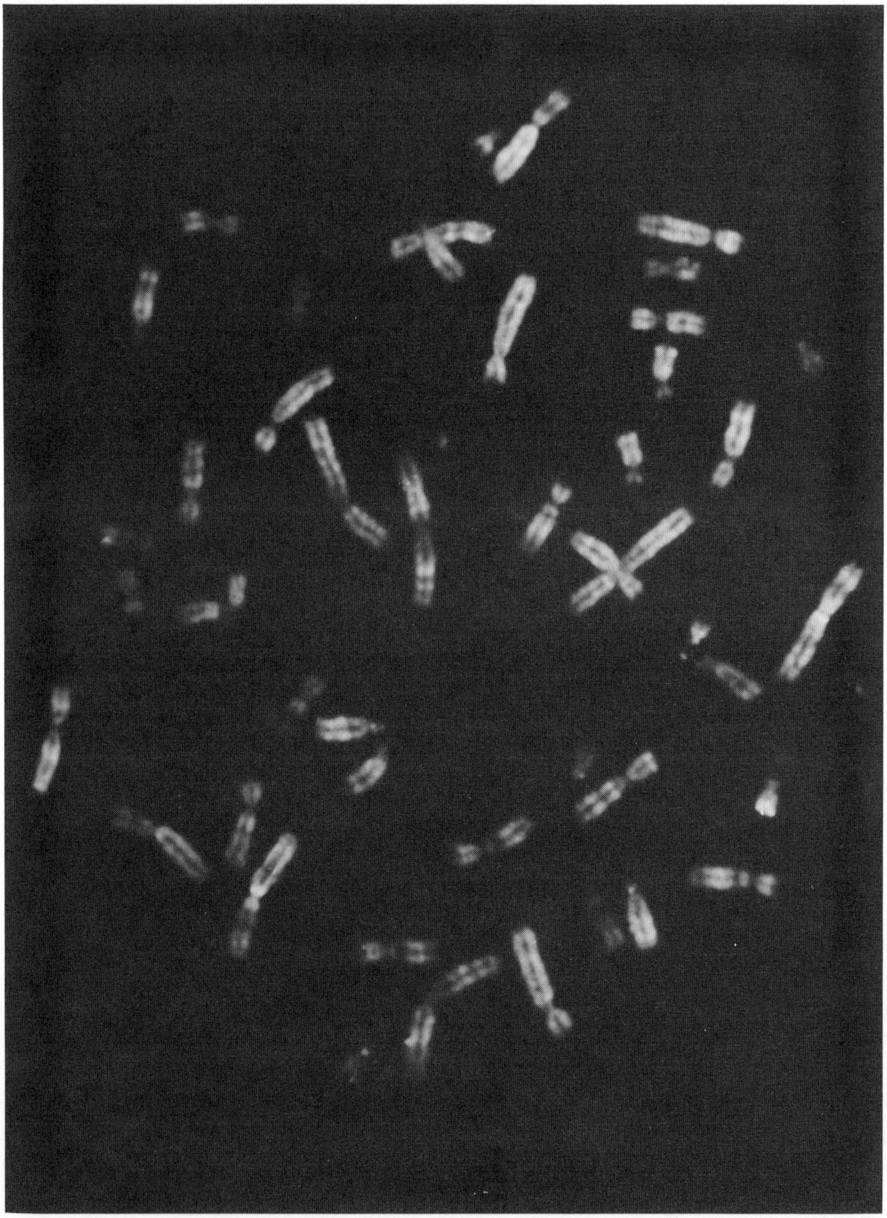

Figure 3.1. Photograph of unmatched chromosomes prior to preparation of a karyotype. Chromosomes will be cut out of a picture and matched to produce the karyotype shown in Figure 3.3. Photo courtesy Dr. M. Evans.

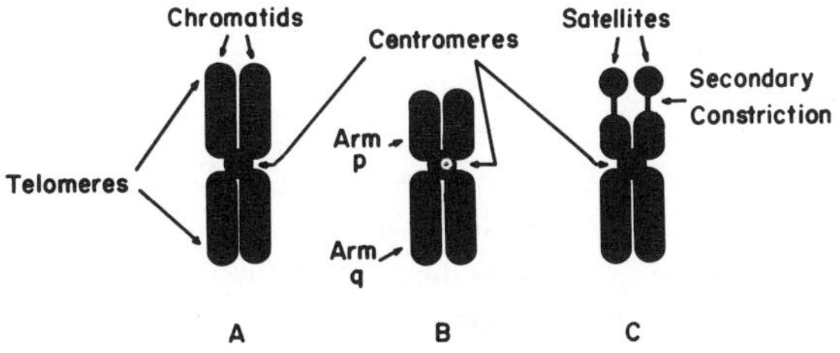

Figure 3.2. Schematic drawing of a chromosome. During mitosis, the threadlike chromosomes become visible and certain physical characteristics become distinguishable, especially the centromere and the arms of the chromosome. Chromosomes are characterized in terms of their size, the position of their centromere, and other distinguishable characteristics such as satellites.

After preparing them in this way, cytogeneticists examine the chromosomes under a microscope and take pictures of them like that in Figure 3.1.

Next, the chromosomes are cut out of the picture and arranged into matching pairs on the basis of their sizes and position of their *centromere,* the part that constricts them (see Figure 3.2).

Arranging chromosomes in terms of size is a basic way of grouping them. There is about a fourfold difference in size between the largest human chromosome (number 1) and the smallest (number 23).

Another way is by matching those of similar size on the basis of the position of the centromere, which varies from chromosome to chromosome. Some chromatids have arms of equal length above and below the centromere; others have centromeres closer to one of the ends. By convention, the short arm is called "p" for "petite" and the long arm is called "q." If the centromere is in the middle, the chromosome is called metacentric; if off center, submetacentric; if near the end, acrocentric; and if at the very end, telocentric. Those with no centromeres are acentric. Another distinguishable area on some chromosomes is the satellite, a region at the end of a chromosome separated from it by another constriction.

The arrangement of chromosomes in pairs in descending order of size just prior to division is called a *karyotype*. Karyotypes are now performed on fetal cells that are shed into amniotic fluid. Examination of these karyotypes allows for the early detection of individuals who will be born with chromosomal disorders. (Methods for prenatal diagnosis of birth defects are surveyed in Appendix A.)

Each mammalian species has its own unique number of chromosomes. Humans have 46. The shorthand nomenclature in Table 3.3 describes this complement.

Autosomes (indicated by "A") are the nonsexual carriers of heredity. Autosomes are

Table 3.3. Shorthand Nomenclature for Describing Normal Human Chromosomal Complement

Sex	Chromosomal complement[a]
Normal male	44A + XY = 46 chromosomes
Normal female	44A + XX = 46 chromosomes

[a] A = autosome.

so-called because each pair of chromosomes is similar to the other in contrast with the single pair of sex chromosomes, which is made up of either a matched pair of homozygous chromosomes for females (XX) or an unmatched (hemizygous) pair for males (XY).

In referring to a chromosomal complement for an autosomal disorder like Down syndrome in which there is an extra 21st autosomal chromosome, the number of chromosomes in the cell is written first, then the sex chromosome complement, followed by the number for the extra chromosome, e.g. (47,XY + 21) or (47,XX, + 21). When a sex chromosome is involved, this is written, for example, as (47,XXY) for Klinefelter syndrome. A karyotype of an individual with Klinefelter syndrome is shown in Figure 3.3.

In this picture, chromosomes have been arranged in pairs on the basis of their lengths and position of their centromeres. There are 23 numbered pairs. Twenty-two are autosomes. The 23rd is a sex chromosome pair. In this case two sex chromosomes are X and one is Y, so this is the karyotype of a male.

The ability to visualize and identify individual chromosomes was a major breakthrough in cytogenetics. Within a few years all of the main human chromosomal anomalies had been identified. An equally important breakthrough occurred in the 1960s when researchers discovered that staining chromosomes with special stains, such as quinicrine hydrochloride or giemsa stain, caused multiple bands (Q bands or G bands respectively) to appear along the arms of each chromosome. This enabled geneticists to match chromosomes even more accurately. Equally important, the accurate identification of single chromosomes paved the way for future studies that showed that these different bands represented different genes. This in turn made gene mapping possible.

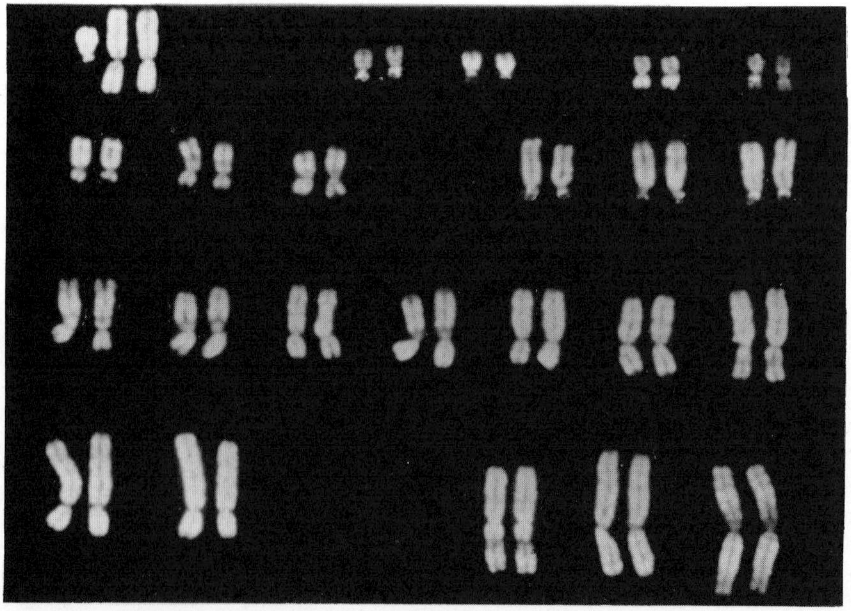

Figure 3.3. Karyotype of a child with a trisomy. Notice the presence of three number 23 chromosomes, the sex chromosome. Such children develop small testes and are infertile. Most have mild to moderate mental retardation. Photo courtesy Dr. M. Evans.

Chromosomal mutations are repositionings of chromosomal segments. These repositionings can take the form of

1. Complete breaks in single chromatids resulting in complete loss of material.
2. Deletion of part of the chromosomal material in which the deleted material may or may not be visible as a fragment. Chromosomal breaks or deletions are generally lethal for the cells in which they occur because they represent lost genetic material or because they cannot subsequently undergo mitosis.
3. Translocations, or breaks involving two or more chromosomes in which whole segments of chromosomes exchange positions with other segments.
4. Inversions, or double breaks in a single chromosome in which the broken piece does a 180-degree turn and reunites so that what was once a sequence like A B C D E F G H is now something like A B C F E D G H.
5. Nondisjunctional events, meiotic or mitotic occurrences, in which one or more chromosomes are not evenly passed on to daughter cells. The result is cells with an abnormal number of chromosomes, a condition called *aneuploidy* (*aneu,* "uneven"; *ploid,* "number"). Most clinical syndromes associated with excess chromosomes are called "somies"—conditions in which an individual has one or more extra chromosomes rather than a pair of similar chromosomes. A "trisomy" refers to a single extra chromosome. The best known and most frequently occurring trisomy (1.25 for every 1,000 live births) is trisomy 21, more commonly known as Down syndrome.

The only way to estimate the incidence of chromosomal mutations among newborns accurately is by doing a karyotype on every single live-born infant. One such analysis, on 56,952 consecutive live births, showed that chromosomal mutations occur in about 6.2 newborns out of every 1,000 live births (Russell, 1985).

The fact that the incidence of chromosomal anomalies at birth is only 6.2 per 1,000 compared with 30% among abortuses (Hook, 1981) indicates that far more children with chromosomal mutations are spontaneously aborted than those that come to term. Apparently, there is an intrinsic selection process against the birth of children with birth defects, especially those due to chromosomal mutations.

Down Syndrome

People have been born with trisomy 21 probably since the beginning of human life, but until Dr. John Langdon Down described the mental retardation and physical features associated with this disorder in 1866, it was not recognized as a syndrome.

Beyond a description of the features associated with this syndrome, there was very little known about it until 1959. In that year the chromosomal abnormality was first identified and found to be due to an extra chromosome in the 21st pair. This chromosome is the smallest of the 23 human chromosomes and contains about 45 million base pairs of DNA, about 1.5% of the three billion total base pairs in the human genome (Patterson, 1987).

Most Down syndrome children are identified on the basis of their facial features. The most common of these are listed in Table 3.4. Other features include decreased immune function, increased susceptibility to leukemia, and premature death. Most, if not all, people with Down syndrome also experience the same neuropathological changes seen in Alzhei-

Table 3.4. Common Characteristics Associated with Down Syndrome

Physical
Flat facial profile
Upwardly slanting eyes
Small mouth
Thick protruding tongue
Dysplastic ear
Webbing at neck
Muscle hypotonia
Dysplastic pelvis
Dysplastic middle phalanx of fifth finger
Simian palmar crease
Absent moro reflex
Hyperflexible joints
Short stature
Behavioral
Subnormal intelligence (IQ 30–50)
Delayed sitting and walking
Delayed speech and language skills
Delayed self-feeding skills
Cheerful
Affectionate
Rarely aggressive

mer disease and both conditions are related to abnormalities associated with chromosome 21 (see Chapter 2).

Behaviorally, IQ ranges from 25 to 70. Most IQs range from 30 to 50. Speech and language skills are most impaired. Hearing is also impaired. Expressive ability is more affected than ability to understand. Performance of simple tasks and rote skills is less affected. Intellectual retardation usually becomes evident at 3 to 6 months of age and becomes more apparent with age. Although retarded, many Down syndrome individuals can learn to read and write and can lead semi-independent lives in sheltered communities. Those raised in their own homes attain higher IQs than those raised in institutions.

In almost all cases of Down syndrome, a *nondisjunction* (failure of separation) occurs in chromosome pair number 21 during meiosis I or II. What happens in one instance is that the chromosomes of the 21st pair do not separate equally into the two daughter cells during anaphase I. Instead, both chromosomes go to the same gamete cell. In the second instance, nondisjunction occurs during anaphase II. The chromatids separate and each heads for a different daughter cell, but there is a lag in movement of the chromatids and the nuclear membrane forms before one of the chromatids can get there. If a germ cell with an extra chromosome combines with a normal gamete, a trisomy results (indicated by XY + 21 for a male and XX + 21 for a female). If the gamete lacking the chromosome combines with another gamete, a *monosomy* results. Monosomies are generally spontaneously aborted and are rarely seen at birth.

About 95% of all Down syndrome conditions result from nondisjunctions or anaphase lag. The other 5% result from a translocation in which only a piece of chromosome 21 attaches to another chromosome. The fragment is then passed on during meiosis.

If nondisjunction occurs when the embryo is dividing mitotically, the abnormal cell line starting from the abnormal segregation will divide along with normal cells. The result will be a person who has both normal cells and cells with abnormal chromosome numbers, a condition called *mosaicism* (see Figure 3.4).

Mosaicism therefore occurs after fertilization. The appearance (phenotype) of such a person will depend on the proportion of abnormal to normal cells. This is one resons why varying degrees of severity occur in Down syndrome and some other genetic syndromes.

An extra chromosome is not abnormal in itself—it is structurally intact and contains all its genes, yet its presence is damaging to the cell. This may be because it affects the amount or the speed of operation of certain enzymes or other important components of the cell.

A major impediment to determining how an extra chromosome like the one in Down syndrome adversely affects human development is the inability to trace developmental pro-

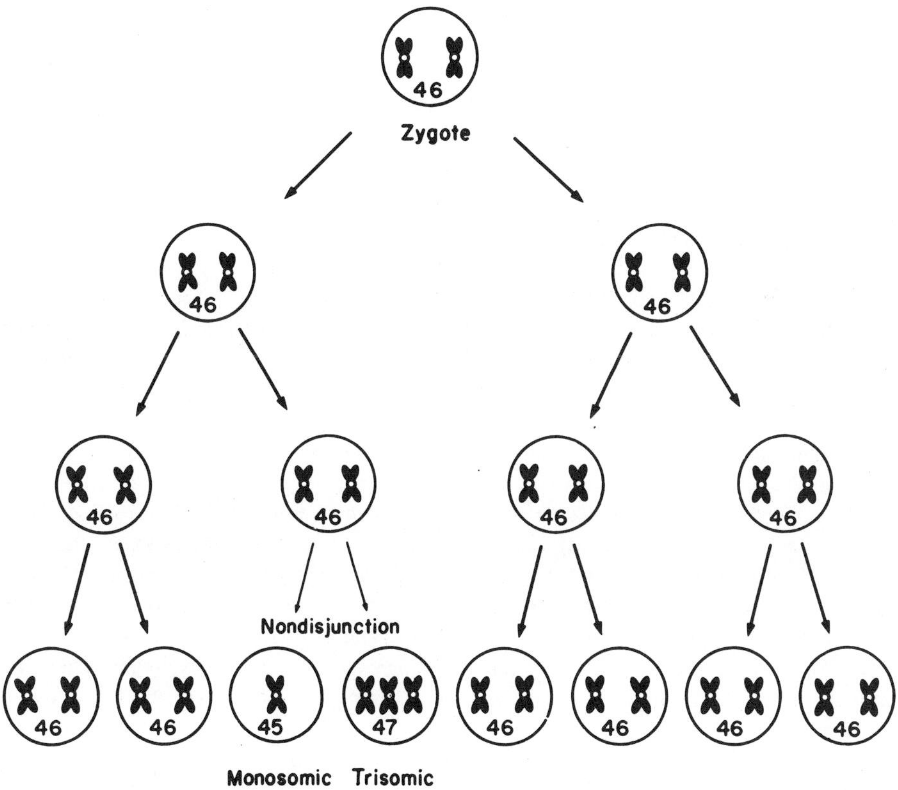

Figure 3.4. Formation of mosaic lines of cells. During the third mitotic division, a nondysfunction occurs in one cell resulting in a line of monosomic and trisomic cells. If these cells continue to divide, the organism will have some cells with the normal complement of chromosomes and some with an abnormal number.

cesses in humans and the unavailability of cells and tissues from various parts of the body, especially the brain. The only viable alternative is to study such processes in animals.

Until quite recently, such studies of genetic anomalies in animals were not possible because no appropriate models were available. However, in recent years, a large number of genetically related developmental disorders have been produced in mice (Epstein, 1986) that are similar to those of humans.

The counterpart to human trisomy 21 in the mouse is trisomy 16. This linkage between human and mouse chromosome is based on identification of three of the same genes at approximately the same part of the chromosome in both humans and mice. The similarities in type and location of genes suggest additional similarities, but these have yet to be identified. Mice with trisomy 16 die *in utero* or shortly after birth, however, so that they cannot be used for behavioral studies.

Despite the inability to evaluate these animals behaviorally, embryological studies have found that nearly all have heart disease and defects in their auditory and visual systems. Facial anomalies are also present including the flat face and skull and low-set ears seen in people with Down syndrome. Developmental studies indicate that as early as gestation day 9, these mice have fewer normal cells in the brain and development of this and other organs is delayed. By gestation day 18, the most severely affected areas include cerebellum, midbrain sensory relay nuclei, and areas of the cortex (Kornguth, 1987).

Sex Chromosome Anomalies

Phenotypic and behavioral disturbances due to sex chromosomal anomalies are much less severe than those associated with autosomal anomalies and are usually not identified until later childhood or adulthood. A number of long-term prospective longitudinal studies have followed individuals with sex chromosome anomalies through childhood into adolescence and beyond. The most recent summary of these studies is contained in a monograph edited by Ratcliffe and Paul (1986). Much of the data presented here are taken from this source.

Abnormal sex chromosome combinations are not necessarily accompanied by mental retardation but other kinds of behavioral abnormalities may be present. A girl with *Turner syndrome* has 44 autosomes but only one X chromosome (designated XO, or 45,X). Such girls are sterile, short, and usually have only partially developed female sexual characteristics. Behaviorally they are characterized by poor emotional expression, narrow or limited interests, low aggressiveness, and passivity. Intelligence varies widely. Some girls have IQ scores as low as 67; others have scored 130. Surprisingly, these girls often are superior in verbal IQ and comprehension. Earlier reports of perceptual-cognitive dysfunction have not been corroborated. A summary of characteristics associated with Turner syndrome is presented in Table 3.5.

In *Klinefelter syndrome,* a male has one or more extra X chromosomes so that his total complement is XXY, XXXY, XXXXY, etc. The karyotype of an XXY male is shown in Figure 3.3. Males with this syndrome usually have small testes. Onset of puberty is normal but they have feminine features, which may cause them problems with sexual identity. Characteristics are summarized in Table 3.6.

Table 3.5. Main Physical and Behavioral Characteristics Associated with Turner Syndrome

Physical
Short stature
Infantilism
Various malformations
Ovarian dysgenesis
Webbed neck
Abnormal secondary sex characteristic (clitoral hypertrophy, pin-point nipples, delayed development of pubic hair, undeveloped breasts)
Behavioral
Wide range of IQ from subnormal to normal
Normal or above average verbal IQ
Lower performance IQ
Poor spatial orientation
Poor emotional expression
Passivity
Low productivity
Narrow or limited interests
Low aggressiveness
No psychopathology
Lack of flexibility

Table 3.6. Typical Characteristics Associated with Klinefelter Syndrome

Physical
Small testes
Sterility
Gynecomasia
Late development of pubic hair
Taller than normal
Behavior
Verbal IQ lower than performance IQ
Stuttering
Antisocial behavior
Immaturity
Shyness
Hypoactive
Passiveness
Prone to violent outbursts
Increased incidence of schizophrenia

Table 3.7. Relation of Extra X and Y Chromosomes to Average IQ[a]

Karyotope	IQ
45, X	93
46, XX	100
47, XXY	75
47, XXX	54
48, XXXY	55
48, XXXX	60
49, XXXXY	36
49, XXXXX	20–50
46, XY	100
47, XYY	84
48, XXYY	62

[a]From Barlow, 1973.

Most people with Klinefelter syndrome are not mentally retarded, but the greater the number of extra chromosomes, the more severe the mental retardation (see Table 3.7). Verbal IQ tends to be lower than performance IQ.

Boys with Klinefelter syndrome tend to be passive and less active than other boys their age and have impaired language skills including impaired reading ability.

In a recent study, Bender and his co-workers (1986) monitored 40,000 consecutive newborns in Denver. Of these, 20 XXY boys were identified but three of the boys died in infancy. Parents of 14 of the remaining children agreed to allow their children to participate in a long-term study. Control subjects were brothers or cousins.

The full IQ range for the 14 Klinefelter subjects ranged from 71 to 122; only four had IQs below 85. The Klinefelter subjects did not do as well as controls on two tests of perception but differences were not significant. Likewise, scores on seven language tests were not significantly different, although there was a general trend for Klinefelter subjects to do worse.

In tests of auditory memory for numbers and sentences, however, Klinefelter subjects did considerably worse than controls. These subjects also did significantly worse in reading ability. Thus, although more than half the subjects with Klinefelter syndrome had normal intelligence, they still had impaired auditory memory and reading skills.

Several years ago, men with an XYY chromosome defect (dubbed "supermales") began to receive considerable notice because there appeared to be a higher proportion of them compared with the general population or with less aggressive inmates in institutions housing mentally subnormal criminals (Hook, 1973). Some of the more common characteristics associated with this chromosomal anomaly are summarized in Table 3.8.

XYY boys are taller than normal relative to weight but reach puberty at the normal age. Testicular size in these boys is larger than normal and they tend to become increasingly taller than normal with increasing age. Intelligence scores are normal. Speech development, however, is impaired. These children are not noted for behavioral disturbances in school. Leonard and Sparrow (1986) did not find any specific personality or character

Table 3.8. Characteristics of "Supermale" (XYY Chromosome)

Taller than average
Impulsivity
Mild retardation
Incoordination
Shyness
Normal fertility

abnormalities up to age 17–18 (when the oldest were evaluated) in XYY boys. Although some had quick tempers, none had antisocial, delinquent, or criminal tendencies. These findings are at odds with the report of aggression, criminal tendencies, and mental illness.

Genome Mutations

Numerical chromosomal anomalies in which whole sets of chromosomes are involved are indicated by the suffix "ploidy." *Polyploidy* is the general term for an extra set of chromosomes. Two sets of chromosomes is normal. Someone with triploidy (three sets of chromosomes) has 69 chromosomes; tetraploidy involves four sets of chromosomes (92 chromosomes). Individuals with these conditions do not live long after birth. Polyploidies are considered genome mutations and are possibly produced by agents that interfere with formation of spindles during cell division or by agents promoting fertilization by more than one sperm, a condition called *polyspermy*. Any extra chromosomes always mean some form of birth defect, usually serious.

Agents Causing Mutations

Ionizing Radiation

Ionizing radiation is electromagnetic radiation, such as from x-rays and gamma rays, that has very short wavelengths (10^{-11} to 10^{-7} cm), high frequencies, no mass or charge, and produces ions in the tissues through which they pass. Gamma rays emanate from atomic nuclei; x-rays emanate from sources outside an atomic nucleus. Other than origin, x-rays and gamma rays are identical.

Four units of measurement are generally used in describing ionizing radiation. The *roentgen* (R) refers to the amount of radiation in the air; the *rad* refers to amount of absorbed radiation and is defined as the dose depositing 100 ergs per gram tissue. Usually, exposure to one roentgen of x-rays results in absorption of one rad, so roentgens and rads are roughly equivalent. The third unit, called the *rem,* refers to ionizing radiation from particulate matter like radioactive plutonium. A *rem* is generally equivalent to a rad. The fourth unit is the *Gray.* A Gray is equal to 100 rads.

We are constantly being exposed to ionizing radiation. In the United States, each person is exposed to background radiation of 100 to 400 millirads (1 to 4 milligray) per year (Brent, 1976). This background radiation comes from cosmic rays, decomposing radioactive materials, such as radium, in the earth's crust, naturally occurring radionucleotides in all bodies, and residual nuclear fallout.

Patients undergoing radiation treatment for cancer are given around 2,000 times this background radiation (about 200 rads per day), five times a week. The mean acute lethal dose in humans is about 400 rads (Puck & Waldren, 1987). A single dose of about 100 rads will cause radiation sickness. If treatment is for pain relief only, the patient will be given a series of exposures to a maximum of 3,000 rads; if a cure is being attempted, a maximum of 5,000 to 6,000 rads will be administered (Goldstein, personal communication).

For organisms like *Drosophila,* exposures are cumulative, so that 10 exposures of 200 rads will have the same effect as a single exposure of 2,000 rads. In such organisms, there is also a linear relation between administered dose and incidence of mutations; that is, each doubling of dosage results in a doubling of mutations. Thus when dose and lethality are plotted against each other, the line passes through the origin suggesting no safe level of exposure.

In mammals, spaced exposure has a lesser effect than massive exposures to the same amount of radiation. This is because mammals have repair mechanisms capable of eliminating some of this damage (see below). As a result, there are thresholds of exposure below which biological effects are not observable. Embryopathic effects, for example, generally do not occur in animals below exposures of 40 rads—far above the levels of background radiation but far below what patients receive during radiation therapy.

Survivors of the radiation from the atom bombs dropped on Hiroshima and Nagasaki were exposed to about 150 rads if they were about 1.2 kilometers from the bomb blast (about 2 miles). Levels of exposure such as these are rarely encountered except in atomic warfare or in a nuclear reactor accident such as the one at Chernobyl in Russia in 1986.

Twenty years after the bomb blasts in Japan, survivors had a significant increase in somatic cell chromosomal abnormalities, for example, 34% anomalies compared with 1% for controls (Miller, 1969). About 39% of those exposed *in utero* developed chromosomal anomalies compared with 4% of controls (Miller, 1969). Although *in utero* exposure did not result in a higher incidence of physical birth defects (Schull *et al.,* 1981), 33% of those exposed prior to 15 weeks gestation were microcephalic and about 15% were mentally retarded. Less than 10% of those exposed later in gestation were microcephalic and less than 5% were mentally retarded (Wood *et al.,* 1967). About 70,000 newborns and infants conceived after the bomb blasts were examined at birth and at 8 to 10 months of age. No significant increases in malformations, stillbirths, or chromosomal abnormalities were detected, nor were changes in birthweights, physical development, or postnatal mortality noted. Behavioral test results have not been reported as yet (Schull *et al.,* 1981; Neel *et al.,* 1980).

Studies in animals have shown that levels of 10 to 40 rads cause aberrant nerve cell development without killing many cells. After irradiation, cells often appear normal until they enter mitosis. At first, the rate of mitosis is slowed. This is followed by a compensatory increase in rate along with high incidence of abnormally segregated chromosomes, breakages, and rearrangements (Upton, 1982). There is a definite relation between the extent and area of structural damage and behavior (Hicks & D'Amato, 1976; Furchtgott, 1975). However, in cases of *in utero* exposure to x-rays, it is often unclear whether the effect is due to a DNA mutation, some change in spindle formation, an RNA mutation; amino acid incorporation into protein, or some other aspect of development after transcription of DNA. Less ambiguous are effects on offspring resulting from paternal exposure to x-rays because these can only be due to changes in DNA or in the proteins (histones) that regulate DNA regulation.

For example, Kirk and Lyon (1984) irradiated male mice and then bred them to non-irradiated females at various times afterward so as to evaluate effects on postmeiotic stages and spermatogonial stem cells. Irradiation of early spermatids (week 3 after irradiation) resulted in the greatest number of malformed offspring, with dwarfism and exencephaly the two most commonly observed malformations.

When ionizing radiation enters cells, it imparts energy to the electrons in that cell's molecules causing them to be ejected from their orbits. These electrons are captured by other molecules depending on their proximity to the ejected electrons and their electron affinities. Because oxygen has one of the highest electron affinities of all molecules and is present in cells in relatively large proportions, it is a primary candidate for capturing free electrons. Such electron capture often begins with removal of an electron from water because cells are mainly made up of water:

$$(\mathrm{H} \,:\!\ddot{\underset{\cdot\cdot}{\mathrm{O}}}\!:\, \mathrm{H} \rightarrow [\mathrm{H} \cdot\!\dot{\underset{\cdot\cdot}{\mathrm{O}}}\!\cdot \mathrm{H}]^{+} + e^{-}).$$

Free electrons cannot exist free for long and are quickly captured by other molecules such as oxygen creating a free radical, like superoxide (symbolized as $\overline{O_2}$):

$$e- + \; \dot{\underset{\cdot\cdot}{\mathrm{O}}}\!:\!\dot{\underset{\cdot\cdot}{\mathrm{O}}} \rightarrow \dot{\underset{\cdot\cdot}{\mathrm{O}}}\!\cdot\,\dot{\underset{\cdot\cdot}{\mathrm{O}}}\!\cdot$$

When the cellular content of oxygen is low, for example, during periods of anoxia, irradiation produces a much lower mutation rate than when levels of cellular oxygen are normal. The difference illustrates the role of oxygen in radiation-induced mutagenesis.

The term "free radical" refers to any group of atoms that contains one or more unpaired electrons and is capable of independent existence. Ordinarily, the spins of paired electrons in a molecule are parallel and there is no detectable magnetic field. When a radical is formed, its uneven number of electrons causes the affected atoms to become paramagnetic; that is, they become slightly attracted to a magnetic field and as a result their formation can be detected.

Because of their unpaired electron, free radicals are biologically very reactive; if they form chemical bonds with cellular macromolecules like DNA, they can cause biological effects including mutagenesis, carcinogenesis, teratogenesis, and necrosis (Kocsis *et al.*, 1986; Slater, 1984). Among the most pervasive of these radicals are superoxide ($O_2\cdot$) and hydroxyl radical ($HO\cdot$). These radicals have been proposed as intermediaries in x-ray and ultraviolet irradiation and in alcohol toxicity for over half a century (see Table 3.9).

The hydroxyl radical ($HO\cdot$) is the most reactive of all biological radicals and will strip an electron from the nearest molecule to combine with its unpaired electron. For example, in the following reaction, the hydroxyl radical has removed a hydrogen atom and an electron from an organic molecule (RH) creating an organic radical ($R\cdot$);

$$RH + HO\cdot \rightarrow R\cdot + H_2O$$

Hydroxyl radicals can form in a number of ways. One of the most commonly proposed routes starts with generation of superoxide. Once superoxide is formed, it is acted on by an enzyme called superoxide dismutase (SOD), which converts it to hydrogen peroxide in the presence of water.

Table 3.9. Mechanism by which Ionizing Radiation Causes Biological Effects

Ionizing radiation (e.g., cosmic rays, x-rays)
↓
Excitation of electrons
↓
Ejection of electrons from atomic orbit
↓
Free radical
↓
Breakage of chemical bonds in DNA
↓
Chromosomal damage
↓
Biological effects

$$H_2O + \overline{O_2}\cdot \xrightarrow{SOD} H_2O_2 + O_2$$

When both superoxide and hydrogen peroxide are generated, there is the potential for formation of hydroxyl radicals:

$$\overline{O_2}\cdot + H_2O_2 \rightarrow O_2 + H_2O + HO\cdot$$

This reaction is known as the *Haber–Weiss reaction.* Although it is a very slow reaction, it can be catalyzed by metal ions such as iron or it can be increased by oxidative stress.

Normally, iron-mediated catalysis does not occur because metals like iron are compartamentalized in cellular organelles. However, if iron were to become decompartamentalized because of the action of a drug or other agent, the Haber–Weiss reaction would occur much faster creating the potential for cellular damage.

Oxidative stress occurs as a result of (1) increased oxygen concentrations (e.g., hyperbaric oxygenation), (2) activation of large numbers of activated phagocytes that kill bacteria by releasing bursts of oxygen, or (3) chemicals that react with oxygen to form superoxide (e.g., adriamycin, paraquat, alcohol). When oxidative stress occurs, the steady-state levels of either superoxide or hydrogen peroxide can become elevated and hydroxyl radical generation may be increased because the protective function of enzymes like sulfuroxide dismutase and scavengers may be inadequate to handle the additional load.

There are no enzymes in the body to break down hydroxyl radicals. Instead, they are removed by reacting with any molecules in their vicinity. Cellular damage may result from the direct interaction between hydroxyl radical and cellular constituents (e.g., membranes, enzymes, nucleic acids), or between the secondary radicals (e.g., (R·)), created by the interaction of hydroxyl radicals with cellular molecules, which then transform them into radicals. If these radicals approach DNA, they can combine with nucleotide bases to form *adducts* (see below), which can result in germ or somatic cell mutations:

DNA (germ cell) + Radical → DNA adduct (possible inheritable defect)

DNA (somatic cell) + Radical → DNA adduct (possible cancer)

Ultraviolet Radiation

Ultraviolet (UV) radiation is also high-energy radiation but it is nonionizing; that is, it does not produce ions in the tissues it passes through. The main source of UV radiation is sunlight. Although UV radiation does not cause molecules to lose electrons, it does cause the electrons in organic ring compounds like the pyrimidines in DNA to enter a high-energy excited state. If two adjacent pyrimidine nucleotides are irradiated with ultraviolet light, a double covalent bond, called a *pyrimidine dimer,* may form between them. These pyrimidine dimers are cross links that can interfere with DNA replication. The area in which these cross links are located is bypassed and is left to be filled in later. However, the correction is often incorrect, resulting in a mutation in the base sequence.

Chemical Causes of Mutation

Although x-rays were being used to induce mutations in the laboratory in the late 1920s, it was not until 1947 that chemical mutagenesis was demonstrated (using mustard gas).

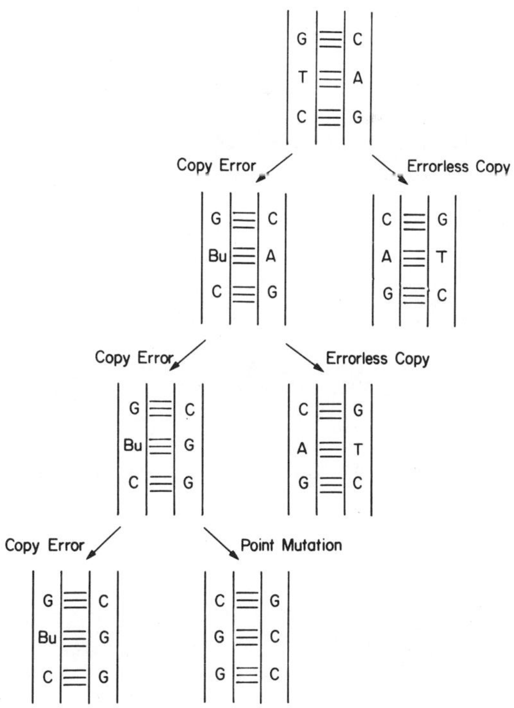

Figure 3.5. Stages in formation of a point mutation due to tautomeric shift caused by 5 bromouracil (BU) substitution for thymine. In the first replication, BU is taken up in place of thymine. The BU undergoes a tautomeric shift and takes the form of cytosine, which attracts guanine. When this change replicates, a new double-stranded sequence is produced.

There are three basic types of chemical mutagens. *Substitutive* mutagens act by virtue of their similarity to nucleotide bases. Chemicals like 5-bromouracil (Brd Urd), for instance, have a structure similar to thymine and may be taken up in its place during replication. However, Bu easily undergoes a spontaneous change (called a *tautomeric shift*) and rearranges its structure so that it now resembles cytosine instead of thymine. This results in its pairing with guanine so that instead of a TA bond, a CG base pair is formed (see Figure 3.5). Such tautomeric shifts represent a major mechanism in generating point mutations.

Additive mutagens represent a second type of chemical mutagen. These cause formation of covalent bonds between themselves or their metabolites and DNA. Such attachments to a nucleotide are called *adducts*.

One example of this class of mutagens is mustard gas, a chemical whose mutagenic actions were discovered in connection with chemical warfare during World War II. Another, whose mutagenic mechanism of action is identical to mustard gas, is cyclophosphamide, currently one of the most widely used for treating cancers.

Mustard gas and cyclophosphamide are alkylating agents; that is, they produce alkylating agents that attach to DNA. Alkylating agents are electrophilic organic molecules, like methyl (CH_3) and ethyl ($C_2H_5^+$). These can be formed when a terminal atom (chloride in the case of cyclophosphamide) splits off an alkene—a hydrocarbon that has the general formula C_nH_{2n+2}. These alkylating agents are very biologically reactive. The most common sites of attachment in DNA are the nucleophilic (electron-rich) atoms such as the nitrogens in the four bases, especially the 7 nitrogen atom in guanine.

In the case of cyclophosphamide, the reaction with DNA first involves its metabolism to a toxic mustard metabolite (phosphoramide mustard), which may then attach to the 7 nitrogen of guanine (see Figure 3.6). This can cause a number of different effects. One is a tautomeric shift so that instead of binding with cytosine, the alkylated base forms a base

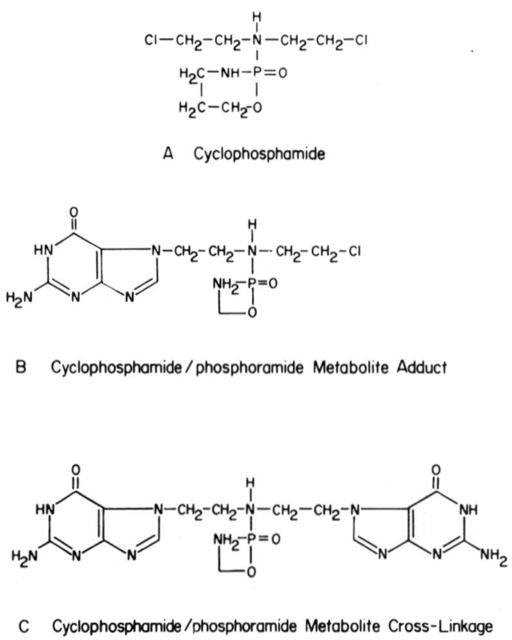

Figure 3.6. A. Formula of cyclophosphamide. The reactive nitrogen mustard moiety has two functional alkylators. B. Formation of adduct between phosphoramide metabolite and number 7 nitrogen atom on guanine. C. Cross-linkage between two guanine bases with phosphoramide adduct providing the linkage.

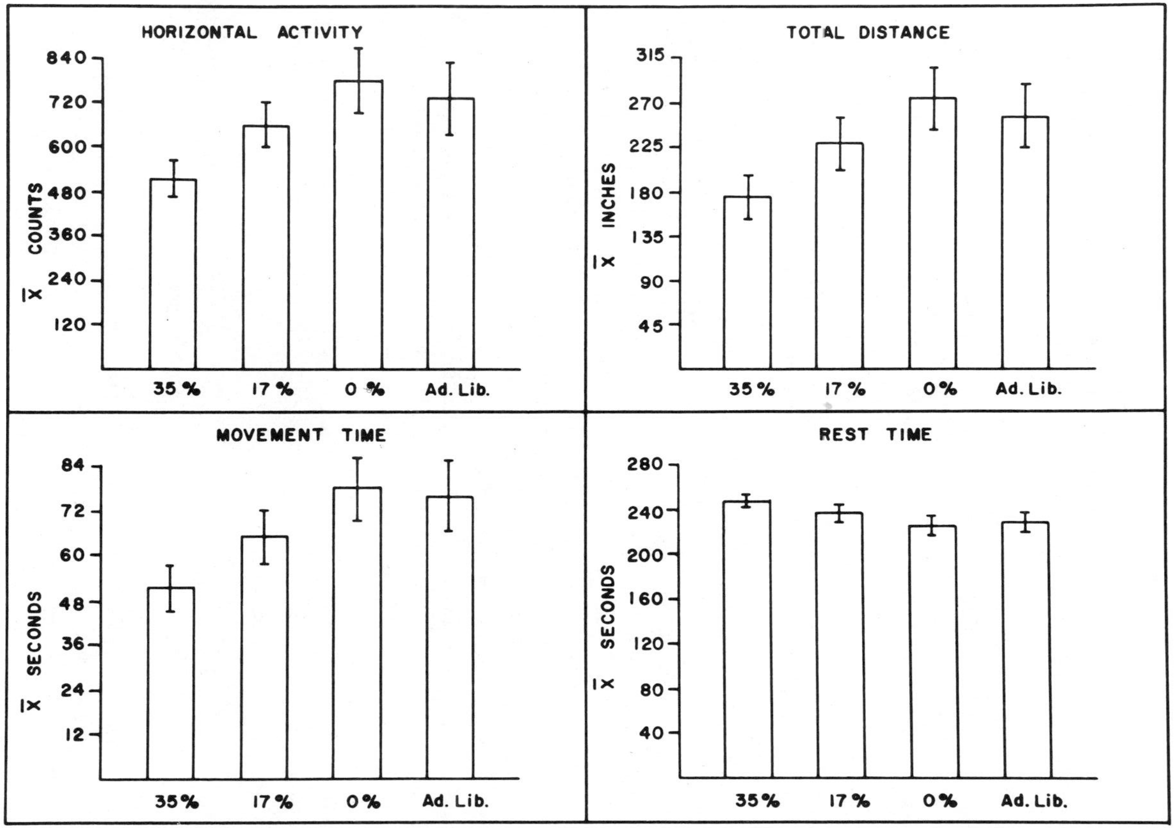

Figure 3.7. Effects of paternal alcohol consumption on activity of offspring in rats. Male rats consumed liquid diets containing varying amounts of alcohol and were subsequently bred to untreated females. Male alcohol consumption produced a dose-dependent decrease in activity. (From Abel and Tan, 1988.)

pair with thymine ultimately resulting in substitution of an adenine-thymine base pair for a guanine-cytosine base pair. A second possible reaction involves a cross linkage in which the other side chain alkylates a second guanine, another nucleotide moiety, or a part of a protein.

Because cyclophosphamide's therapeutic value as a cancer chemotherapeutic agent resides in its ability to interfere with mitotic activity and function of rapidly proliferating cancer cells, it is not surprising the drug has mutagenic effects. The deleterious effects of paternal cyclophosphamide exposure on offspring (Adams *et al.*, 1981; 1982), for instance, probably results from formation of DNA adducts in postmeiotic spermatids and spermatocytes (Adams *et al.*, 1985). The drug also has teratogenic effects but these occur at a higher dose than its mutagenic effects (Novotna & Jelinek, 1986).

Another antitumor alkylating agent, ethylnitrosurea (ENU), has now been shown to induce phenylketonuria in mice. ENU is a potent mutagen in mouse spermatogonial cells, producing an average of 1 mutation in 1,500 per DNA base (Justice & Bode, 1986). Bode and McDonald (1988) mutagenized males with ENU by injecting them with 250 mg/kg in weekly intervals over a period of 3 to 4 weeks. They then mated them to untreated females and tested their offspring for high concentrations of blood phenylalanine using the Guthrie test. Out of 105 male offspring, one carrier of hereditary PKU was produced—a relatively high rate of mutagenicity.

As previously noted, the presence of free iron may generate hydroxyl radicals by catalizing the Haber–Weiss reaction. Since alcohol causes iron to be released from binding sites in cells, it is possible that alcohol increases formation of hydroxyl radicals that form adducts with DNA. Little and Sing (1986) found that newborns whose fathers were regular drinkers weighed about 180 grams less than those whose fathers were occasional drinkers. This difference occurred independently of their mother's alcohol, tobacco, and marijuana use and could not be explained by differences in placental weight, height, parity, age, race, weight gain, or any other reproductive variable they examined.

Paternal alcohol exposure can also affect the behavior of offspring. Abel and Tan, (1988) placed male rats on liquid alcohol diets. After they had consumed alcohol for over 50 days, these males were bred to nontreated females. Offspring of these males were less active than controls and the more alcohol the fathers had consumed the less active were their offspring (see Figure 3.7).

A third type of chemical mutagen is *destructive*. This is a relatively rare mutation and involves cleavage of moieties in nucleotides, for example, removal of amine from adenine or cytosine.

Some Laboratory Tests for Mutagens

Numerous laboratory methods have now been developed for identifying possible mutagens. Most of these are based on the pioneering work of Bruce Ames and his co-workers (see below). Only some of the more common tests can be mentioned here. Additional information about these and other tests can be found in Venitt and Parry (1984).

A positive outcome in an *in vitro* test does not necessarily mean a substance is a mutagen in humans. For an agent to cause damage to successive generations it must penetrate to the reproductive cells of the body and it must produce mutations in them. Although ionizing radiation is capable of penetrating to all cells, not all chemicals do so because of peculiarities of absorption or distribution. Some chemicals may also be active only when metabolized. Metabolism of some substances will not readily occur in an *in vitro* test unless mammalian enzymes are added to the test system. When positive results are obtained, such results may suggest a possible *in vivo* action for a mutagen, but even then the metabolite may be short-lived.

Another major problem associated with many mutagenicity tests is that they use test organisms that lack mechanisms capable of repairing mutagenic damage. These organisms are deliberately selected to maximize the possibility that a potential mutagen will be detected. As a result, a substance identified as a mutagen in these test organisms may be a very weak human mutagen or may not be a mutagen at all in humans.

A procedural problem associated with mutagenicity testing, especially in tests using cell populations in which the marker is located on chromosomes essential for cell reproduction, is that the mutagenized population is examined only after it has multiplied back to its original numbers (see below). Because mutagenized cells are often growth retarded (due to delayed G_2 synthesis), populations of cells (numbers of animals) with the mutation may be less numerous than populations of normal cells. If the genetic marker is confined to a particular site, a large amount of random genetic damage may have to occur before changes in the test population are seen. To increase the likelihood of this occurring, large amounts of the mutagenic agents are typically administered. For example, most studies with x-rays,—the model mutagen—use test exposures in the range of 600 to 3,000 rads, (Puck & Waldren, 1987), whereas the average lethal exposure for humans is about 400 rads (see above). Although we are interested in levels of exposure far below these test levels, such levels are needed to produce the mutations of interest out of the thousands of random mutations. Alternatively, hundreds more test organisms need to be exposed at lower levels than if high-dose levels are used.

Because it is less costly and time consuming to use high doses rather than large numbers of test organisms, the convention is to use the high doses and extrapolate to lower exposures. However, the way such extrapolation is performed has caused considerable disagreement. If the dose response relation is linear, some researchers contend that extrapolation must be linear; others contend that extrapolation should involve a threshold region to determine levels of exposure that have no effect. Most regulatory agencies endorse the threshold argument. The difficulty of extrapolation is compounded because many dose relationships are not linear. In some instances, the relation is convex; that is, low doses are more effective per unit dose than higher doses. Alternatively, the relation may be concave; that is, low doses are less effective per unit does than higher doses.

Cytogenetic Tests

Any break in a DNA strand may result in some abnormal effect. If the break appears in both strands, a complete chromosomal break will occur. X-rays can induce double-strand breaks directly; chemical mutagens may or may not do so indirectly via a metabolite and therefore are usually seen only after *in vivo* procedures.

The basic procedures used to identify mutations in chromosomes are identical to those previously described for performing a routine karyotype (see above). In experiments, rats, mice, or Chinese hamsters are treated for varying periods of time and cells (e.g., lymphocytes) are removed and examined for cytogenic damage. Although more biological data are available for mice and rats, Chinese hamsters are often preferred for this kind of analysis because of their smaller chromosome complement ($2n = 22$) compared with the rat ($2n = 42$) or mouse ($2n = 40$).

Sister Chromatid Exchange (SCE)

SCEs are generally regarded as one of the ways of increasing genetic variability rather than as a mutation but because they involve breakage and rejoining of sister chromatids, they can also be considered mutations.

To test for agents that increase sister chromatid exchanges, cells from an exposed animal are cultured in a growth medium. Twenty-four hours later the chemical bromodeoxyuridine (BrdUrd), an analogue of the thymidine nucleotide, is added. This chemical promotes SCE. After exposure to BrdUrd, the cells are stained with a BrdUrd-sensitive dye and examined during metaphase for reciprocal exchanges.

The baseline rate for SCEs using this technique is about 5 to 10 per cell and the rate of SCE is examined against this baseline. When *in vivo* SCEs are evaluated, an animal is treated with BrdUrd and SCEs are examined in bone marrow or spermatogonia cells. However, BrdUrd itself can cause SCEs. Consequently, any test using this compound is actually a test of synergism.

Ames Test

The most commonly used method for screening possible point mutagens is the "Ames Test" (Ames *et al.*, 1975). This is also known as the "reverse mutation assay" because the goal of the test is to see if a test substance can reverse or suppress a preexisting mutation. The rationale for the test is as follows: A cell population with a specific mutation is selected from the *Salmonella typhimurium* bacterium. Several test strains of the bacteria are used, each of which contains a mutation in the histidine operon. Without a dietary supplementation of histidine, these mutants are unable to grow. The mutant bacteria are placed together with a suspected mutagen on a test plate that contains only a little histidine. This small amount allows all bacteria to grow and divide from a few generations but those with the mutation (his −) are unable to form large populations. The percentage of cells that have reverted to their original premutation state (his +), that is, that are now able to make their own histidine, is determined and the percentage is regarded as a measure of mutagenic action in the cell population. Because the preexisting mutation has been reversed, the test substance is considered to be a mutagen.

The shortcoming of this test is that only survivors are scored for mutations. Cells killed by the test agent, that is, lethal mutations, are not taken into account. This is an important consideration because an agent that causes a lethal mutation for the chromosome with the marker gene may produce mutations at other chromosomes in other organisms without being lethal. For example, in several chromosomal disorders such as Down syndrome, the chromosome containing the marker gene (number 21) undergoes nondisjunction. Two daughter cells are then produced, one with the missing chromosome and one with the extra. Generally, cells with the missing chromosome die, whereas many cells with the extra chromosome are viable. However, because these viable cells contain the marker, the mutation will not be detected by this method (Puck & Waldren, 1987).

Host-Mediated Assay

Another shortcoming of any test using bacteria is that they lack the metabolic systems that activate some agents to their reactive state. The host-mediated assay, developed by Michael Gabridge and Marvin Legator, was designed to overcome this problem. In this test, the test bacteria are injected into the peritoneal cavity of a rat or mouse followed by the test agent. If the agent has to undergo metabolic changes, it can do so by making use of the host's metabolic systems. The bacteria are then recovered from the host and are cultured on media that will indicate mutagenicity.

Dominant Lethal Mutations

This test for dominant lethal mutations examines DNA damage in gametes, which is lethal for zygotes. In rodents, the defective embryo does not implant or dies soon after implantation. In the case of a postimplantation death, the aborted implantation site may leave a deciduum or "mole," which can be easily recognized by inspection if the abortion occurred relatively late or by staining the uterus with a dye if inspecting for early resorptions.

Although dominant lethal effects can be assessed in either males or females, males are primarily tested because treatment could affect nongenetic mechanisms in females (e.g., hormones), which could subsequently affect implantation. Males are also preferable because they can be bred repeatedly with many females, whereas females can be examined for only one pregnancy at a time.

Detection involves administering the test agent to the animal for the entire period of spermatogenesis. In mice, this is about 8 weeks (Dixon & Hall, 1981); in the rat, it is about 10 weeks (Dixon & Hall, 1981). By breeding the male to untreated females after every week of treatment, different stages of germ cell development can be assessed (see Table 3.10). The female is sacrificed shortly before term and the uterus is examined.

Behavioral Mutagens

In the 1960s, Benzer and his co-workers (1973) initiated a series of studies in which they induced mutations in single genes to examine the genetic basis of different behaviors. Some of these studies involving sexual behavior and learning were mentioned earlier in this chapter. At present, the fruit fly represents the quintessential species for such studies because it is possible to identify the actual type and location of genetic damage in fruit flies better than in any other species. However, these types of studies have not as yet been used to screen for possible mutagens.

Table 3.10. Germ Cell Stages in Mice and Rats

Time	Mouse	Rat
Week 1	Epididymal sperm	Epididymal sperm
Week 2	Late spermatid	Late spermatid
Week 3	Early spermatid	Late spermatid
Week 4	Late spermatocyte	Early spermatid
Week 5	Early spermatocyte	Late spermatocyte
Week 6	Mature spermatogonia	Immature spermatogonia
Week 7	Stem cell	Immature spermatogonia
Week 8	Stem cell	Immature spermatogonia
Week 9		Stem cell
Week 10		Stem cell

DNA Repair Mechanisms

Fortunately, not all mutations are expressed. Remember that genes come in pairs. For a mutation to be expressed, the gene on which the mutation occurred would have to be dominant or the mutation would have to be present on both genes (i.e., recessive). In addition, for the mutation to be passed on to successive generations it would also have to be nonlife-threatening. Because mutations usually offer no advantage, most are life-threatening.

Potential mutations may also never be expressed because the damage is repaired before it becomes converted to a final mutation. For instance, even though an agent may form an adduct with DNA, the adduct need not cause permanent damage. Many cells are able to protect themselves against potential damage by repair mechanisms. Such repair has been observed in human cells and occurs in one of two ways:

Excision Repair

Excision repair is an error-free repair process in which a covalently attached adduct is removed prior to cell replication and the damaged base is replaced by a new base. The closer the damage to the G_1 phase of cell division, however, the less likely it will be repaired and the more likely it will be passed on to subsequent daughter cells.

The fact that DNA replication has a prolonged period in spermatogonial stem cells and resting oocytes means that there is an extended period of time available for such repair. As a result, these cells are generally resistant to mutations from chemical mutagens that are often damaging to other cells.

The first step in excision repair of an adduct involves an enzymatic excision of the damaged area and adjacent nucleotides. This is followed by a nucleotide polymerization by another enzyme to fill the gap with new bases. The final step is the rejoining of the strand breaks by a third enzyme. Such excision repair has been seen after damage from x-rays, ultraviolet light, alkylation, and the like.

There are two types of DNA excision repair. "Short patch" repair involves replacement of only a few nucleotides and usually occurs after damage from agents that produce single-strand breaks (e.g., x-rays and alkylation by methylating and ethylating agents). "Long patch" repair involves replacement of as many as 100 adjacent nucleotides and usually occurs after damage from agents that cause double-strand breaks, such as ultraviolet radiation, or agents with large ring systems such as polynuclear aromatic hydrocarbons.

Postreplication Repair

Postreplication repair is an error-prone repair process that takes place after lesions have formed in DNA. The discovery of this repair mechanism followed experiments with UV-irradiation in which newly synthesized DNA was found to be shorter than normal but eventually returned to normal size after time. The inference was that gaps in the DNA originally produced by irradiation were filled in by newly synthesized DNA. The gap is filled by using the other strand to code for the correct replacement nucleotides. Because the other strand may also have been damaged, the repair is not always error free. This kind of damage is a major source of mutation.

SOS Repair

A third process operates to rescue cells from possible death. This emergency system, dubbed SOS, involves several different cellular processes that increase the efficiency of the two other systems. In protecting the cell from death, however, the SOS system also increases the likelihood of a nonlethal mutation, especially during postexcision repair. This system has been clearly identified in bacteria, but its existence in mammals is still conjectural.

Several methods are available for detecting DNA repair. One of these, called *unscheduled DNA synthesis* involves placing radiolabeled thymidine ([3^H] thymidine) in a cell culture and measuring the radioactive thymidine in the DNA when the cell is *not* replicating. Any incorporation of thymidine at this time reflects repair of DNA, a process called *unscheduled DNA synthesis.*

Although repair of DNA damage can occur, the extent of such repair depends on the amount of mutagenic exposure. If the repair system is severely taxed, repair mechanisms may be unable to keep up with the changes and more cells would be damaged than could be repaired, resulting in cell death or mutation.

Appendix A. Prenatal Testing

A number of prenatal tests are now available to determine if a fetus has a birth defect. These are maternal serum alpha-fetoprotein (AFP) screening, ultrasound, amniocentesis, and chorionic villus sampling.

Alpha-Fetoprotein Screening

Alpha-fetoprotein screening (AFP) is the main test for diagnosis of open fetal neural-tube defects in early pregnancy. AFP is a protein that all fetuses normally produce as the fetus develops. This protein is excreted by the fetus into the amniotic fluid and back into the mother's blood. If a fetus has an open neural-tube defect (an opening in the spine or brain), it will excrete a higher than normal amount of AFP. By measuring the amount of AFP in a mother's blood around 16 to 20 weeks of pregnancy it is possible to detect an open fetus with a neural-tube defect long before birth. However, levels of AFP vary during pregnancy and an error in estimating gestational age for a pregnancy can result in an apparently abnormal value that becomes normal when corrected for actual gestational age. To make sure that the fetus is abnormal further tests may be performed involving ultrasound or amniocentesis.

Ultrasound

Ultrasound is a method similar to submarine sonar whereby sound waves are sent into the mother's abdomen. These sound waves bounce off the fetus and are picked up by a receiver that uses a computer to translate them into a gray and white image of the fetus. Very often it is possible to see neural tube defects in these pictures. The ultrasound image also lets us make measurements of the length of the fetus that can then be used to date the pregnancy. This will also enable us to evaluate AFP results that suggest the fetus has a neural tube defect.

Amniocentesis

Amniocentesis is probably the best-known method of prenatal diagnosis. Around the 16th to 20th week of pregnancy, a needle is inserted through the mother's abdomen into the amniotic fluid that surrounds the fetus. A sample of fluid is removed. The fetal skin cells in the fluid are grown and cultured in the laboratory and subsequently examined for abnormalities.

Chorionic Villus Sampling

Chorionic villus sampling (CVS) is a recent alternative to amniocentesis in which a syringe is attached to a catheter that passes through the vagina and the cervix. A small amount of tissue called the chorionic villi, which will develop into the placenta, is removed and sampled. Because the placenta is part of the fetus, it has the same chromosomal number and structure as the fetus and can be used to detect any abnormalities. CVS is almost painless and has the advantage over amniocentesis of allowing detection of abnormalities much earlier (9th to 12th week of pregnancy rather than the 16th to 20th for amniocentesis). Results are also available within days for CVS compared with about three weeks for amniocentesis.

References

Abel, E. L., and Tan, S. E. Decreased activity and shock avoidance learning in progeny of male rats consuming alcohol prior to mating. *Neurobehavioral Toxicology and Teratology,* 1988, *10*, 187–192.

Adams, P. E. Cyclophosphamide teratogenesis: A review. *Teratogenesis, Carcinogenesis, and Mutagenesis,* 1985, *5*, 75–88.

Adams, P. M., Fabricant, J. D., and Legator, M. S. Cyclophosphamide-induced spermatogenic effects detected in the F_1 generation by behavioral testing. *Science,* 1981, *211*, 80–83.

Adams, P. M., Fabricant, J. D., and Legator, M. S. Active avoidance behavior in the F_1, progeny of male rats exposed to cyclophosphamide prior to fertilization. *Neurobehavioral Toxicology and Teratology,* 1982, *4*, 531–534.

Ames, B. N., McCann, J., and Yamasaki, E. Methods for detecting carcinogens and mutagens with the salmonella/mammalian-microsome mutagenicity test. *Mutation Research,* 1975, *31,* 347–364.

Bastock, M. A. A gene mutation which changes a behavior pattern. *Evolution,* 1956, *10*, 421–439.

Bender, B. G., Puck, M. H., Salkenblatt, J. A., and Robinson, A. Dyslexia in 47 XXY boys identified at birth. *Behavior Genetics,* 1986, *16*, 343–354.

Benzer, S. Genetic dissection of behavior. *Scientific American,* 1973, *229* (Dec), 24–33.

Bode, V., and McDonald, J. D. HpH-1: A mouse mutant with hereditary hyperphenylalaninemia induced by ethylnitrosurea mutagenesis. *Genetics,* in press.

Brent, R. L. Environmental factors, radiation. In R. L. Brent and M. I. Harris (Eds.), *Prevention of Embryonic, Fetal and Perinatal Disease*. Washington, D.C., Department of Health, Education and Welfare, 1976, pp. 179–197.

Dixon, R. L., and Hall, J. L. Reproductive toxicology. In A. W. Hayes (Ed.), *Principles and Methods of Toxicology*. New York: Raven Press, 1981, pp. 107–121.

Epstein, C. J. Developmental genetics. *Experimentia,* 1986, *42*, 1117–1128.

Furchtgott, E. Ionizing radiation and the nervous system. In G. E. Guall (Ed.), *Biology of Brain Dysfunction.* New York: Plenum Press, 1975, pp. 343–379.

Hicks, S. P., and D'Amato, C. J. Effect of ionizing radiation on developing brain and behavior. In G. Gottlieb (Ed.), *Early Influences.* New York: Academic Press, 1978, pp. 36–72.

Hook, E. B. Behavioral implications of the human XYY genotype. *Science*, 1973, *179*, 139–150.

Hook, E. B. Epidemiology and population monitoring in genetic risk assessment. In F. J. De Serres and R. W. Pero (Eds.), *Individual Susceptibility to Genotoxic Agents in the Human Population*. New York: Plenum Press, 1984, pp. 21–30.

Justice, M. J., and Bode, V. C. Induction of new mutations in a mouse t-haplotype using ethylnitrosurea mutagenesis. Unpublished manuscript, 1986.

Kato, H., and Shimada, H. Sister chromatid exchanges induced by mitomycin C: A new method of detecting DNA damage at chromosomal level. *Mutation Research*, 1975, *28*, 459–464.

Kirk, K. M., and Lyon, M. F. Induction of congenital malformations in the offspring of male mice treated with x-rays at pre-meiotic and post-meiotic stages. *Mutation Research*, 1984, *125*, 75–85.

Kocsis, J. J., Jollow, D. J., Witmer, C. M., Nelson, J. O., and Snyder R. (Eds.) *Biological Reactive Intermediates III*. New York: Plenum Press, 1986.

Kornguth, S. Neural development of trisomy in 16 fetal mice. *Birth Defects*, 1987, *23*, 341–350.

Latt, S. A., Allen, J., Bloom, S. E., Carrano, A., Falke, E., Kram, D., Schneider, E., Schrock, R. and Wolff, S. Sister-chromatid exchanges: A report of the gene-tox program. *Mutation Research*, 1981, *87*, 17–62.

Leonard, M. F., and Sparrow, S. Prospective study of development of children with sex chromosome anomalies: New Haven study. IV. Adolescence. *Birth Defects: Original Article Series*, 1986, *22*, 221–249.

Little, R., and Sing, C. F. Association of father's drinking and infant's birth weight. *New England Journal of Medicine*, 1986, *314*, 1644–1645.

Miller, R. Delayed radiation effects in atom-bomb survivors. *Science*, 1969, *166*, 569–574.

Neel, J. V. Search for mutations affecting protein structure in children of atomic bomb survivors. *Proceedings of the National Academy of Sciences*, 1980, *77*, 4221–4225.

Novotna, B., and Jelinek, R. A comparison of mutagenic and embryotoxic effects of cyclophosphamide on the chick embryo. *Environmental Mutagenesis*, 1986, *8*, 241–252.

Patterson, D. The causes of Down syndrome. *Scientific American*, 1987, *255*, 52–60.

Perry, P., and Evans, S. New Giesma method for the differential staining of sister chromatids. *Nature*, 1974, *251*, 156–158.

Puck, T. T., and Waldren, C. A. Mutation in mammalian cells: Theory and implications. *Somatic Cell and Molecular Genetics*, 1987, *13*, 405–409.

Quinn, W. G., Sziber, P. P., and Booker, R. The Drosophila memory mutant amnesiac. *Nature*, 1979, *277*, 212–214.

Ratcliffe, S. G., and Pual, N. (Eds.) *Prospective Studies on Children with Sex Chromosome Aneuploidy*. New York: A. R. Liss, 1986.

Reimer, D. L. and Singh, S. M. Cyclophosphamide-induced in vivo sister chromatid exchanges (SCE). *Genetics*, 1983, *105*, 169–179.

Russel, D. *Mental Handicap*. New York: Churchill Livingstone, 1985.

Schull, W. J., Otake, M., and Neel, J. V. A reappraisal of the genetic effects of the atomic bombs: Summary of a thirty-four year study. *Science*, 1981, *213*, 1220–1227.

Slater, T. F. Free-radical mechanisms in tissue injury. *Biochemical Journal*, 1984, *222*, 2–15.

Upton, A. C. Radiation injury: Past, present and future. In R. B. Hill and J. A. Terzian (Eds.), *Environmental Pathology*. New York: A. R. Liss, 1982, pp. 9–34.

Venitt, S., and Parry, J. M. (Eds.) *Mutagenicity Testing*. Oxford: IRL Press, 1984.

Watson, J. D., and Crick, F. H. C. Molecular structure of nucleic acids; a structure for deoxyribose nucleic acid. *Nature*, 1953, *171*, 737–738.

Wolff, S. Chromosome aberrations, sister chromatid exchanges, and the lesions that produce them. In S. Wolff (Ed.), *Sister Chromatid Exchange*. New York: Wiley, 1982, pp. 41–57.

Wood, J. W., Johnson, K. G., and Omori, Y. In utero exposure to the Hiroshima atomic bomb. An evaluation of head size and mental retardation: Twenty years later. *Pediatrics*, 1967, *39*, 385–392.

CHAPTER 4

From Gene to Brain

At the moment of ejaculation, muscular spasms propel a grayish viscous fluid from a male's body into a female. The fluid is called *semen* and can contain more sperm than there are people in the United States (about 250 million). Semen is another of nature's marvelous formulas. Only 2% is sperm. Another 90% is water. The remaining 8% contains various chemicals. Some neutralize the foreign acidity of the female's vaginal fluids making the new environment more hospitable for its sperm; others trigger spasms in the walls of the vaginal tunnel catapulting sperm onward. Still others bathe sperm in enzymes to give them energy for the arduous race ahead. And it is arduous, make no mistake about it.

Measuring only 1/500 of an inch, a sperm is the smallest cell in the human body. It has three parts: a head, a midpiece, and a tail. By lashing its tail back and forth, the average sperm can swim about five inches in an hour. But some swim in circles, some swim straight for a few inches and then veer off to the right or left. Nearly all who start the race never finish. Most die as soon as they enter the vaginal channel. Of the 250 million that may start out together, as few as five may reach their journey's end (Doak *et al.*, 1967).

If the timing is not opportune, when a sperm gets to the narrows of the cervix, it will encounter a thick plug of cervical mucus packed with hungry white blood cells. These blood cells live for the chance to gobble up intruders like these invading sperm. But if the time is auspicious, if the female's ovary has just ejected an egg cell, the mucus covering becomes more hospitable. Its white blood cell count plummets and its viscosity thins. Tiny canals form through it. A sperm must still swim between a rock (remaining white blood cells) and a hard place (the thick mucus on either side of the canals), but hundreds, perhaps thousands, manage the treacherous passage.

Negotiating the cervical passage is but one of many challenges. Still before the voyagers is the seemingly endless abyss of the uterus. Then comes a crucial decision. At the apex of the uterus the passage branches into two fallopian tubes. One houses the lady; the other, the tiger.

Wisked into one of the tunnels from its ovary by hairlike structures called *fimbria*, the egg cell drifts toward the approaching suitors at a leisurely rate of about 1/16 of an inch an hour. In about three days it will reach the uterus. If intercourse occurs just before its entry into the tube, 100, or perhaps 10, or even only one of the original 250 million may be waiting to intercept it, but the chance for a pregnancy during any individual menstrual cycle is only 25% (Wilcox *et al.*, 1988). This low probability of conception is due in part to opportune timing and in part to yet another rite of passage the favored sperm must endure.

The egg is surrounded by several protective layers. The outer corona radiata contains substances that release enzymes from the sperm's head allowing it to forge a pathway through to the next layer, the zona pellucida. Without the release of these enzymes, further entry will be denied. The zona pellucida is thick and contains receptors. To gain further entry into the egg a sperm must bind to one of these receptors. Only sperm from the same species have a chance. This is nature's way of making interspecies fertilization impossible. If binding occurs, another enzyme is released from its head to get it through the zona pellucida to the third and last barrier, the vitelline plasma layer.

If the gametes are sea urchins, the sperm's contact with this last layer opens channel pores in the membrane that normally keep sodium ions out. The entry of these ions detonates an ecstatic shock of about 80 mv, which triggers the pouring of enzymes into the vitelline layer making it impenetrable to other sperm. Small extensions in the vitelline membrane envelop the head of the winner in a fertilization cone, which then contracts and pulls the sperm into the egg leaving behind its midpiece and tail in the vitelline layer. A similar block to polyspermy does not occur in humans because there are so few sperm to compete for entry into the egg. As a result, polyspermy occurs in about 5% to 10% of all human conceptions (Edwards, 1986). Most of these triploid embryos are spontaneously aborted, although some do survive to term.

Like sleeping beauty, the egg remains in a state of limbo unless fertilized. For at least 14 years, ever since it was first formed during its sixth week of embryonic existence, the egg has been waiting to complete its second meiotic division (see Chapter 1). The sperm's princely presence within its cytoplasm wakes the egg from her slumber and she completes her cell division. One of the new daughter cells disintegrates; the other lives. About 10 hours after the sperm's entry into the egg, their chromosomes marry.

Preimplantation

Centuries ago, scientists believed sperm cells contained miniature people who were transferred to the ovum where they were nurtured until birth. We now know that nothing is preformed. (Taken to its logical extreme, *preformation* would mean that every human on earth was contained in a single ancestral sperm cell.) Instead, we all begin as a hodgepodge of chemicals and minerals blended by a formula contained in the unique collection of genes united at the moment of fertilization.

Whether fertilized or not, an egg travels down the fallopian tube toward the uterus. If fertilized along the way, it is called a *zygote*. A day later, the single-celled zygote divides to form a two-celled embryo. It continues to be called an *embryo* from the time of its first cell division to the eight week of its existence, after which it is called a *fetus*. The embryo undergoes a number of cell divisions called *cleavage*, which takes it from a relatively large, single, fertilized cell into a cluster of two cells, then four cells, etc. Compared with other species, cleavage in humans is very slow. The first cleavage doesn't occur until about 24 hours after fertilization and subsequent cleavages occur every 12 hours. By the fifth day of its existence, the embryo consists of 12 to 16 cells and is called a *morula* (the Latin word for "mulberry"); by its 10th day, it is a collection of 120 cells (Fowler & Edwards, 1973).

In some species like the sea urchin and frog, early cleavage is not controlled by the embryo but by maternal mRNA or proteins stores in the unfertilized egg (Davidson, 1976).

In mammals like the mouse, the embryonic genome begins to control early development beginning at the two-cell stage of development (Bensaude *et al.*, 1983) and drugs, which block gene transcription, can arrest embryonic development (Braude *et al.*, 1979).

Newly formed embryonic cells are called *blastomeres*. The cluster of blastomeres is the same size as the original zygote but instead of a single cell it is comprised of several cells (see Figure 4.1). Initially, these blastomeres are loosely compacted. By around the eight-cell stage, a sudden change comes over them and they squeeze against each other forming a compact ball.

Before compaction, the early collection of cells is porous and uterine fluid is able to enter it. The formation of "tight junctions" between the cells in the outer layer marks the loss of this porousness and the formation of a barrier between the developing embryo and the uterine fluids. Soon after, the cells within separate into three parts—an outer layer called the *trophoblast* (which eventually forms into the placenta), an inner layer called the *embryoblast* (which becomes the embryo), and a cavity called a *blastocele*. When these three divisions occur, the embryo is called a *blastula* or *blastocyst*. Up to the morula stage, there is no cellular growth and the size of each new cell is smaller than the ones before it. Despite considerable variation from species to species, all animals undergo this transformation from single cell to blastocyst without hardly any change in size at this early stage of development.

But other changes occur. Because the embryo is like a graft of tissue from another body, the mother's immune system would ordinarily reject it. To keep this from happening the blastocyst secretes a number of substances such as chorionic gonadotropin and early pregnancy factor that suppress her immune system (Smart *et al.*, 1981).

Another change at this time depends on the sex of the embryo. In males, the Y chromosome causes all its cells to become "imprinted" with a characteristic called an *H-Y antigen*. All genetic males have this characteristic, genetic females do not. Castration does not cause males to lose it and early injection of testosterone does not cause females to acquire it.

The presence of the H-Y antigen was discovered during skin grafting studies in the 1950s. Using inbred mice (which are genetically identical), researchers attempted to graft pieces of skin from one animal to another. Skin taken from females was not rejected when grafted to males but when skin was taken from males and grafted to females, it was rejected. Subsequent studies determined the H-Y antigen was responsible for this rejection. Although the H-Y antigen is found only in male cells, both male and female gonadal cells contain receptors for H-Y antigen. This antigen is secreted early in development and causes development of testes from undifferentiated tissue (Zenzes *et al.*, 1978).

If the embryo is female, it will have no Y chromosome but it will have two X chromosomes. Because the Y chromosome does little else but imprint cells with H-Y antigen, the burden of gene expression for the sex chromosomes is delegated to the X by default. If it gets the job done perfectly on its own, why does the female need two Xs.?

The answer is that, with the exception of ovarian development, it doesn't. In fact, for a somatic cell, two Xs is one too many. The X chromosome contains the codes for over 100 genes but too much of a good thing is sometimes a bad thing. It can create confusion. An extra number 21 chromosomes is a case in point. To protect itself from this genetic overload, the early embryo "lyonizes" one of its Xs. Lyonization, named after Mary Lyon who first hypothesized its occurrence, is the inactivation of one of the X chromosomes. Like H-Y antigen imprinting in males, lyonization occurs around the eight-cell embryo stage in females (Epstein *et al.*, 1978). The inactivated chromosome disintegrates and when

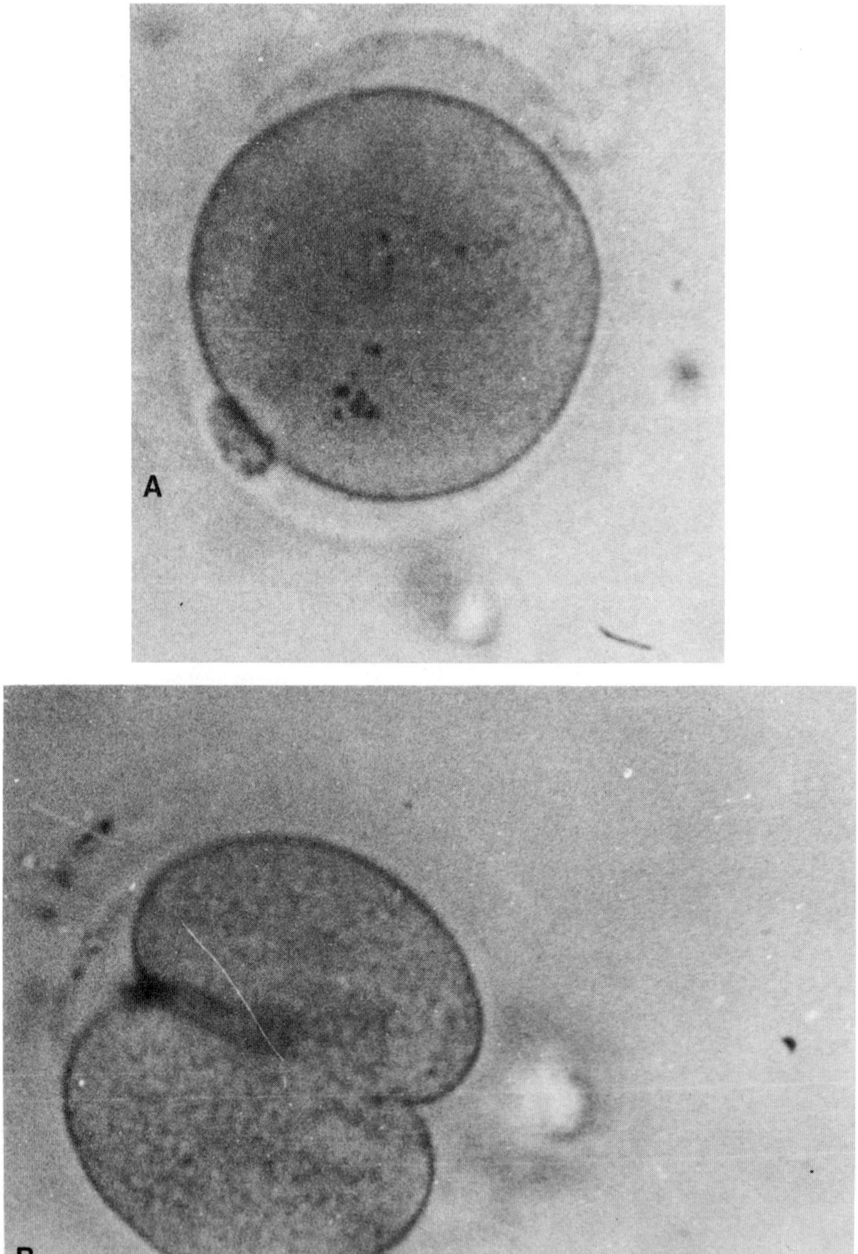

Figure 4.1. Early cleavage in newly fertilized human egg cell. A. One-celled zygote. B. First cleavage. C. Second cleavage. D. Third cleavage. Photos courtesy Dr. A. Sacco.

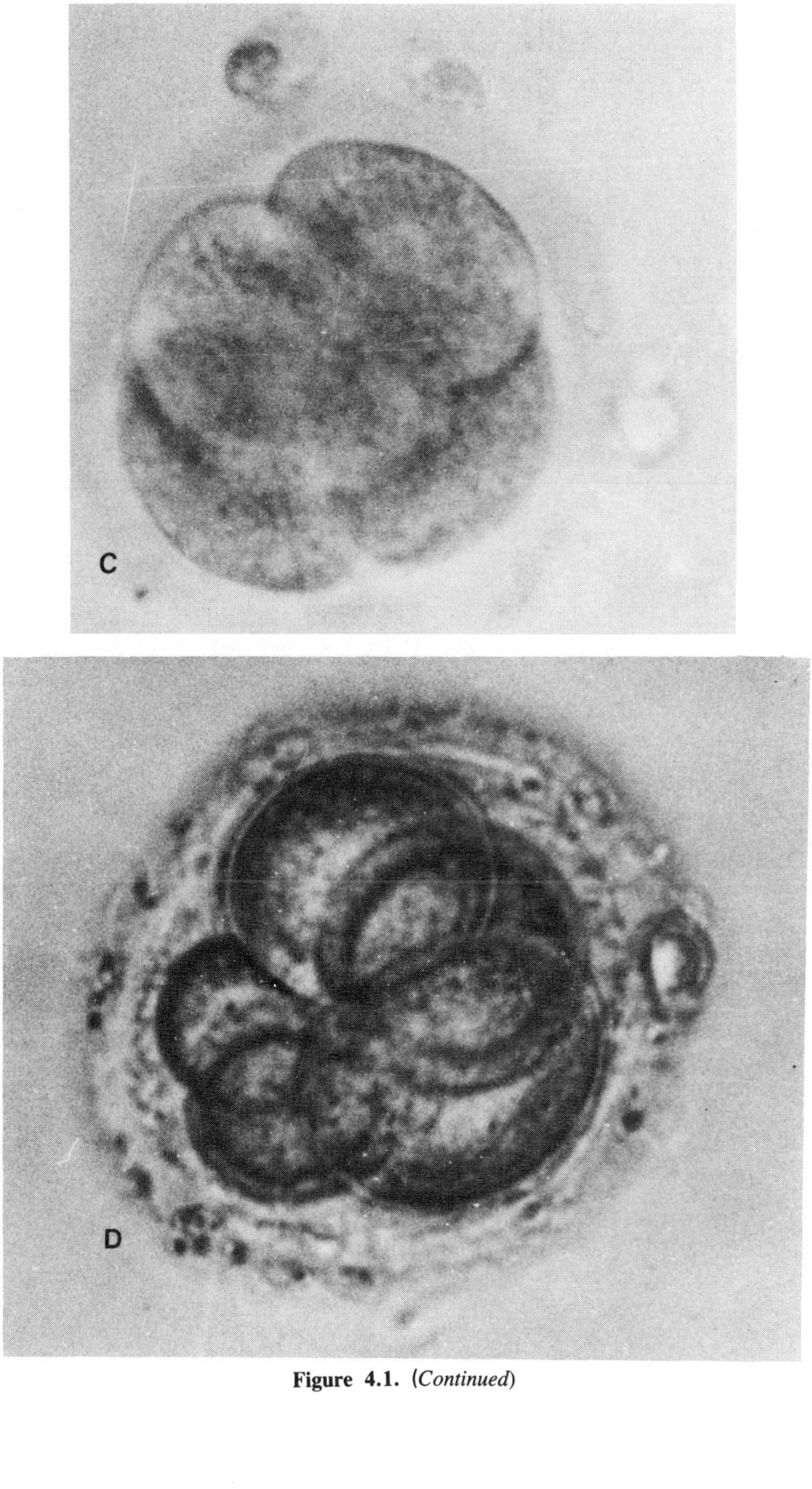

Figure 4.1. (*Continued*)

cells from a female are stained and put under a microscope, it looks like a dark clump called a *Barr body*, named after Murray Barr, the neuroanatomist who first identified it.

After the embryo passes through the fallopian tunnel, it reaches the area of the uterus where it spends about four more days floating in the uterus's fluids. At the end of this waiting period (the first week after fertilization), it implants in the uterine wall.

Right from the start, the embryo can encounter problems. About a third of all human embryos are abnormal (Shiota *et al.*, 1987). Most of these are either monosomic or trisomic (Edwards, 1986) and degenerate before implantation. Some do not cleave properly. For some, mitosis goes awry. When these or other problems occur, abnormal embryos are spontaneously aborted. About 22% of all pregnancies are aborted by the time of implantation (Wilcox *et al.*, 1988). In nearly all of these cases, a woman does not even know she is pregnant. After implantation, another 30% of all pregnancies are spontaneously aborted (Wilcox *et al.*, 1988). Depending on when these abortions occur, a woman may or may not become pregnant.

Implantation

During the second week of life, implantation occurs if the uterus has been properly prepared and the embryo is still viable.

Implantation involves three important events. First, the trophoblast (placenta-forming) cells attach themselves to the uterine wall. About three days later, these cells start invading the walls of the uterus trying to get closer to the important maternal blood vessels that will connect them to the mother's blood. Simultaneously, the embryo burrows itself further into the uterine wall. By the 12th day of its existence, it is completely embedded in the uterus. All mammalian embryos undergo these events leading up to and including implantation. A comparison of the timing of these events up to birth in different species is shown in Table 4.1.

For implantation to occur, the uterus must be prepared to accept the embryo. Such preparation depends on synchronized priming by estrogens and progesterone from the ovary. If the timing in the release of these hormones is improper, the uterus will be a less hospitable host. Impaired blood flow to the uterus will also decrease the embryo's ability to

Table 4.1. Timing of Events from Fertilization to Implantation in Various Species

	Mouse	Rat	Rabbit	Monkey	Human
Fertilized egg appears in Fallopian tube (days)	½	1	¼	1	1
Two-cell stage	1	2	½	2	2
Four-cell stage	2¼	3	1	3	2½
Eight-twelve cell stage	2½	3¼	2	4	3
Blastocyst in uterus	4	5	4	7½	5
Implantation begins	4½	6	7	9	6
Organogenesis					
Start	7.5	9	7	20	20
End	16	17	20	45	55
Birth	20–21	21–22	30–32	164–170	260–280

implant; poor utero-placental perfusion after implantation will cut off the embryo's food supply and decrease growth or cause early embryonic death.

Usually, human embryos implant on the posterior or anterior wall of the uterus. Occasionally, an errant embryo will implant far down in the uterus near the cervix. As the placenta grows it covers this opening, a condition called *placenta previa*.

The embryo can also implant outside the uterus. This condition is called an *ectopic pregnancy*. Some ectopic pregnancies occur when the embryo implants in the fallopian tube. In such cases the embryo dies by about the second month of pregnancy.

Prior to implantation, the embryo draws upon nutrients within it and from those in the fluids of the fallopian tube and uterus. Other substances may also be present in these fluids, but there is little evidence as yet that these substances influence development of the embryo. After implantation, the embryo draws upon nutrients from the maternal circulation and from the amniotic fluid in which it is suspended. Most of the amniotic fluid comes from maternal blood. The fetus also contributes to the volume of fluid by urinating into it. In the course of its existence in the womb, the fetus swallows as much as 400 ml of this fluid a day. Whatever is in this fluid will enter the fetus's body.

Gastrulation

While implantation is occurring, a number of cell migrations take place within the embryo called *gastrulation*. Gastrulation rearranges the undifferentiated blastomeres into a three-layered embryo. In some animals, this rearrangement is associated with an unequal distribution of cytoplasm within the egg at the time of fertilization. In the amphibian zygote, for instance, the entry of a sperm repels a portion of the egg's cytoplasm to the opposite site of the cell and this heavier cytoplasmic region is called the *vegetal pole*; the sperm's point of entry, which now has less of this cytoplasm, is called the *animal pole*. Because of the difference in cytoplasm, one of the two sides of the cell has polarity. The vegetal pole will eventually become the dorsal side of the embryo; the animal pole, the ventral side.

Gastrulation begins when cells from the dorsal area of the blastula sink inward. These cells then spread further inward toward the animal pole pulling other cells behind them. This sequence of invagination and movement of cells into the interior is responsible for the three-layered embryo. The upper layer is called the *ectoderm*. This is the layer from which most of the nervous system will develop. The lower layer is called the *endoderm*. Cells from this layer will develop into the lining covering the stomach, intestines, lungs, liver, and other organs. The middle layer is called the *mesoderm*. From it will come muscle, the heart, kidneys, genitals, and blood vessels.

Although it seems reasonable to assume development involves some complex interplay between operonlike gene mechanisms, there are simply not enough genes in the genome to account for the complexity of the development. For instance, the human brain contains about 15 billion cells. The organization of these 15 billion cells into a brain requires body planning (called *pattern formation*), which at minimum involves the genesis, multiplication, adhesion, migration, and differentiation of nerve cells, including their axonal growth, formation of synapses, death, and rearrangement of axonal connections (see Table 4.2).

An operon's "one-gene, one-enzyme" mechanism can explain a wide variety of cellular events like blue eyes or brown hair but there are simply too many individual changes during development for each to have its own operon. Take synaptic organization, for in-

Table 4.2. Major Stages in Brain Development

Stage	Characteristics
Induction of neural plate	Commitment of ectodermal tissue to formation of neural tube and nerve cells
Proliferation of nerve cells	High rate of nerve cell formation in different areas of neural tube
Migration of nerve cells	Movement of nerve cells from birthplaces to other regions of the developing brain
Aggregation	Nerve cell aggregation into identifiable regions
Differentiation	Development of neurons into cells with characteristic shapes and functions
Synaptogenesis	Growth of axons and dendrites to form connections between neurons
Nerve cell death	Selective demise of certain neurons
Modification of connections	Reorganization of original connections, stabilization and formation of new connections

stance. There simply isn't enough room in the genome for each of the thousands of billions of synapses in the human brain to have its own operon.

The genetic solution to this impasse is to establish a hierarchy of control akin to federal, state, and local government. The federal authority sets the guidelines for the whole organism including the early stages of development like gastrulation, the state authority implements these guidelines for different organs after gastrulation, and the local authority works through a series of basic stages like those outlined in Table 4.2 to complete the fine points of development. These stages are not mutually exclusive. Nor is there any specific moment when one stage begins or ends, but they do occur in a particular sequence and are irreversible; each transition is decisive.

Induction

The first stage is induction. A dramatic example of this local mechanism is the classic experiment with newts in which Hans Spemann transplanted mesoderm tissue from the dorsal side of an early gastrula to the ventral side of another. The transplanted tissue caused the host tissue to develop another nervous system on its ventral side (Spemann, 1938). Subsequent experiments showed that if ectoderm tissue is isolated and placed in a culture medium where it can grow, it will produce only epidermal tissue. However, if mesoderm tissue is put into the medium as well, the ectoderm forms a neural structure. These experiments and many others like them, indicate that the ectoderm's "ground state" is epidermal tissue but during gastrulation, the mesoderm layer induces ectoderm layer to form a nervous system. But how?

Despite a great deal of intensive searching, the inducing factor has not as yet been identified. It appears to be chemical in nature because a nonporous barrier like cellophane placed between inducing and responding tissue prevents induction. If the barrier is porous, induction occurs.

Induction implies that at an early stage of development, cells are unspecialized and indifferent as to their futures. Their fates are not yet sealed; they can become any type of cell. This capability is called *competence* or *totipotentiality.*

Competence is ephemeral. The first stage narrowing a cell's potentiality or fate is called *primary induction.* This is the period when the inducer causes responding tissue to lose its freedom of development, thereby committing it to a particular fate. Spemann's demonstration of nerve tissue formation in response to transplanted mesoderm cells is an example of this stage of induction.

Secondary induction occurs when inducing tissue acts on responding tissue that is already committed to some degree, causing it to complete its differentiation. An example comes from another study in which Spemann removed tissues from the part of the eye that develops into the lens as a result of contact with facial epidermis from the belly. The lens still formed, reflecting a general influence from epidermal tissue. The second major phase involving acquisition of form and function, comes later (see below).

The amount of time involved in primary and secondary induction is also different. For tissue like amphibian muscle, about 1 to 2 hours of inducer contact is needed before a response will occur, the extent of induction is greater if contact lasts up to 5 hours (Gurdon *et al.*, 1985). For secondary induction, contact of only an hour or less may be required (Smith, 1987).

The response to the inducer does not occur in all the competent cells capable of responding but is restricted to only a few cells located in close proximity to the inducing tissue. By gradually reducing the size of the responding tissue, embryologists discovered a critical mass of cells is needed for inducton to occur. Whereas single cells from a blastula will differentiate in accord with their surrounding cells, differentiation is facilitated by being with cells of one's own kind, a "communitylike" effect.

Critical Period

Totipotentiality is a transient property of neurons. It occurs only during the "critical period" of embryogenesis before differentiation. In the study in which Spemann transplanted dorsal mesodermal cells from one embryo to the ventral area of another, the host cells developed a second nervous system only if the transplant occurred when the host tissue was in its gastrula phase.

Totipotentiality is completely dependent on the developmental stage of the responding not the inducing tissue. An interesting example of critical periods for totipotentiality involves transplantation of embryological neural tissue into adult brain sites. For such transplantation to be successful, the transplanted tissue must survive, grow, differentiate, and become integrated in the host's brain. The success of transplantation varies with the age of embryological tissue because it contains more or fewer differentiated cells. Transplants taken from gestation day 15-, 16-, or 17-day old embryos have a better survival rate than those taken from older embryos (Das, 1983). Adult nueral or neonatal tissue cannot be successfully transplanted because it has already undergone differentiation. If transplanted, such tissue rapidly degenerates (Das, 1983).

The concept of critical periods in development is fundamental to teratology and accounts for the fact that the same amount of a drug like thalidomide can cause missing arms in some newborns, missing legs in others, or no damage at all. If exposure occurs during the critical period in pregnancy when an arm or leg is being formed, thalidomide will interfere with the development of that arm or leg. If drug exposure occurs before or after the critical period for arm or leg formation, no damage to these limbs will occur.

Brief exposure to a teratogen is also unlikely to disrupt development. Because primary induction takes hours, exposure to a teratogen would have to be continuous and in an amount great enough to be present at the induction site.

Different regions of the brain have different critical periods. The auditory system and the cortex are continuously organizing throughout development; therefore they have a prolonged critical period and can be affected at any time. Some areas of the hippocampus rapidly undergo organization at a specific time and then stop. Such regions have only a single critical period. Other areas of the hippocampus organize slowly at first, then become very active, and then slow down again. These cells have an extended critical period but are most vulnerable during the period of most rapid growth.

The Beginning of a Brain

When the human brain is finally developed, its 1,400 cubic centimeters contain about 15 billion nerve cells. Not all of these 15 billion cells are identical. Instead, there are several hundred different types of cells, each with distinctive morphological, physiological, and biochemical characteristics. Although there are as many cells in the digestive system and as many different kinds of cells in the immune system as in the brain, no organ in the body has as complex a pattern of cell connections as the brain.

The period in which organs are formed is called *organogenesis*. Organogenesis of different bodily structures like the brain requires exact organization of different cells into specific compartments. Although no organs of the body resemble any other, they all rely on the same mechanisms of genesis, proliferation, migration, differentiation, and cell death for their final form and function. Different organs rely on differences in location, extent, and timing of these various events. Abnormalities in development arise when one or more of these processes fails.

Brain development begins relatively early compared with other organs and is over before many other organs have formed. Nearly all the main parts of the brain are in place within the first few weeks after fertilization. If anything untoward is going to happen to the structure of the brain it will most likely happen during this time.

The major stages in brain formation were outlined in Table 4.2. Brain development begins with induction. The first phase is called *neurulation* and is marked by the appearance of a structure in the dorsal surface of the mesoderm called the *notocord*. The presence of the notocord is a primary inducer, causing cells immediately above it in the ectoderm to form a *neural groove*. This groove gradually widens and cells on either side thicken and form a flat layer of cells called the *neural plate*.

The edges of the neural plate become very distinct and bend upward whereas cells in the middle invaginate. The ends of the plate come closer to each other and eventually fuse to form a *neural tube* underneath the overlying ectoderm (see Figure 4.2). This neural tube will develop into the central nervous system, whereas neural crest cells situated between the neural tube and the overlying ectoderm will develop into the peripheral nervous system, Schwann cells, which insulate peripheral neurons, and the adrenals. Another area in the neural crest located toward the anterior of the neural tube thickens into a cluster of cells called *placodes*, which develop into sense organs in the head.

While the neural tube is closing, segmented blocks called *somites* appear on either

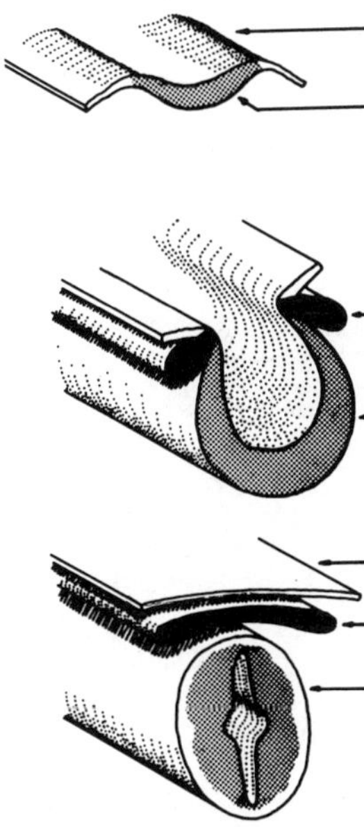

Figure 4.2. Schematic depiction of formation of neural tube. The neural plate first appears as a thickened sheath in the ectoderm. The plate invaginates and a groove forms. The edges of this groove move together to form a tube.

side of it in the mesoderm layer. The somites eventually develop into vertebrae, connective tissue, and muscles. As each area in the ectoderm and mesoderm becomes committed to a particular type of tissue, sequential inductions bring about the eventual regionalization of the brain and spinal cord.

Meanwhile, the mesoderm continues to affect neurulation. Neurulation does not occur simultaneously throughout the ectoderm. Instead, the three stages of neural groove, plate, and tube can all be present in different areas of the ectoderm. The middle part progresses from plate to tube first, followed by the cephalic head and caudal tail areas (see Figure 4.3).

The two ends of the neural tube differ in appearance. In the cephalic area where the brain will develop, the tube is wide and swellings appear that will become the forebrain vesicle (prosencephalon), midbrain vesicle (mesencephalon), and hindbrain vesicle (rhombencephalon). These vesicles eventually develop into the cerebral hemispheres, midbrain, and hindbrain, respectively. The other end of the tube is narrow and tapering. This is where the spine will develop. The two open ends of the neural tube corresponding to the head and spine are called the *anterior* and *posterior neuropores*, respectively. These neuropores allow fluid to pass through the neural tube but eventually close off.

If the posterior neuropore fails to close, a condition called *spina bifida* will develop. If the anterior neuropores fails to close, anencephaly will occur. These malformations are discussed more fully in Appendix A. Basic functional neuroanatomy is described in Appendix B.

Proliferation

When development is through, the human brain will contain billions of nerve cells. There are as many neurons in the human brain as there are the stars in our galaxy. Some

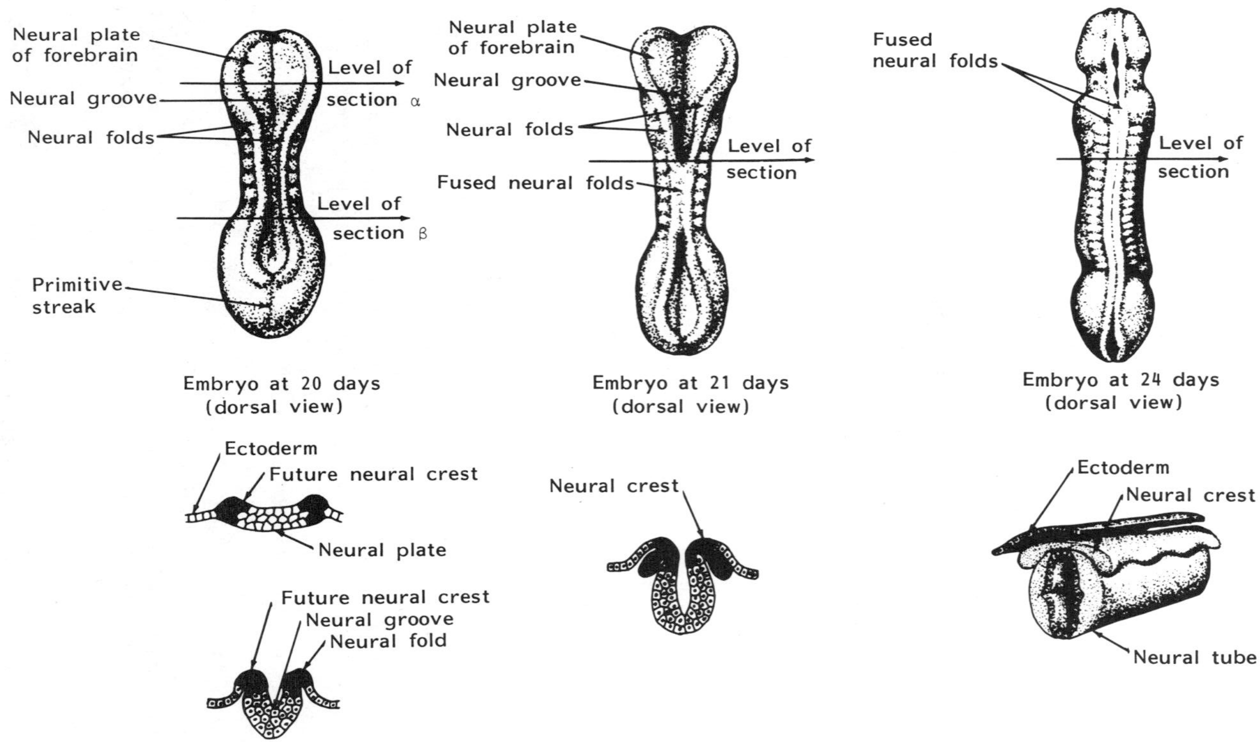

Figure 4.3. Origin of various cells from neural tube and neural crest.

texts put the number of neurons and stars at 10 to 15 billion. Nobody really knows. Let's be generous and give ourselves and our galaxy the 15 billion. To create a brain with this many cells, a single cell has to double itself only 34 times. Because all our brain cells come into existence during the first twelve weeks of life, the developing brain must produce about 150,000 nerve cells . . . a minute! (Purves & Lichtman, 1985).

This explosive proliferation of nerve cells begins as soon as the neural tube closes. Initially, the wall of the neural tube contains only a single germinal neuroepithelial layer of cells. This layer undergoes a sudden change from a single layer to a multicellular layer of rapidly dividing neurons. During nerve cell division, a cell in the neuroepithelium (also called *ventricular*) zone loses its processes and its nucleus migrates to the inner surface of the epithelium of the neural tube. It then replicates its DNA, develops new processes, and moves back to the ventricular area. After a number of such replications, a neuron loses its ability to synthesize DNA (a developmental landmark called its "birthday") and it migrates past the ventricular zone to the ventricular surface layer, where it and others form a new layer called the *intermediate zone* or the *mantle*. This layer contains young neurons (called *neuroblasts*) that will never divide again and precursor glial cells with no such limitations. Neuroblasts form connections with one another in the mantle and send out axons away from the neural tube creating a marginal zone.

These three zones characterize early cell proliferation throughout the brian. Glial cells subsequently cover these axons with myelin. Because of the whitish appearance of these myelinated axons, the marginal layer is called the *white matter*, whereas the neuroblasts in the mantle are called the *gray matter*.

The number of cells that will occupy a particular area of the brain depends on how long its proliferation period lasts. In some areas, the proliferative period lasts for several weeks, in others, only a few days. A second factor is the length of the cell cycle in the proliferating zone. This can be as short as a few hours in the early days of proliferation or as long as a few days in older embryos. Yet another factor is the size of the precursor pool. The smaller its overall size, the fewer the cells that can arise from it.

Neurons in different areas of the neural tube have their birthdays at different times. Typically, large neurons whose axons project over long areas like those in the visual system are "born" earlier than smaller neurons whose processes are more modest in length, but there are exceptions. Cells with the same birthdays also eventually settle in the same area of the brain after their migrations (see below). If you know a cell's birthday, you can predict its subsequent "address." For instance, a cell whose residence is in the deeper layers of the cortex will have an earlier birthday than a cell whose address is in a more superficial layer.

If proliferation is compromised, areas of the brain will be diminished or missing. Such losses can be general or highly localized. For instance, decreased proliferation could result in microcephaly, which results in a small brain, or a normal-sized brain but one with decreased cells in a particular layer of the cortex.

Agents that block DNA synthesis were among the first antiproliferative agents to be identified as behavioral teratogens (Rodier, 1986). Some of them such as colchicine, vinblastine, and urethane arrest mitosis by inhibiting cell spindle formation or function. These substances do not attack DNA directly but rather interfere with its expression after it has been synthesized.

As noted in Chapter 1, colchicine or its derivative colcemid are routinely used by cytogenetists to stop mitosis at metaphase. When colchicine is administered to pregnant animals, it produces a high incidence of spontaneous abortions if exposure occurs early in pregnancy; later in pregnancy, it produces structural abnormalities in the brains of offspring and behavioral anomalies such as increased fearfulness and hypoactivity (Petit & Isaacson, 1976). During the latter half of gestation, colchicine also produces intrauterine death by greatly increasing uterine contractility to the point that fetuses are crushed (Theirsch, 1958).

Another antiproliferative agent with a different mechanism of action is methylazoxymethanol (MAM). MAM causes damage to DNA by selectively methylating guanine bases. This results in denaturation of DNA. In animals, prenatal exposure to MAM results in decreased thickness and disorganization of cells in the cerebral cortex, corpus collosum, and hippocampus, fewer and shorter dendrites, and decreased spine density on surviving neurons (Dambska *et al.*, 1982). The behavioral consequences of such damage include hyperactivity (Rabe & Haddad, 1972) and impaired maze learning (Haddad *et al.*, 1969).

Cell Migration

Whereas cells in the medulla retain their early three-zone pattern of cell proliferation, cells in the higher regions of the brain leave their birthplace in the mantle of the neural tube's outer surface, detach themselves, migrate through the white matter, and aggregate in a new zone called the *neopallial cortex* or *cortical plate*, which eventually gives rise to the six-layered cortex and other cortical structures.

How do these cells know where to go?

Position and directionally are provided by pathways formed by radial glial cells. Generations of cells born in the same area of the proliferating zones migrate along the same radial pathway and come to rest in the same area in a column of cells. This migration is facilitated by a glycoprotein called *neural cell adhesion molecule* (NCAM), which is produced by a single gene (Edelman, 1984). Where neurons finally come to rest is very significant because function depends on connections between neurons. Therefore, factors affecting the migration of cells are of particular interest to neurobiologists and teratologists.

Neurons in certain areas of the brain such as the cerebral cortex and the cerebellum are arranged in layers. Those within a given layer share basic morphological features and connections. The cerebral cortex has six different layers of cells arranged in an inside-to-outside pattern with the deepest layers forming first (Rakic, 1988). Cells in the dentate gyrus of the hippocampus are arranged in an outside-to-inside sequence (Schlessinger *et al.*, 1975).

The final organization of the brain is the product of myriads of cellular events culminating in an exact spatial arrangement and connection of nerve cells. Because different areas of the brain undergo nerve cell proliferation at different time, teratogens may interfere with the formation or organization of one particular area of the brain and not another, just as they can cause damage to a leg and not an arm depending on the time of exposure.

Disorders of neuronal migration are the most common of the cerebral malformations. *Lissencephaly* occurs as a result of decreased neuronal migration around the 4th month of gestation; instead of a six-layer cortex only four layers are formed. As a result, very few sulci develop and the brain has a smooth appearance. *Heterotopias* are nodules of gray matter that bulge into the ventricles. These disorders are due to localized defects in neuronal migration. *Polymicrogyria* is an abnormal formation of many small gyri in the cortex due to abnormal migration of cells to one area of the cortex. It is often associated with certain kinds of infection during gestation.

Although mercury has many direct and indirect effects on the developing nervous system (Chang & Annau, 1984), one of its major effects is delayed migration of neurons, especially of cells destined for the cerebellum (Reuhl *et al.*, 1981). The most dramatic examples of prenatal mercury toxicity are the methylmercury poisonings that occurred in Minimata, Japan, and in Iraq in the 1960s and 1970s respectively. The Minimata tragedy occurred after pregnant Japanese woman consumed fish contaminated by mercury (methyl mercury in fish can reach levels of 20 mg/kg). In Iraq, the poisoning occurred as a result of eating home-made bread from seeds treated with a fungicide containing mercury.

Brains of infants who died as a result of prenatal mercury exposure were characterized by decreased thickness of the corpus callosum and cerebellum, narrowing of gyri, disor-

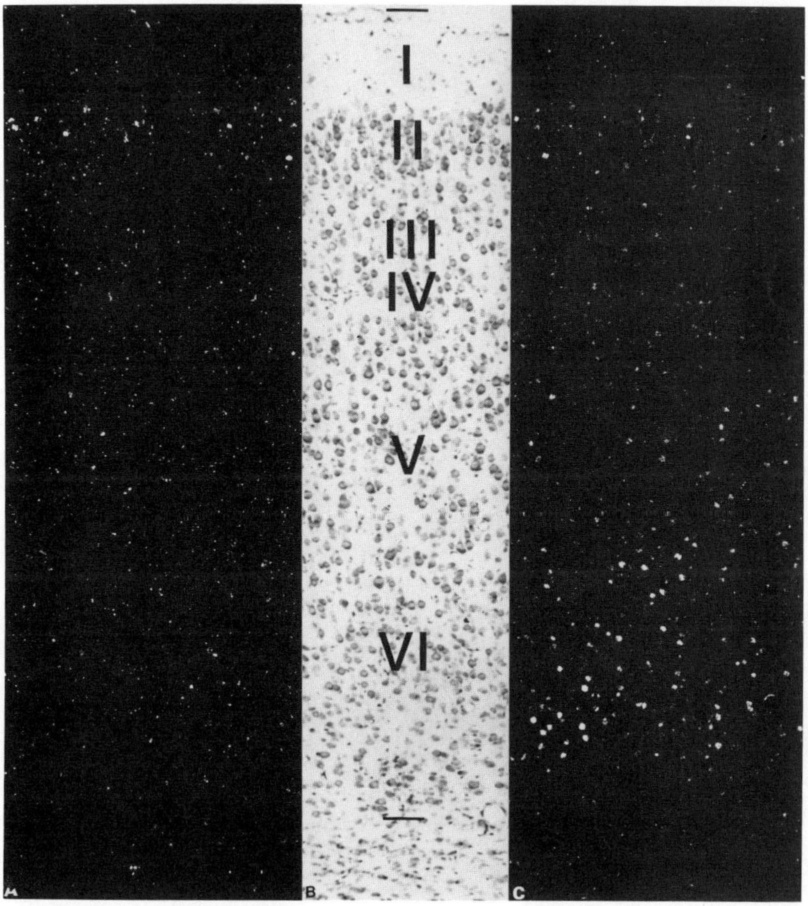

Figure 4.4. Effects of prenatal alcohol exposure on nerve cell migration in the motor cortex of 3-month-old rats. Figures A and B represent a pair-fed control under dark and light field light microscopy. Figure C is an animal exposed to alcohol on gestation days 6 to parturition. Note that layer II is more densely packed in control animals whereas layer VI is more densely packed in alcohol-exposed animals. This indicates that migration of cells in the cortex has been inhibited. (From Miller, 1986; photo courtesy Dr. M. Miller.)

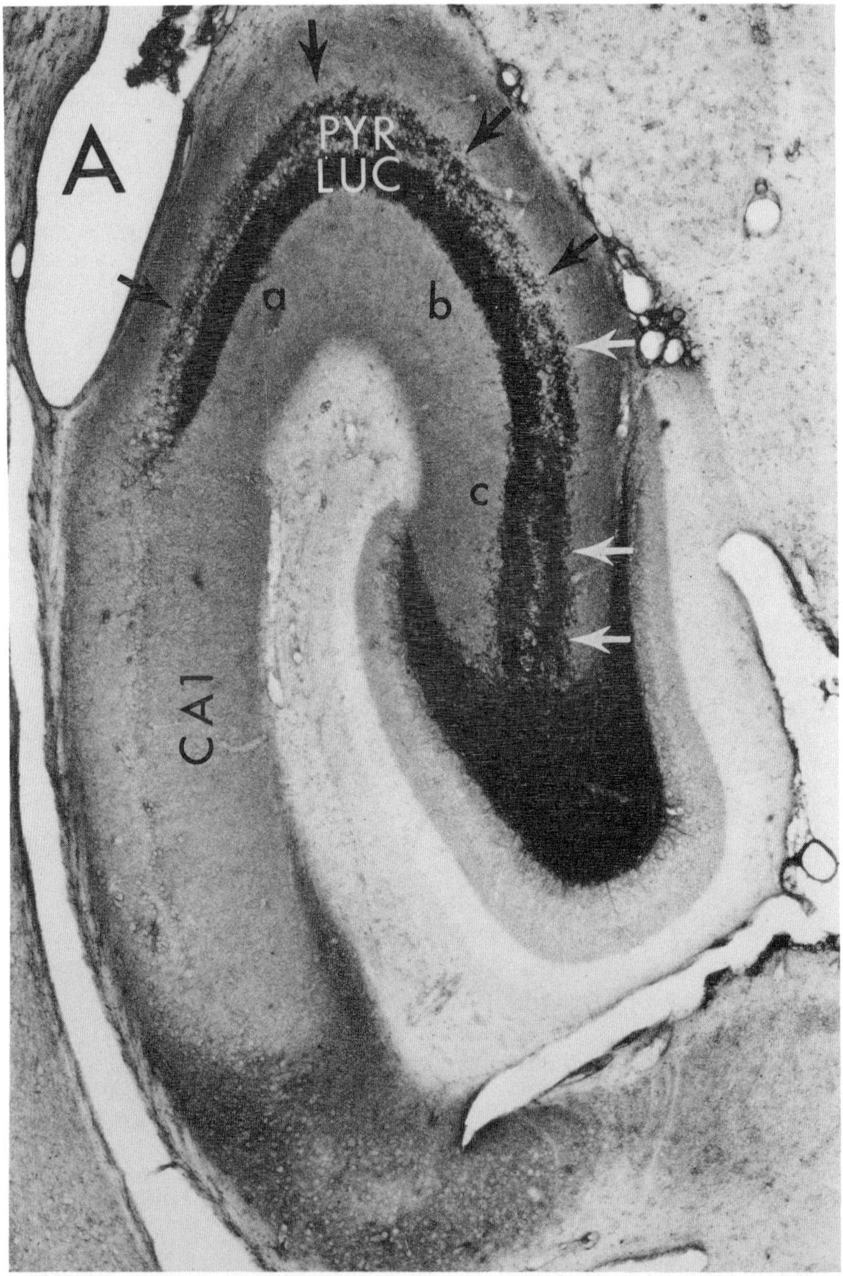

Figure 4.5. A. Section from an adult rat prenatally exposed to 35% ethanol-derived calories during days 1–21 of gestation. B. A section similar to that in A, but of an adult rat from a pair-fed control dam. White arrows indicate normally occurring mossy fibers. Black arrows indicate abnormal darkly

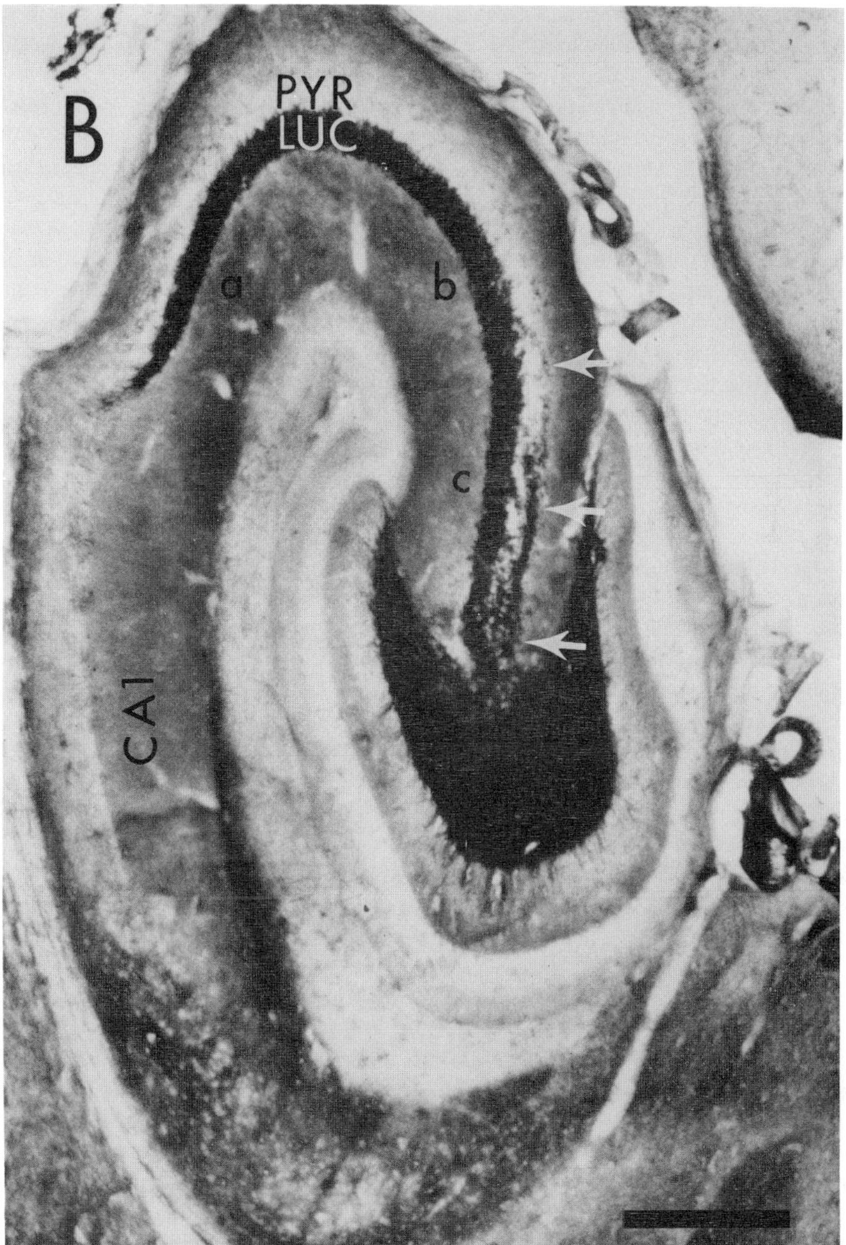

stained fibers. PYR = pyramidal cell layer; LUC = stratum lucidum; CA1 = hippocampal field CA1; a,b,c = subfields of hippocampal field CA3. (From West *et al.*, 1981; photo courtesy of J. West.)

dered cerebral cortical layers and irregular neuronal alignment (Choi, 1983). This kind of damage is different from the localized destructive lesions that occur in adults poisoned by mercury. Less severe mercury poisoning during gestation results in mental retardation, speech disturbances, and cerebral palsy (Amin-Zaki *et al.*, 1976; Snyder, 1971). In some instances, infants appear normal at birth and later develop psychomotor impairments (Amin-Zaki *et al.*, 1976). Animals prenatally exposed to mercury likewise have impaired coordination (Spyker & Smithberg, 1972) and learning performance (Eccles & Annau, 1982).

Alcohol also has many direct and indirect effects on the developing nervous system. Like mercury, alcohol causes delayed and aberrant migration of neurons. Fetal and neonatal brains from children with fetal alcohol syndrome who died shortly before or after birth are characterized by lissencephaly and disturbances in cortical laminar organization (Clarren *et al.*, 1978) not seen in neuropathies in adult alcoholics (Freund, 1983). In rats prenatally exposed to alcohol, neurons that normally appear in the motor cortex by gestation day 13 do not appear until gestation day 14. In normal rats, histogenesis is complete by gestation day 13, whereas in alcohol-exposed fetuses it is not completed until gestation day 23 (Miller, 1986). Alcohol exposure also interferes with the normal inside-to-outside pattern. Instead of the deepest layers forming first, many neurons migrate to the outer layers first (see Figure 4.4).

Alcohol-induced aberrations in neuronal cell migration are reflected not only in changes in the distribution of cells in different layers of the cortex but in changes in the circuitry of mossy fiber nerve cell tracts in the hippocampus (see Figure 4.5). Exposure to high levels of thyroxine during the critical period when the intrinsic circuitry of the hippocampus is forming produces a similar effect on mossy fiber tract formation.

Although some agents such as alcohol affect migration of nerve cells and tract formation, the mechanism(s) underlying these aberrations are not necessarily the same. There is currently no unequivocal explanation for alcohol's effects on aberrant hippocampal circuitry. It may be that alcohol produces this effect by altering neuronal cell migration,but it is also possible that rearrangement may be due to other causes such as altered rates of cell death or loss of synaptic connections (see below).

Pattern Formation

Body planning is an essential feature of development. Unless the various organs of the body are always positioned exactly alike in every member of a species, no member would have any resemblance to any other. The spatial transformation of cells into a functionally organized unit is called *pattern formation.* In humans, this organization occurs between the 2nd and 9th week of gestation. What an animal or human eventually looks like and what it will be able to do depends to a much greater extent on pattern formation, that is, how cells are spatially arranged, than on the different kinds of cells in its body (humans have about 200 different types of cells). Fingers projecting from the palm are a lot more useful than fingers projecting from the elbow.

Cells first acquire their positions; after that they acquire their unique characteristics. Because position is a prerequisite for differentiation, a cell must know its geographical location in the body. It has to know the boundaries of its territory relative to other cells, its orientation or polarity so that it recognizes the dimensions of anterior–posterior, dorsal–ventral, and proximal–distal, and it has to have some quantitative way of measuring that position (Wolpert, 1969). One way this coordinating system is achieved is through the embryonic field.

An *embryonic field* is a collection of no more than 100 cells taking up about one millimeter of embryonic tissue. It takes several hours for an embryonic field to become established but once formed, cells derived from it develop into a whole structure. Embryonic fields were discovered by experiments in which embryologists removed groups of cells at various times during development and observed the resulting course of development. Complete removal of the neural plate during embryonic development prevented development of the nervous system; partial removal caused the remaining parts to take over the function of the missing parts and the whole nervous system formed.

One hypothesis accounting for the embryonic fields is that they are formed through the diffusion of a chemical ''morphogen'' from an originating cell to surrounding cells. If there is a fixed amount of morphogen, a concentration gradient could provide position information. Cells closer to the source would be exposed to a higher concentration than those farther away. Above a certain threshold, cells in the field would develop one way, below it, another.

One such morphogen is retinoic acid, a derivative of retinol (Vitamin A). Gradients of retinoic acid have been implicated in the formation of the chick wing along with a morphogen receptor with high affinity for retinoic acid (Petkovich *et al.*, 1987). The retinoic acid receptor also resembles receptors for steroid and thyroid hormones. Apparently, morphogenes diffuse through cell membranes to receptors within the cell. This enables them to enter the nucleus, where they bind with regulatory genes to affect subsequent pattern formation, the mechanism for which is still unknown.

Cell lineage studies, using a variety of techniques such as marking primordial cells with dyes, have shown that during pattern formation cells remain with restricted boundaries and their further development takes place in these compartments. The fruit fly has five compartmental boundaries. One boundary separates anterior and posterior compartments, another separates dorsal and ventral compartments, etc. Cells destined for one compartment do not cross into other compartments. How these boundaries are created is still unknown, some kind of cell recognition mechanism is probably at work.

Bizarre mutations that transform structure characteristics of one part of the body into those appropriate for other parts of the body indicate that such cell recognition is genetically controlled. In insects like *Drosophila* whose bodies are divided into segments, mutation called *opthalmoptera* causes wings instead of eyes to sprout from the head. Another mutation called *antennapedia* causes a leg to form in place of an antenna. Yet another, called *engrailed*, is less dramatic but far more interesting because it affects the posterior areas of wings, legs, and the eye antenna, making them develop into anterior parts. Because a mutation in the engrailed gene affects more than one structure, it is possible that the engrailed gene codes for a general property like ''posterior position.''

Such mutations indicate that certain genes control temporal and spatial pattern formation. Presumably, there genes initiate, monitor, direct, and stop development and are arranged in a hierarchical network. If so, a gene like engrailed might operate in an on–off mode at various decision points. In its ''on'' mode, it might signal ''posterior segment'' development in all posterior segments and would be inactive in all anterior segments. If, however, it became mutated, it could also signal ''posterior segment'' in regions otherwise destined to be anterior (Garcia-Bellido *et al.*, 1979).

Collections of genes like engrailed genes are typically part of families of about 180 genes called *homeoboxes*. A homeobox appears to be a set of regulatory and structural genes that instruct specific segmental units to unify into a specific body part. Homeoboxes are highly conserved. This means that the same sequence of 180 genes that controls segmentation in *Drosophila* is also present in mice and in many other species including humans (See, for example, Levine *et al.*, 1984).

Two such homeoboxes located about 20 kilobases apart on mouse chromosome 6 control development of the central nervous system (Toth *et al.*, 1987). If one of the genes in these homeoboxes undergoes a mutation, the homeobox that directs formation of one kind of segment may transform into another segment. Some malformations may thus be due to mutations in single genes homeoboxes. In many cases, like the engrailed gene, the same gene operates in more than one homeobox to signal a general pattern of development. A mutation in such single genes could thus result in multiple malformations.

Developmental pathways taken by various parts of the body may depend on the activity of specific sets of homeoboxes and those homeoboxes that contain specific regulatory genes involved in pattern formation. However, as development proceeds, the area of the embryonic field becomes much smaller relative to the whole and with further development regulatory homeobox genes "delegate" increasing influence to local events.

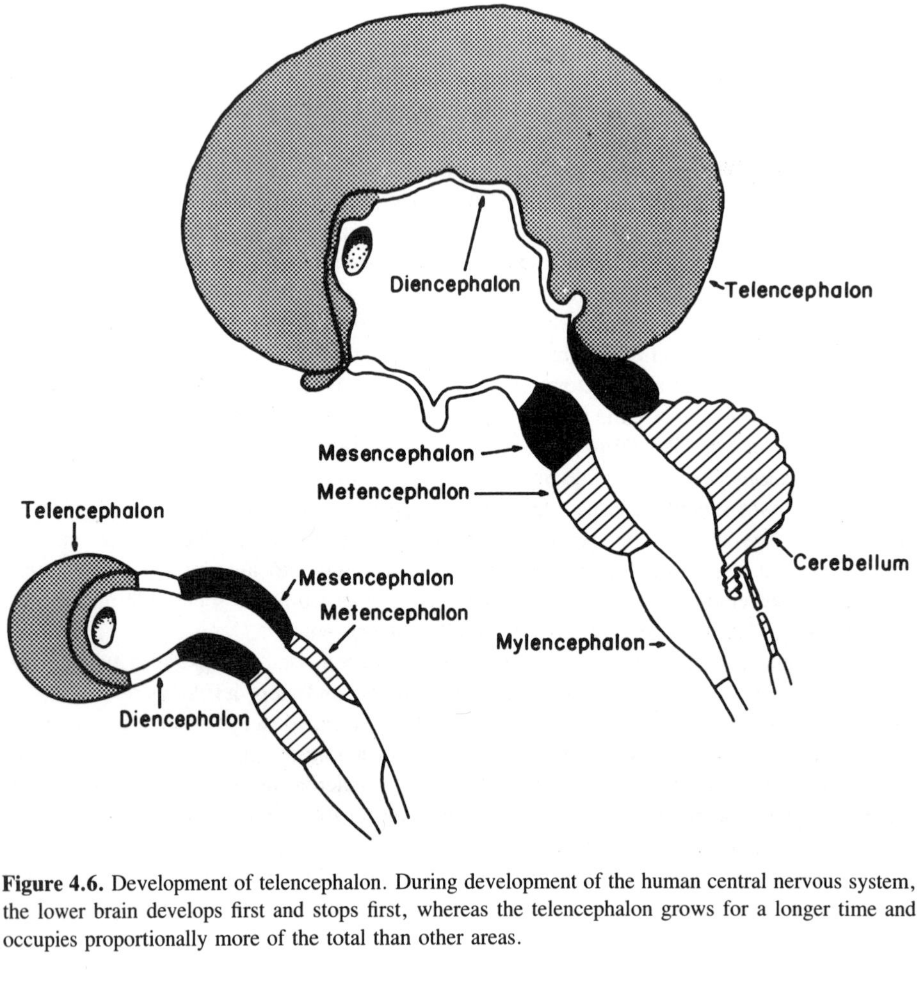

Figure 4.6. Development of telencephalon. During development of the human central nervous system, the lower brain develops first and stops first, whereas the telencephalon grows for a longer time and occupies proportionally more of the total than other areas.

Differentiation

All of the different cells of the central nervous system and their connections are formed from identical precursor cells in the neural tube. Newly formed neurons are relatively small and densely packed. They develop their characteristic shapes and functions after their arrival in a particular part of the brain. As in real estate, the three most important features that sell a house or characterize a neuron's shape and function are location, location, location.

The second growth spurt in the brain begins about the final 2 months before birth. Because all the neurons have been born, this growth period can only be due to the sprouting of axons and dendrites, the proliferation of glial cells, and the myelination of new axons. The migration of cells to the human brainstem is over by about the end of the second month after birth but cells continue to flood into the cerebral cortex until the 5th month after birth causing it to increase in size relative to other areas (see Figure 4.6). Final maturation is not over until about the second year of life.

Relationship of Animal Brain Development to Human Brain Development

The timing of the brain growth spurt is different in every species. In humans, the growth spurt occurs around birth. The growth spurt of the rat, one of the most commonly used animal models in the behavioral sciences, occurs during the first week after birth, whereas the monkey experiences its growth spurt about two weeks prior to birth (Dobbing & Sands, 1979). Although the brain spurt period is often regarded as a period of increased vulnerability to insult, the brain growth spurt represents an overall surge in growth and does not represent the growth of individual regions of the brain.

Comparisons of the timing of the brain growth spurt in different species are also based on arbitrary adjustments for life span and choice of intervals of time (Dobbing & Sands, 1979); there is no way of knowing if a day in the life of a rat is equivalent to a month in the life of a human. Even if development were equal at one period of life, it does not mean that it is equivalent at another. The main value of charting brain growth is to describe brain development in terms of pre-, peri-, and postnatal development within a species. Such comparisons are not necessarily a good rationale for administering teratogens in animal models. If percentage of adult brain weight achieved by various species at birth were chosen as the basis for comparison, the rat would be a better model for human brain development than the monkey.

The Neuron

Once they have differentiated, neurons do three things. They receive information, they process information, and they transmit information to other cells. Neuronal differentiation facilitates these jobs by creating four main structural parts—dendrites, a cell body, an axon, and a terminal button (see Figure 4.7).

Dendrites are the receiving part of the neuron. They come off the neuron in branches. These branches in turn have branches that also have branches. In fact, the word "dendrite" comes from *dendron*, Greek for "tree." No matter how conscientious, an illustrator cannot accurately portray a neuron, especially its dendrites. There are just too many of them to

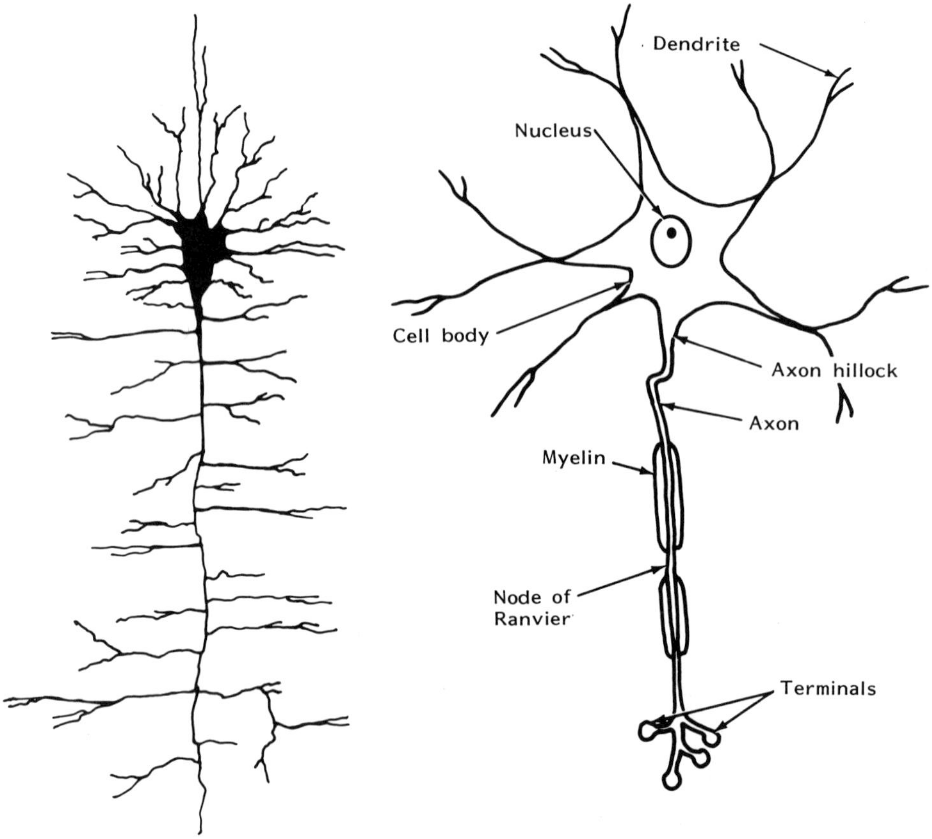

Figure 4.7. Actual and schematic depiction of a neuron from the motor cortex.

distinguish. To simplify the image of the neuron the artist generally attaches only a few dendrites to each nerve cell body instead of its thousands of branches. The picture looks neat but it robs the neuron of one of its most significant characteristics—its ability to receive impulses from thousands of other neurons.

Branching patterns and arrangements also vary from nerve cell to nerve cell allowing a single neuron to receive input from as many as 10,000 other neurons (see Figure 4.8).

Several conditions and agents affect dendritic development. Increased phenylalanine levels associated with PKU, for example, will reduce dendritic length and decrease aborization (Hogan & Coleman, 1981) as will hypothyroidism.

The final branches on dendrites are twigs, some of them as tiny as a millionth of an inch. These twigs, called "spines," are the actual contact points for dendrites. There are three different kinds of spines. Type 1 spines are long with thin necks and small end bulbs, Type 2 are short with thin necks and large end bulbs, and Type 3 are short with thick necks and large end bulbs.

Mental retardation in humans has been linked to decreased dendritic aborization and spine density (Huttenlocher, 1974). In rats, the number and proportion of spines are both

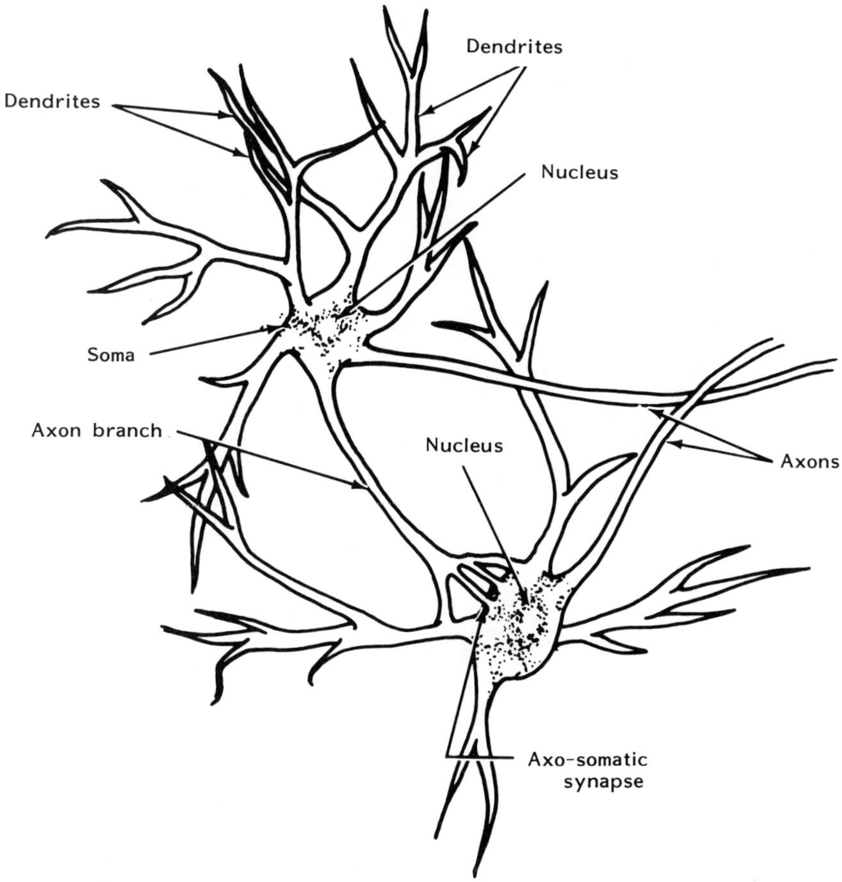

Figure 4.8. Schematic depiction of two interrelated neurons. Each neuron in the brain can receive thousands of contacts with other neurons.

altered by prenatal exposure to teratogens such as alcohol (see Figure 4.9) and this alteration is associated with impaired learning ability (Abel *et al.*, 1981). Conceivably, the mental retardation associated with fetal alcohol exposure may in part be due to alcohol's effects on dendritic spines.

The *soma* or cell body contains the nucleus and is the integrator part of the neuron. Although a neuron can have thousands of dendrites, it usually has only one *axon*. The axon is the long, thin part of the neuron that conducts nerve impulses from the cell body to the terminal button. Some axons are several feet in length and are covered with a fatty sheath called *myelin*. Myelin acts like an insulator on an electric wire to increase the speed of a nerve impulse along the axon by about tenfold. In multiple sclerosis, the myelin around nerves degenerates resulting in much slower transmission of nerve impulses and consequent loss of ability.

Very few neurons make direct physical contact with each other. Instead, there is a tiny gap between neurons called a *synapse*. The *terminal button* is the end part of the axon

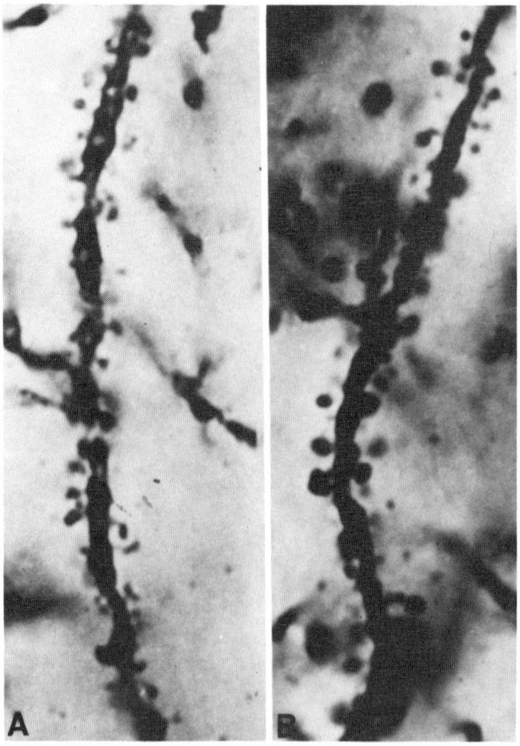

Figure 4.9. Pyramidal neurons from CA1 region of hippocampus. A is pair-fed control; B is section from a 3-month-old rat prenatally exposed to 3 g/kg of alcohol twice daily throughout gestation. Alcohol-exposed animals had 30% fewer total spines. The proportion of Type 1 thin spines was lower in alcohol-exposed animals (58% vs. 74%), higher for Type 2 round spines (36% vs. 20%), and unchanged for Type 3 intermediate spines (6% vs. 5%). (From Abel *et al.*, 1981.)

(see Figure 4.9) and it secretes a biochemical neurotransmitter into the synapse that links neurons (see below).

Although there are a wide variety of neuronal shapes, nearly all possess these four basic structures. If removed from an embryo and cultured, hippocampal neurons will develop these basic structures without the need for any input from any other cell or any environmental cue. If an axon of a young neuron is cut, the original axon may regenerate or one of its dendrites may develop into an axon (Dotti & Banker, 1987). Basic neuronal structure is thus controlled by a neuron's genes; subtleties of structure are controlled by the cellular environment in which development occurs.

The Nerve Impulse

Prior to firing, the nerve cell has an electrical potential across its membrane due to a higher distribution of positive sodium ions outside compared with inside the cell making the inside more negative than the outside.

If a week current is generated in a nerve cell, sodium ions enter and the resting potential becomes less negative. This decrease in negative potential is called a *depolarization.* If the stimulus causing the depolarization is increased in strength, it may cause depolarization beyond a critical threshold level. When this happens, the cell membrane suddenly stops holding sodium back and many sodium ions rush in, causing the interior of the nerve cell to go from negative to positive.

This change, called *an action potential*, lasts for only a sliced instant. Following an action potential, the neuron needs time before it returns to its resting potential. This is called the *refractory* period. The duration of this refractory period determines how many nerve impulses can be initiated in a given time. This is how the nervous system codes for intensity—only very strong stimuli can excite nerves during their refractory period. Neurons can outfire the fastest machine gun. Bursts up to 200 times a second are not unusual and some cells in the spinal cord have been clocked at over 1,000 responses a second. By comparison, the Uzi and Ingram machine guns can only fire 200 bullets a second.

How fast do nerve impulses travel? It depends on the diameter and insulation of a nerve's axons. Sensory and motor nerve axons have the biggest diameters and are insulated with myelin. They transmit impulses at a rate of 120 meters a second. This allows a signal to travel from your toes or fingers to your brain and back again in less than a second. By contrast, the autonomic nervous system contains some nerve cells that transmit impulses at rates of only 1 meter a second.

Neurons do not directly cause other neurons to fire. Instead, they release a neurotransmitter from their terminal button that may produce a local change in the membrane of the postsynaptic neuron. Some nerve cells receive input from only one terminal button, but it discharges so much neurotransmitter that it nearly always drives the nerve cell to its threshold. Other neurons are innervated by thousands of terminal buttons. If some of these decrease the local potential, it makes the neuron more likely to fire. This change in potential is called an *excitatory postsynaptic potential* (EPSP). If there is an increase in the local potential, the neuron is less likely to fire. The increased potential is called an *inhibitory postsynaptic potential* (IPSP) because an excitatory impulse acting on the neuron will not be able to depolarize it to the threshold level required to trigger an action potential. It is the amount of change in the membrane potential that determines if a nerve will fire. EPSPs and IPSPs are how neural activity in the brain is fine-tuned, allowing for the interplay of many neurons in the final occurrence of a response.

Synaptogenesis

Synaptogenesis is the formation of synaptic junctions between neurons. When neurons influence one another by way of neurotransmitters, they discharge these neurotransmitters into a ten-millionth-of-a-millimeter-wide gap between them called a *synapse*.

The most active period of synaptogenesis in the human cerebral cortex occurs after the 28th week of gestation, but formation of new synapses never stops. Synaptic connections between most neurons are formed after they have migrated to their target areas. They involve the growth of axons and dendrites as a result of an intrinsic mechanism within neurons (Globus & Sheibel, 1967) and by induction toward a particular area.

The specialized area in neurons where axons and dendrites grow is known as the *growth cone*. Although several studies suggest growth cones can be induced to develop in a particular direction by neurotrophic factors in target tissues such as *nerve growth factor* (NGF), this is unlikely because the neurons in these studies had already innervated their targets (Davies *et al.*, 1987). NGF is not even present in some target tissue until more than a day after the arrival of axonal processes (Davies *et al.*, 1987). The role of NGF seems instead to be that of protecting neurons from cell death (see below).

Once growth cones arrive at a particular site, they must still establish synaptic contact with specific neurons in an area containing many different kinds of neurons. There must be a recognition process for establishing appropriate synaptic contacts. This recognition process may be mediated in some cases by neurotransmitters.

Neurotransmitters as Inducers

About 50 different neurotransmitters are present in the brain (Black *et al.*, 1984) and neurotransmitter maps have been created that trace neuronal pathways mediated by some of them (Hokfelt *et al.*, 1984). Neurotransmitters are synthesized in axons or in nerve cell bodies from amino acids. They are then transported down an axon to its terminal portion, the axonal button, where they are stored in synaptic vesicles.

When a nerve impulse occurs, calcium ions enter the terminal button from extracellular fluid. The entry of calcium ions causes some of these vesicles to migrate to the terminal button and fuse with the cell membrane. The vesicle then bursts, discharging its contents into the synapse.

Once in the synapse, some of the neurotransmitter comes in contact with receptors located on the adjacent postsynaptic nerve cells. If the transmitter fits into the receptor, it will produce a local change in the membrane potential. Although a single neuron typically releases only one kind of neurotransmitter, the dendrites on a single neuron may have receptors for several different kinds of neurotransmitters.

After a response is triggered, the neurotransmitter's activity is ended by enzymes that break it down in the synapse or postsynaptic membrane or that cause it to be taken back up into the presynaptic terminal buttons.

Neurotransmitters have been divided into five basic groups:

1. *Cholinergic neurotransmitters.* The prototype of the cholinergic neurotransmitter is *acetylcholine* (ACh). ACh is the most abundant neurotransmitter in the central nervous system and is believed to be intimately involved in learning and memory.

2. *Biogenic amine neurotransmitters. Biogenic amines* are among the first neurotransmitters to appear in the developing brain. The three main biogenic amine neurotransmitters are serotonin, norepinephrine (NE) or norephinephrine (E), and dopamine (D). These neurotransmitters are located in regions of the brain involved in hunger, arousal, sleep, and learning.

3. *Amino acid neurotransmitters.* One of these, *gamma-aminobutyric acid* (GABA), has been recognized as a possible inhibitory neurotransmitter for many years. There is now good evidence that other amino acids such as *aspartate, glutamate*, and *glycine* can also act as neurotransmitters. GABA is located in relatively large amounts in the cerebellum and the hippocampus. Glutamate and aspartate have excitatory actions and act as neurotransmitters in part of the hippocampus.

4. *Neuropeptide neurotransmitters.* Neuropeptides are chains of two or more amino acids. The best known of these are the endorphins, which have morphinelike effects. Receptors for morphine were not put there for people to stimulate them with morphine but because they react to endogenous opiates whose structures are very similar to morphine. The highest density of endorphins in the brain occurs in the hypothalamus and in the limbic system.

5. *Purine neurotransmitters.* The final category of neurotransmitters is the purines such as adenosinetriphosphate (ATP). Although ATP appears to have a neurotransmitterlike function in the peripheral nervous system, its role as a neurotransmitter in the brain is still speculative.

Neurotransmitters are present very early in brain development. They not only mediate synaptic communication but also influence cell migration and differentiation, closure of the neural tube, and maturation of neuronal axons. Since many psychoactive drugs, such as haloperidol, diazepam, reserpine, and alcohol, act directly on neurotransmitter mechanisms, they may adversely affect brain maturation by disturbing their intrinsic evocator actions.

Neurotransmitter systems develop sequentially rather than concurrently and influence development of their own chemical circuitry. First to appear are serotonin-containing neurons, followed by neurons containing dopamine and then norepinephrine (Seiger & Olson, 1973; Nobin & Bjorklund, 1973). Serotonin is a key evocator in nerve cell migration in the brain (Yamamoto *et al.*, 1980). In cell cultures where the extracellular environment of actively growing neurons can be controlled, serotonin causes an immediate and premature inhibition of nerve growth cones in some (but not all) neurons, resulting in decreased synaptogenesis in affected neurons (Haydon *et al.*, 1987). Drugs that interfere with neurotransmitter activity will more than likely suppress or inhibit further synaptogenesis, resulting in behavioral anomalies in offspring.

Long-Range Induction

During early development, brain organization relies on short-range induction such that one group of cells (the inducers) causes a change in an adjacent group of responding cells prompting them to differentiate in a particular way. This short-range induction subsequently results in the differentiation and organization of cells into various structures of the brain. Later on in development, short-range induction is replaced by a long-range induction using hormones to complete the organizational process. The best-known example of hormonal induction is metamorphosis.

Frogs and toads start off life as aquatic tadpoles. Under the influence of thyroid hormones, almost every organ in their bodies undergoes a major transformation and they develop into terrestrial animals that bear no physical resemblance to their earlier forms. One of the dramatic changes that takes place during this transformation from tadpole to frog is a loss of tail. This loss is due to thyroid-hormone-induced death of tail cells, which are programmed to die when they are exposed to these hormones. If cells from a part of the eye are transplanted to the tail region, they will remain intact even through the tail degenerates (Gilbert, 1985). Tail degeneration also occurs in humans during the fourth week of gestation (Fallon & Simandl, 1978).

In their role as inducers, hormones also affect morphological, histological, physiological, and neurological mechanisms underlying behavior. The role of the sex steroids on brain development is an example of this influence.

Initially, the primordial gonadal cells in males and females are identical. The presence of H-Y antigen (see above) causes these cells to develop along the lines of a male; absence of the male chromosome results in differentiation as a female. Once formed, the primordial testis produces two types of hormones. The first type, the androgens, act on receptors in nearby structures to masculinize other primordial gonadal cells into internal and external male genitalia. These receptors are also present in female embryos; if testosterone is administered to a female embryo, they will become masculinized. The second hormone, called an *anti-mullerian duct factor*, causes degeneration of tissue that would otherwise develop internal and external female genitalia, e.g., the uterus and vagina. The Y chromosome determines genetic sex, but the male hormones determine sex-related physical characteristics.

After differentiation of the genitalia, brain organization is likewise affected by the presence or absence of androgens during development. Female guinea pigs prenatally injected with testosterone do not act sexually like females when they are adults and if given testosterone during adulthood they behave sexually like males (Phoenix *et al.*, 1959). Likewise, males castrated at birth or chemically castrated *in utero* by injecting their mothers with cyproterone acetate or flutamide do not act sexually like males when adults; if given estrogen when adults, they behave sexually like females (Ward, 1972). Castrated males given testosterone replacement therapy exhibit normal male sexual behavior when challenged with testosterone as adults. These effects on sexual behavior have been observed in animals with short gestations like the rat (Ward, 1972) and rhesus monkeys, which have long gestations (Phoenix, 1974).

Mammalian brain development normally proceeds along female lines unless subverted by male androgens. When testosterone reaches the fetal brain, local enzymes convert it to a metabolite, dihydrotestosterone, and to the female hormone, estradiol. There are an equal number of androgen and estrogen receptors in the fetal brain but because of higher testosterone levels, more of these receptors are occupied in the male brain (McEwen *et al.*, 1988). Occupancy of these receptors determines the two primary aspects of sexual differentiation: masculinization and defeminization. Masculinization, i.e., enhancement of characteristic male traits such as androgen-associated aggressiveness and sexual behavior, is related to androgen receptors; defeminization, i.e., suppression of characteristic female traits such as cyclic hormonal surges and sexual behavior, is related to occupancy of estrogen receptors as a result of conversion of testosterone to estradiol (McEwen *et al.*, 1988). In rats and mice, masculinization of the brain occurs prenatally, whereas defeminization occurs postnatally.

The critical period for sexual differentiation of the brain varies widely in different species (MacLusky and Naftolin, 1981). In the rat, which has a relatively immature brain at birth, the critical period extends from gestation day 18 to postnatal day 5. For species like guinea pigs and primates, which are more fully developed at birth, the critical period is completed prior to birth (MacLusky & Naftolin, 1981).

Human research in this area is obviously constrained but a number of ''natural experiments'' are available from which to make comparisons. In the human fetus, the testes produce testosterone around the 8th to the 10th week of gestation and levels rise to their maximum around the 20th week of gestation, (Siiteri & Wilson, 1974). After the 20th week, levels decrease until birth. Peak testosterone levels in the male are about nine times higher than in the female fetus (Abramovich, 1974).

In certain congenital disorders such as adrenogenital syndrome, an excess of androgen is produced by the fetal androgens. Females with this syndrome are born with masculinized genitalia (Ehrhardt & Money, 1967) and engage in masculine psychosocial behaviors—they prefer boys as playmates, they have little interest in playing with dolls or dresses, and they are more ambivalent about marriage and maternal activities (Ehrhardt *et al.*, 1968). In males, a mutation called *androgen in sensitivity syndrome* is characterized by absence of functional testosterone receptors. Although they have testes and secrete testosterone, males with the syndrome cannot respond to testosterone and they develop many female sex characteristics like breasts.

If estrogen is a female hormone, why aren't females defeminized? Normally, the female fetus produces very little estrogen and alpha fetoprotein acts as a ''back up'' mechanism to keep levels low. By binding to estrogen, alpha fetoprotein protects the female brain from masculinization. Pharmacological doses of estrogen, however, will masculinize females. [Interestingly, an increase in alpha fetoprotein levels in maternal blood or amniotic fluid indicates failure of the neural tube to close during fetal development (see Chapter 2)].

Females are also protected from possible estrogenic influences from the mother by the placenta, which does not allow natural estrogens to pass from the mother's circulation to the fetus. However, the placenta is a less efficient barrier to synthetic estrogens like diethylstilbesterol (DES). Furthermore, alpha fetoprotein does not bind synthetic estrogens like DES. As a result, many women prenatally exposed to DES became defeminized (Ehrhardt *et al.*, 1985).

After puberty, sexual hormones produce changes in body structure reminiscent of metamorphosis (Gilbert, 1985) and act as "energizers" for sexual behavior. In adults, castration causes a profound decrease in sexual motivation, which can be restored by sex-appropriate replacement hormone therapy. Sexual motivation and behavior in castrated males will be restored only by replacement with testosterone. Estrogens will not cause a castrated male to exhibit female sexual behavioral patterns or vice versa.

The main areas of the brain that undergo sexual differentiation are the preoptic and ventromedial areas of the hypothalamus, the corpus callosum, and the amygdala (MacLusky & Naftolin, 1981). Following testosterone-induced activation of androgen and estrogen receptors in the developing brain, neurons in these areas increase in size and send out neurites to establish increased synaptic contacts with other cells. In men, the sexually dimorphic cell group in the preoptic area of the hypothalamus is about 2.5 times larger than in women (Swaab & Fliers, 1985). The preoptic area of male monkeys contains about 20% more dendritic branches per neuron than the same area in females. Males also have a higher percentage of intermediate and spiny dendrites in this area, whereas females have a higher percentage of sparsely spined dendrites. Differences in dendritic length, however, are not significant (Ayoub *et al.*, 1983). Such differences in size and structural organization of different brain regions in males and females are probably responsible for differences in sexual behavioral patterns and sexual differences in nonsexual behaviors (Quadagno *et al.*, 1977).

It is also possible that sex differences in response to many psychoactive drugs are also related to sexually dimorphic differences in brain structure and function, especially differences in neurotransmitters and their receptors. Many of these differences are reversed in conjunction with early hormonal manipulations (see review by McEwen *et al.*, 1988). Although sex differences in drug responsiveness are often related to metabolic effects, differences in brain mechanisms may also be involved.

Testosterone thus has an inductive and organizational function in the development of the brain. Males behave sexually as males because they are exposed early in life to testosterone; females behave as females because they aren't exposed to testosterone early in life.

Because testosterone is a vital evocator for sexual differentiation of the brain, substances that block testosterone's synthesis or release *in utero* could interfere with sexual differentiation. One such substance is alcohol. Male rats prenatally exposed to alcohol have significantly smaller than normal volumes of the sexually dimorphic nucleus (Rudeen *et al.*, 1986) and exhibit femalelike saccharin consumption and maze-learning patterns (McGivern *et al.*, 1984) as well as female sexual patterns, such as lordosis (Hard *et al.*, 1984). Neuroanatomical studies support these observations because these males also.

Other compounds such as nicotine (Lichtensteiger & Schlumpf, 1985) and barbiturates (Reinisch & Sanders, 1982) have also been found to interfere with organization of sexually dimorphic behavioral patterns. These observations demonstrate that an important behaviorally teratogenic effect can occur as a result of the actions of drugs on hormonal systems during development.

Glia and Myelination

The other basic unit in the nervous system besides the neuron is the *glial cell* (from the Greek word for ''glue''). Glial cells outnumber neurons by a ratio of 10 to 1. This means that there are about 150 billion glial cells compared with the human brain's 15 billion neurons. Glial cells surround neurons and keep them from sloshing around inside the brain, but they are much more than just a filler. Glial cells also furnish a supportive function for nerve cells. They provide the material from which myelin develops, they affect the metabolic activities of nerve cells, and they may be involved in some memory function.

Myelination occurs after histogenesis begins and lasts longer. In the peripheral nervous system, myelination is performed by Schwann cells; in the brain it is performed by glial cells called *oligodendrocytes*.

Different tracts in the brain undergo myelination at different times and rates (Gilles *et al.*, 1983). In humans, myelination begins around the 4th month of gestation but mostly takes place postnatally during infancy and early childhood. The areas of the brain that become myelinated first are those in the medulla, pons, and mesencephalon, which control more primitive functions, such as breathing and reflex development. Myelination in the forebrain occurs later in gestation (about 30 to 40 weeks), continues postnatally, and is not completed until about two years of age (Dobbing & Smart, 1973). In the rat brain, myelination occurs primarily after birth, occurs maximally around day 20, and is completed by about 25 to 30 days (Norton & Poduslo, 1973). In the human, myelination also occurs primarily after birth, and some areas such as the reticular formation and association areas of the cerebral cortex are not fully myelinated until about 10 years of age (Yakovlev & LeCours, 1967).

Myelination can be adversely affected by any substance that affects myelin synthesis or causes disorganization of myelin or demyelination. Undernourishment during development will reduce the lipid pool needed for synthesis or myelin and could thereby retard or irreversibly affect the course of myelination. The critical period for such an effect of undernourishment on myelination is birth to day 14 in the rat (Wiggins & Fuller, 1978); equivalent undernourishment after day 21 produces only an ephemeral effect (Wiggins & Fuller, 1978). Irreversible deficits in myelination are associated with at least a 20% decrease in body weight (Fuller *et al.*, 1984).

Prenatal alcohol exposure is also capable of retarding the rate of myelination. Samorajski *et al.* (1986) found that at 14 days of age, rats prenatally exposed to alcohol had 20% to 30% fewer myelinated axons in the optic nerve compared with controls and the ratio of nonmyelinated to myelinated nerves was 60% higher in alcohol-exposed animals. At 28 days, differences in number of unmyelinated axons were no longer significant but the ratio of nonmyelinated to myelinated axons was still 30% higher.

Another class of compounds that depresses the rate of myelination is the anticonvulsants, e.g., diphenylhydantoin, sodium valproate, and phenobarbital (Patsalos & Wiggins, 1982). Some industrial solvents and anesthetics and metals like tin cause disorganization of myelin after it has formed (Wiggins, 1986).

Demyelination is a third way in which myelin can be adversely affected. The prototypic demyelinating agent is lead, which causes myelin to disintegrate in segments (Fullerton, 1966). This demyelination is independent of axonal degeneration and is directly associated with decreased nerve conduction velocity. Prior to the 1960s, eating lead-based paint flakes and chips was the major source of lead-induced brain damage in children. Recognition of this problem resulted in removal of lead from paint. However, lead is still used extensively in industry. Consequently, inner-city dust and garden soil, especially near freeways and lead smelters, are highly contaminated with lead. As a result of contamination of air with lead, lead exposure is inevitable.

Modification of Synaptic Connections

During embryological development, certain structures appear at an early stage and then disappear. One example is the appearance and then disappearance of a rudimentary tail in humans—undoubtedly a phylogenetic remnant of our primate ancestry.

In the central nervous system, nerve cell death removes errors and serves to fine-tune matchups between pre- and postsynaptic neurons. If a dye is placed in the pyramidal tract of newborn rats, for instance, it diffuses along these neurons and is carried throughout the cortex; if placed in the same area of 3-week-old rats, it does not appear in the visual cortex (Stanfeld *et al.*, 1982) indicating that these tracts no longer innervate the visual cortex. Apparently, elimination of neurons after they are in place by nerve cell death is a more economical way of ensuring accurate synaptic connections than predetermined synaptic linkages.

In some cases the extent of cell loss is considerable. In the hamster retina, about 50% of the cells are lost (Sengelaub & Finlay, 1981). Cell death also occurs in the neocortex, where different lamina undergo different amounts of cell death (Finlay & Slattery, 1983).

A dramatic example of cell death occurs in connection with imprinting in birds. Baby ducks and goslings normally follow their mothers away from their nest within a day of hatching. If, however, they hatched away from their mother and you were the first moving creature they encountered, they would waddle along after you and follow you everywhere. Even if their own mother crossed their path, they would treat her like a stranger. And when these birds become mature, the males would court you.

Imprinting is a special learning process that takes place during a critical period of development in the first few hours after hatching. It can occur not only to visual stimuli but also to auditory and olfactory stimuli, that is, sounds or odors acquire special significance and animals will follow objects that emit these sounds or odors. Imprinting has two important biological consequences. The first is the formation of a permanent social attachment to an individual, usually the mother. Second is recognition of appropriate mating partners. Other learning processes differ from imprinting in not having a known critical period and having the possibility of forgetting and modification of associations.

To determine the neuroanatomical consequences of imprinting, Wallhausser and Scheich (1987) compared brains from 7-day-old imprinted chicks and controls. Imprinted chicks had over 22% fewer dendritic spines than controls. Whereas traditional learning is generally accompanied by growth of neurons and formation of new synapses, imprinting resulted in an irreversible loss of connections between neurons. This loss ensures fixation on the imprinted object, a biologically important event in the life of birds. Neuronal cell death is a natural occurrence in the central nervous system. Interference with this process could result in an excess of connections to some target areas, which could have a negative impact on behavior.

Nerve growth factor (NGF) selectively controls some aspects of neuronal death. Administration of NGF during the time of neural death causes fewer neurons to be lost, whereas anti-NGF antiserum increases neuronal loss (Davies *et al.*, 1987). The amino acid structure of the NGF protein is now well known and the gene that codes for this protein has been identified (Davies *et al.*, 1987).

Another fine-tuning mechanism involves elimination of synaptic contacts. In some areas of the brain, thousands of axons may converge onto a single cell. In the cerebellum of the newborn rat, for example, each Purkinje cell is innervated by several fibers from the inferior olive; in the adult, each Purkinje cell is innervated by only one neuron from the inferior olive (Crepel *et al.*, 1976). Likewise, neurons from one cerebral hemisphere synapse with neurons on the other side through the corpus callosum at birth but no longer do so at maturity (O'Leary *et al.*, 1981). Elimination of synaptic contacts is primarily due to retraction of axons back into their nerve cell bodies rather than to presynaptic cell death. However, though synaptic contact from a particular neuron may decrease, axons that remain in contact with a target cell increase their synaptic contacts with that cell so that they are able to exert greater control over its activity.

One mechanism behind the rearrangement of synapses may involve segregation of inputs to specific cortical layers. For example, at birth, neural input from the retina of the kitten is diffusely distributed throughout the lateral geniculate. In the course of development, these axons become channeled to specific layers in the geniculate (Shatz, 1983). Another mechanism involves neural activity such that patterns of activity in a particular area affect which synaptic contacts are maintained and which are eliminated (Purves & Lichtman, 1980).

Appendix A. Malformations of CNS Development Due to Failure of Neuropore Closure

Malformations of the central nervous system often occur because the neural tube does not close, a condition called *dysraphia.* An excellent review of neural tube defects and agents that cause them can be found in Morrissey and Mottet (1980).

Anencephaly is a dysraphic condition in which the cephalic area of the neural tube does not close. As a result, the cerebral hemispheres do not develop and at birth most of the cortex is missing or dramatically reduced in size.

Anencephaly occurs about once in every 1,000 live births and affected children die soon after birth. *Exencephaly* is considered an early stage of *anencephaly* and is characterized by failure of the skull to close so that the brain develops outside the skull. This condition also leads to death shortly after birth.

Spina bifida is like exencephaly except it occurs in the spine. Literally it means "cleft spine." During development, the spinal column remains open until the 12th week of gestation. In spina bifida, one or more of the vertebrae do not close leaving a cleft in the vertebral column. Because of this cleft, part of the spinal nerves can slip out like a hernia.

The mildest form of spina bifida is called *spina bifida occulta.* In this condition, the cleft is barely perceptible and there are no associated symptoms. Often the only visible evidence is a dimple in the backbone over the defect or a small tuft of hair covering the defect.

In its more serious forms, a sac (called a *cele*) containing portions of the spinal cord bulges through an opening in the spine. Such openings usually occur in the lumbar area of the spine (see Figure 4.10). If the sac contains only some of the coverings of the spinal nerves, the problem is called a *meningocele* (pronounced me-nin-go-seal). If the coverings and the spinal nerves both bulge through, the condition is called a *meningomyelocele* (pronounced me-nin-go-my-e-lo-seal). The latter is the more common and the more serious of the two.

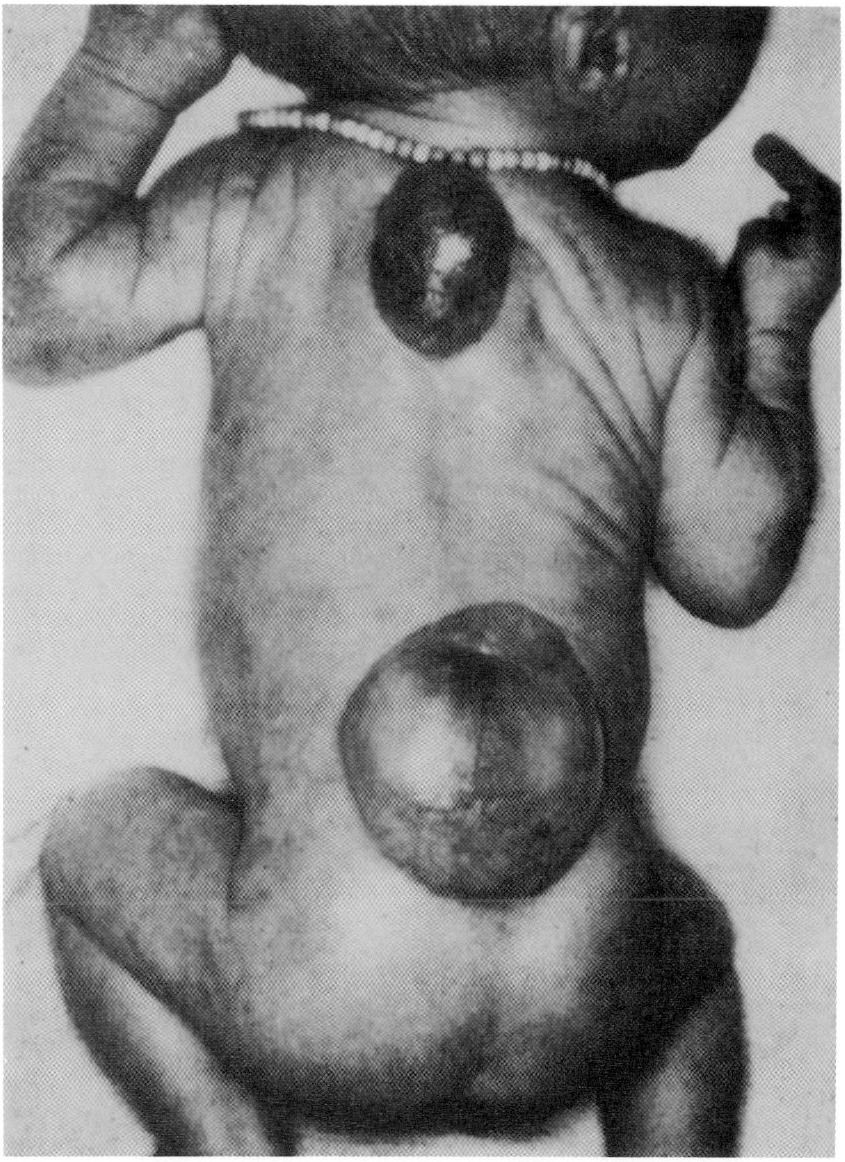

Figure 4.10. Infant with meningomyelocele. Photo courtesy Dr. M. Evans.

The degree of impairment associated with spina bifida depends on the location and extent of the spinal lesion. Because most lesions occur in the lumbar area (the lower part of the spine), the legs are affected most. This may result in weak muscles and poor skin sensation or total leg paralysis and complete lack of sensation in the legs.

Because the covering of the sac is very thin and easily torn, the baby's movements can make it raw and infection can occur. Consequently, surgical repair is generally performed as soon as possible. Surgery usually keeps the condition from getting worse but once the functions are lost, they will not be restored by surgery.

Whenever a neural tube defect occurs, some brain anomaly is generally associated with it. Meningomyelocele is almost always associated with a similar malformation inside the brain. As a result, part of the brain stem and cerebellum may bulge out of the structures usually confining them. This causes a compression of the ventricles and a disorder called *hydrocephalus*. Hydrocephalus is an enlargement of the head due to abnormal cerebral spinal fluid collection in the ventricles of the brain.

Another consequence of failure of the neural tube to close is *agenesis* (lack of development) of the corpus callosum, the bundle of nerve fibers that connects both sides of the cerebral hemispheres.

In spina bifida and hydrocephalus, spinal fluid builds up in the spinal sac or cavities of the brain. The effect of the increased pressure is neural damage from leakage of cerebrospinal fluid into white matter. The result is irreversible axonal degeneration and demyelination.

Another major type of malformation is the diverticulation disorders. *Diverticulation* refers to a branching off of fibers to form nerve tracts such as the olfactory and optic tracts. Diverticulation begins around 5 to 6 weeks of gestation just after the neural tube starts forming brain segments. Disorders of diverticulation include arrhinencephaly, absence of the olfactory tracts, and anophthalmia, incomplete development of the optic tracts. Holoprosencephaly is another diverticulation disorder in which the first two segments of the brain (the prosencephalon) do not develop. This results in failure of the cerebral hemispheres to form and failure of the diencephalon to divide into two thalami.

Appendix B. Basic Functional Neuroanatomy

The nervous system is divided into two basic components, the central nervous system (CNS) and the peripheral nervous system. The CNS includes the spinal cord and the brain. The peripheral nervous system includes the sensory receptors that detect stimuli and nerves whose cell bodies are outside the brain and spinal cord. These include afferent neurons, which bring sensory information to the CNS, and effector (motor) neurons, which carry information to muscles and glands.

The spinal portion of the CNS consists of a long bundle of nerve cells from the base of the brain to the end of the backbone. At various areas of the spinal cord, afferent fibers enter and synapse with interneurons, the integrators of the nervous system. The interneurons synapse with other interneurons or with effector neurons that leave the spinal cord. The simplest arrangement is the reflex arc consisting of one afferent (sensory) neuron, an interneuron, and an effector neuron (motor response). The interneuron acts as an arbiter, enhancing or attenuating the relationship between afferent and efferent neurons. In most cases, however, information from afferent neurons travels up spinal neurons to higher centers for more arbitration than is capable at the spinal level. Once arbitration has occurred, signals are passed from the brain back down the spinal cord in efferent neurons. Afferent and efferent neurons that have their interneurons in the spinal cord are called *spinal nerves*; those that have their interneurons in the brain are called *cranial nerves*.

The Brain

The brain is subdivided into the *forebrain* (or cerebrum), *midbrain*, and *hindbrain* (or brainstem). Each of these areas is further subdivided. The forebrain includes two cerebral hemispheres that make up the cerebrum. The surface of the cerebrum contains the cerebral cortex. Also included in the cerebrum are the basal ganglia and the limbic system. The other major division of the forebrain is the diencephalon, which contains the hypothalamus and thalamus. The midbrain lies between the pons and the thalamus and contains the tectum and tegmentum. The hindbrain is composed of the medulla, pons, and cerebellum.

Formation of these brain structures occurs when the anterior region of the neural tube begins to swell into the three areas that will become the forebrain, midbrain, and hindbrain. A bending then develops in the neural tube and creases appear that correspond to the boundaries of the future brain cavities. The forebrain area then undergoes changes that result in the formation of the telencephalon, which contains areas involved in the final processing of sensory information, complex control of motor activities, and "higher mental functions" such as learning and memory and the diencephalon. The diencephalon is another relay station channeling information between the telencephalon and the rest of the brain. It also controls the pituitary gland, which in turn controls many of the body's endocrine functions.

The cavities in the telencephalon are called the *lateral ventricles*. The diencephalon has one cavity called the *third ventricle*, which is situated in its middle. The midbrain also has a cavity, which is relatively narrow, called the *aqueduct*; the cavity in the hindbrain is called the *fourth ventricle*.

Forebrain. The forebrain is divided into the diencephalon and the telencephalon. The *diencephalon* is made up of four main areas: the thalamus, hypothalamus, subthalamus, and epithalamus. The primary role of the *thalamus* is the relaying of incoming sensory and outgoing motor information. It contains three specific relay interneurons that process hearing signals (the medial geniculate nuclei), visual signals (the lateral geniculate nuclei), and somatosensory signals (the ventralis posterior nuclei) in addition to other sensory nuclei. The only sense that does not have relay nuclei in the thalamus is smell. The thalamus also contains association nuclei that receive input from these relay nuclei and from the cerebellum and send this information to sensory association areas in the cortex that integrate sensory and motor input and act as a "middle manager" for the cortex.

The *hypothalamus* contains many nuclei that regulate hunger, thirst, body temperature, sexual activity, and glandular activities. Regulation of glands occurs via hypothalamic hormones called *neuroendocrines*, which affect the pituitary and through it many other glands. The *subthalamus* contains structures such as the pineal gland, which is involved in homeostatic function.

The *telencephalon* is the most unique part of the human brain. It is made up of the cerebral hemispheres, which contain three basic structures; the limbic system, the basal ganglia, and the neocortex, the most highly developed portion of the brain.

What makes the cerebral hemispheres so unique is their highly developed surface area. Other animals like the elephant have larger brains than people. But no other animal has as much surface area relative to overall brain size. This was accomplished by inner folding of the surface area creating hills (called *gyri*) and valleys (called *sulci*) of nerve cells. Within this surface area called the *cortex* are upwards of six layers varying in thickness, type, and density of cells. From these layers and the connection between the neurons in them, come the ability to walk on the moon or to ask questions like "what has four wheels and flies?" (Answer: a garbage truck).

The forebrain of mammals as primitive as the rat and as complex as the human is divided into four lobes: the occipital, frontal, temporal, and parietal. The *frontal lobes* receive input from all areas of the brain but their function is still unclear. Although this area of the brain was believed to be the "seat" of intelligence, loss of the frontal lobes has little effect on intelligence test scores. The *occipital lobes* receive and interpret visual information from the thalamas and midbrain; the *temporal lobes* receive and interpret auditory information; and the *parietal lobes* receive somatosensory information. Areas in these lobes that receive sensory input directly are called *sensory projection areas;* for example, primary auditory cortex is that part of the cortex that first and mainly receives auditory signals. Parts of the cortex next to the sensory projection areas receive information from the primary sensory areas and integrate them into perceptions and memories. The other area in the cortex is the motor cortex. This area originates signals that control such movement as speech and chewing and running and jumping.

Another area of the cortex integrates sensory and motor information and is called *association cortex.* The association cortex in humans is relatively large compared with the size of sensory and motor cortex, whereas in animals like the rat, the relative amount of association cortex is considerably smaller. It is the size of the human association area cortex that makes complex learning, memory, thought, and language possible.

The association cortex is also unique in that it is not evenly distributed between left and right cerebral hemispheres. The sensory and motor cortex are evenly represented in both hemispheres. The association cortex is asymmetrical. The left hemisphere is dominant (in most people) for control of language. Damage to the left temporal lobe causes speech disturbances called *aphasia.* The right hemisphere appears to contain areas of the brain that control spatial abilities.

Areas in the two hemispheres are connected to one another by bundles of nerve tracts that cross the midline, the most prominent of which is the 200-million-fiber *corpus callosum.* Although the two hemispheres look almost alike, each has specific functions not possessed by the other to the same degree or at all. The left hemisphere is dominant for language—speech, writing, reading; the right is dominant for mathematical, visual, and spatial abilities. The anterior commissure joins the anterior temporal and prefrontal lobes. The third main group of connecting nerve tracts are called *projection fibers.* These fibers link the diencephalon with the cerebral hemispheres and the latter with the brainstem and spinal cord.

Limbic System. The limbic system contains several structures that modulate learning, motivation, and emotional behavior. If animals have electrodes implanted in these areas and are taught to press a lever so as to self-administer electric impulses to this area, they will do so for hours.

One of the major areas of the limbic system especially of interest to behavioral teratologists is an S-shaped structure called the *hippocampal formation.* The hippocampal formation has an inner area called the *dentate gyrus,* an outer area called the *hippocampal gyrus,* and a midarea called the *subiculum.* The hippocampus is connected to the thalamus by a bundle of fibers called the *fornix.* Behavioral teratologists are especially interested in the hippocampus because of its involvement in learning and short-term memory. Another major structure in the limbic system is the *amygdala.* The amygdala modulates visceral activities that underlie emotional expression.

Basal Ganglia. The basal ganglia are a collection of motor centers (the putamen, caudate nucleus, and globus pallidus). These nerve centers receive input from the cortex and cerebellum and send input to motor areas via the red nucleus and substantia nigra and other neuronal tracts. Degeneration of cells on the basal ganglia can lead to Parkinson's disease—a disorder in which movement is slow and awkward and difficult to inhibit once it is started.

Hindbrain and Midbrain. The hindbrain and midbrain contain ascending and descending nerve tracts that relay sensory information to the forebrain and process motor signals from the forebrain to the spinal nerves. The hindbrain also contains centers that control the vital reflex functions of the body including breathing, sucking, swallowing, and blood pressure. In addition, it contains nerve cells that connect input from the cerebellum (which coordinates voluntary movement and equilibrium) to other parts of the brain. The center core of the brain system contains a dense bundle of nerve fibers called the *reticular activating system.* The reticular system receives input from all sensory neurons as well as various centers in the cortex. It modulates the activities of the nervous system by integrating this input from sensory, somatic, and visceral receptors and by acting as an arousal mechanism influencing the alertness and attentiveness of various nerve centers.

The midbrain is a continuation of the hindbrain and contains the tectum, groups of nuclei called *colliculi,* which relay information. The superior colliculi relay visual and somatosensory input to the forebrain and portions of the reticular activating system, whereas the inferior colliculi relay auditory input. The other major part of the midbrain is the tegmentum, which contains the red nucleus and the substantia nigra, parts of the motor system linking motor information from the cortex to the spinal cord. The midbrain also contains the periaqueductal gray matter—an area of neuronal cell bodies that surround the aqueduct and that are involved in control of movement.

Appendix C. Research Techniques in Neuroanatomy and Neurophysiology

Current appreciation and understanding of the prenatal determinants of behavior relies as much on neuroanatomical, neurophysiological, and neurochemical information as on behavior itself. In fact, it was not until behavioral teratologists were able to demonstrate brain damage associated with teratogen-induced behavioral disorders that behavioral teratology emerged from its infancy.

Histological Techniques

The oldest and most widely used technique for detecting structural damage to neuronal pathways involves histologically staining neurons. Because neurons that control various functions tend to be bunched together in specific layers and areas of the brain, their organization can be examined. Proof that nerve tracts do indeed control specific functions can be obtained by electrically stimulating them and watching what happens.

Histological techniques involve removing the brain and slicing it into very thin pieces and then dyeing them to make them visible under the microscope. Different dyes or stains are used to bring out specific areas. The earliest dye to be used to visualize neurons is a nerve-specific silver stain known as a *Golgi stain.* This stain impregnates neural tissue with silver. Then by a process like developing photographic film, neurons that take up the silver become darkened and their dendrites and axons can be distinguished. Some dyes, like the Nissl stain, stain cell bodies but not axons. This enables the neuroanatomist to tell where certain kinds of cell groups are located. Other stains are used specifically to stain the myelin sheath on axons. By following these stained sheaths the neuroanatomist can trace nerve tracts through the brain. Although stain techniques were originally developed for light microscopy, these similar procedures have been adapted for the electron microscope, allowing researchers to examine nerve cell structure with exquisite detail.

Another method for tracing nerve pathways involves injecting the enzyme horseradish peroxidase (HRP) into nerve tissue. This enzyme is taken into neurons spontaneously and is distributed toward and away from the nerve cell body allowing pathways and cell bodies to be visualized.

Neurons that contain certain neurotransmitters can be identified because of the fluorescence of these neurotransmitters. Catecholamines, for example, fluoresce when exposed to ultraviolet light. A similar principle is involved in the use of fluroescent markers, which are injected into nerve tissue and are taken up by neurons.

Using innate fluorescent properties to trace neuronal pathways is called *histochemistry*. A variant of this method, called *immunohistochemistry,* involves making antibodies for certain cell antigens and then adding anti-antibodies treated with fluorescent markers or HRP, which binds with the antibodies.

Quantitative autoradiography is another technique for localizing neurons. In this procedure, animals are injected with a radioactively labeled compound that is taken up by the neurons as HRP. The brain is sectioned and samples from specific areas are exposed to a photographic plate, which creates a high resolution image showing the localization and density of the labeled materials. Two such procedures are the 2-deoxyglucose technique and tritiated thymidine.

2-Deoxyglucose

The 2-deoxyglucose technique is used to determine metabolic activity in areas of the brain as a result of experimental treatments. The procedure involves injecting animals with deoxyglucose, a structural analogue of glucose. After injection, deoxyglucose is transported into brain cells along with glucose and phosphorus is added to it by an enzyme (hexokinase). After it is phosphorylated, deoxyglucose is trapped inside the cell. If the radiolabeled deoxyglucose is injected, the amount of radioactivity in localized brain structures can be determined in specific brain sections. As previously pointed out, glucose is the brain's only source of energy. The amount of uptake of the radiolabeled deoxyglucose is proportional to uptake of glucose and therefore accumulation of radiolabeled deoxyglucose provides an index of localized energy metabolism in specific regions of the brain. Increases in such metabolism are related to increased neuronal activity. Conversely, decreased uptake of 2-deoxyglucose suggests decreased metabolism and decreased neuronal activity.

An example of this technique is the study by Vingan and co-workers (1986) in which rats were prenatally exposed to alcohol. At 90 to 100 days of age they were tested for avoidance learning following which they were injected with 2-[^{14}C] deoxyglucose. Animals were then sacrificed, brains were removed, prepared for sectioning, and regional ^{14}C concentrations were determined by autoradiography.

Animals prenatally exposed to alcohol made only 6.1 avoidances per day compared with 23.4 avoidances for controls demonstrating decreased functional activity in the brains of alcohol-exposed offspring. The 2-deoxyglucose results showed that glucose utilization was decreased in all sensory systems except for parietal cortex, most motor systems, hippocampus, and other limbic structures.

Tritiated Thymidine

The tritiated thymidine procedure is used to identify the "birthdate" of neurons. Radioactive thymidine is injected into actively dividing nerve tissue and is incorporated into the DNA of neuroblasts and precursors of glia cells.

Computer Assisted Tomography

In contrast with other methods that require removal of the brain and therefore are primarily suitable for animal studies, a number of painless techniques are available for looking inside human brains. One such technique is the CAT scan or brain scan, which records x-ray data from several different directions. This information is computer processed and converted into a three-dimensional image of the brain. Data from CAT scans can indicate brain damage such as cerebral atrophy and ventricular enlargement. A shortcoming of CAT scan data is that it is unable to quantify differences between structures in different brain areas. However, once structured anomalies have been identified by the CAT scan, quantitative evaluations can be made by position emission tomography.

Positron Emission Tomography

Positron emission tomography, commonly called the PET scan, is another technique for studying possible brain damage in people. PET is very closely related to the 2-deoxyglucose method used in animal studies. A patient is injected with a small amount of radiolabeled 2-deoxyglucose and the PET scan, which is a radiation detector, quantifies the amount of radioactivity in brain areas. This data allows the investigator to quantify rates of glucose metabolism and hence neuronal activity in specific areas of the brain.

Neurophysiological Procedures

The brain is always active and generates electrical activity on its own. When nerves are induced to fire because of some stimulus, such as light, or noise, the brain responds with changes in electrical activity. Electrodes can record such spontaneous or evoked activity. Neuroanatomical procedures provide information about whether cells appear intact and if they seem to be making the appropriate connections. But neurons and nerve tracts may look normal yet be functionally abnormal. One way of finding out if this is so is by neurophysiological procedures.

EEG. During synaptogenesis, cortical circuits form between neurons, and electrical impulses within the brain become detectable by means of electroencephalography (EEG). Over the last 25 years, normal patterns of bioelectrical maturation have been determined and the EEG can be used to assess maturational development and neurological dysfunction.

In humans, EEG electrodes are placed on the scalp. In animals, portions of the scalp are usually surgically removed and electrodes are placed directly into the cortex.

EEG testing involves recognition of recurrent patterns of distinct types of electrical rhythms emanating from the cerebral cortex. These are expressed in terms of cycles per second or Hertz (abbreviated Hz). The four main frequencies are

Delta	3 Hz or less
Theta	4–7 Hz
Alpha	8–13 Hz
Beta	14 Hz or higher

In addition to frequency, bioelectrical activity has been examined in terms of amplitude, wave form, voltage, and distribution in different areas of the brain and symmetry between the two halves of the brain.

The normal full-term infant's bioelectrical activity is characterized by well-differentiated sleep cycles and organized patterns of electrical activity. When the infant is awake, continuous theta frequencies are seen along with some delta frequencies. During active sleep ("rapid eye movement" or REM sleep) activity similar to the awake state occurs with predominantly theta waves and more delta wave activity. Newborns with less mature brains have activity characterized by discontinuous, undifferentiated, sleep-wakefulness states with mixed frequencies and sharp changes to other frequencies.

An example of brain EEG used to identify brain damage in children that appear physically normal is a study by Chernick and his co-workers (1983). EEG recordings were obtained for 90 to 120 minutes at 3 days after birth immediately after feeding. The study was designed to see if drinking four or more alcoholic drinks per day by the mother during pregnancy produced changes in the bioelectrical activity of newborns. Infants were matched for race, sex, socioeconomic class, and maternal smoking history. The infants born to the heavy alcohol drinkers had much higher amplitudes of electrical brain activity in all stages of sleep than controls, whereas infants born to smokers did not differ significantly from controls.

Evoked Potentials. In contrast to the ongoing, spontaneous bioelectrical activity of the cortex that EEG measures, sensory evoked potentials are specific transient electrical responses of the brain to auditory (e.g., clicks or tones), visual (e.g., flashes or pattern reversals), or somatosensory (e.g., touch) stimuli. These responses are buried within the background EEG (considered to be "noise") and special computer-averaging techniques are used to decrease the background "noise." By recording from different areas of the brain, different sensory systems can be studied and investigators can determine if damage to cortical relay stations, e.g., inferior colliculus or pathways, has occurred.

Evoked potentials are frequently categorized as short-, intermediate-, and long-latency evoked potentials. The brainstem auditory evoked potential (BAEP) is a short-latency evoked potential occurring within 10 msec poststimulus, and reflecting activity from peripheral and brainstem auditory structures. The midlatency auditory evoked potential (MAEP) occurs within 10 to 50 msec poststimulus, possibly reflecting activity from midbrain, diencephalon, and primary cortical auditory structures. The cortical auditory evoked potential (CAEP) is a long-latency evoked potential, reflecting activity from the cortex, hippocampus, and other structures.

Brainstem auditory evoked potentials can be detected in humans as early as 26 weeks of gestation but have larger latencies, smaller amplitudes, and a greater response to changes in stimulation than at birth. At 34 weeks gestation, a dramatic change occurs toward latencies characteristic of full term. This is a very reliable change and has diagnostic value in assessing maturational state. It probably reflects maturation of neurotransmitter systems and myelination (Hecox, 1981).

For testing auditory evoked potentials, brief repetitive (usually 10 to 100 per second)

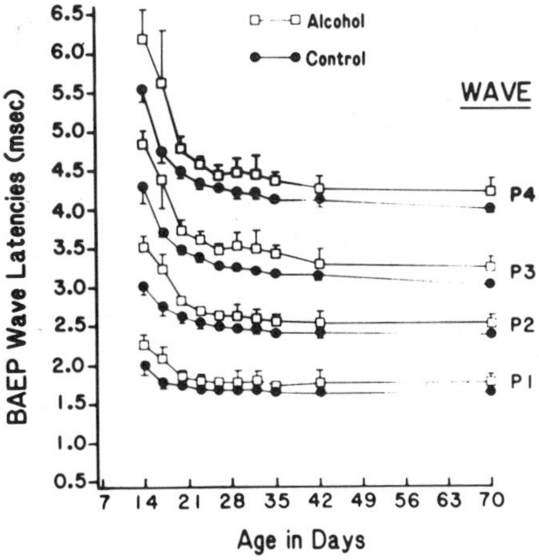

Figure 4.11. Mean (± S.D.) brainstem auditory evoked potential (BAEP) latencies as a function of age for control and alcohol rat pups. The latencies of all four BAEP components for alcohol-exposed pups were prolonged in comparison with control pups throughout postnatal development. The latency differences between these two groups were greatest early in development and narrowed somewhat between days 17 and 23 of the postnatal period. Latencies include a 0.7 msec acoustic transit time from the earphone to the animal's ears. (Data from Church *et al.*, 1987; photo courtesy Dr. M. Church.)

clicks or tones are delivered through earphones into one ear. A recording electrode is placed on the center of the scalp and a reference electrode is placed on the other ear. Electrodes record responses elicited by the stimulus and ongoing "noise." Electronic filtering of electrical activity allows the evoked response to be separated from the EEG "noise." Other sources of "noise" are reduced by signal averaging that examines many responses, determines what is the normal latency of a response to a click, and ignores random activity.

In older children, seven waves are typically generated in the first 10 milliseconds after a click (Jewett *et al.*, 1970) reflecting successive activation of auditory pathways from the acoustic nerve to the thalamus. By examining these wave forms it is possible to identify the site of auditory damage from the ear to the thalamus. Although there is no agreement as to how the waves are generated, there is general agreement that the chief generators of the various BAEP waves are auditory nerve (I), cochlear nucleus (II), superior olivary complex (III), lateral lemniscus (IV), inferior colliculus (V), medical geniculate body (VI), and thalamocortical radiations (VII). Thus each wave may represent the successive activation of neurons and/or nuclei along the ascending brainstem auditory pathway. In clinical evaluation, the presence or absence of these waves are important indicators of function.

Auditory evoked potentials can also be used to assess delayed postnatal maturation of the auditory system in animals prenatally exposed to teratogens. For example, Church *et al.* (1987) examined brainstem auditory evoked potentials (BAEPs) in animals prenatally exposed to alcohol from gestation day 4 to parturition. In the rat, at least four distinct BAEP waves can be distinguished within a 6-msec interval after an auditory stimulus. The principal neural generators of these waves in the rat are believed to be the auditory nerve (P1), the cochlear nucleus (P2), the superior olive/trapezoid body (P3), and the rostral lateral lemniscus or inferior colliculus (P4).

Animals prenatally exposed to alcohol had longer latencies for all BAEP components (see Figure 4.11). Catch-up occurred by about 21 days of age, but differences remained significant at maturation, indicating permanent change along the peripheral and/or brainstem auditory pathway.

Further investigation indicated that this change in BAEP activity was due largely to a recruitment-type sensorineural hearing loss, a finding later confirmed in children with fetal alcohol syndrome (Church & Gerkin, 1988).

References

Abel, E. L., and Greizerstein, H. B. Relation of alcohol content in amniotic fluid, fetal and maternal blood. *Alcoholism: Clinical and Experimental Research,* 1980, *4,* 209.

Abel, E. L., Jacobson, S., and Sherwin, B. T. In utero alcohol exposure: Functional and structural brain damage. *Neurobehavioral Toxicology and Teratology,* 1981, *5,* 363–366.

Abramovitch, R. D. Human sexual differentiation—in utero influences. *Journal of Obstetrics and Gynecology,* 1974, *81,* 448–453.

Amin-Zaki, L., Ehassani, S., Majeed, M. A., Clarkson, T. W., Doherty, R. A., Greenwood, M. R., and Giovanoli-Jakubczak, T. Perinatal methylmercury poisoning in Iraq. *American Journal of Diseases of Children,* 1976, *130,* 1070–1076.

Ayoub, D. M., Greenough, W. J., and Juraska, J. M. Sex differences in dendritic structure in the preoptic area of the juvenile macaque monkey brain. *Science,* 1983, *219,* 197–1983.

Bensaude, O., Babinet, C., Morange, M., and Jacob, F. Heat shock proteins, first major products of zygotic gene activity in mouse embryo. *Nature,* 1983, *305,* 331–333.

Black, I. B., Adler, J. E., Dreyfus, C. F., Jonakait, G. M., Katz, D. M., LaGamma, E. F., and Markey, K. M. Neurotransmitters plasticity at the molecular level. *Science,* 1984, *225,* 1266–1270.

Braude, P., Pelham, H., Flach, G., and Lobatto, R. Post-transcriptional control in the early mouse embryo. *Nature,* 1979, *282,* 102–105.

Chang, L. W., and Annau, Z. Developmental neuropathology and behavioral teratology of methylmercury. In J. Yanai (Ed.), *Neurobehavioral Teratology.* New York: Elsevier, 1984, pp. 405–432.

Chernick, V., Childiaeva, R., and Ioffe, S. Effects of maternal alcohol intake and smoking on neonatal electroencephalogram and anthropometric measurements. *American Journal of Obstetrics and Gynecology,* 1983, *146,* 41–42.

Church, M. W., Abel, E. L., Dintcheff, B. A., Gerkin, K. P., Gritzke, R., and Holloway, J. A. Brainstem and cortical auditory evoked potentials in rats chronically exposed to alcohol in utero. *Electroencephalograhpy and Clinical Neurophysiology,* 1987, Suppl 40, 452–460.

Church, M. W., and Gerkin, K. P. Hearing disorders in children with fetal alcohol syndrome: Findings from case reports. *Pediatrics,* 1988, *82,* 147–154.

Clarren, S. K., Alvord, E. C., Suni, S. M., and Streissguth, A. P. Brain malformations related to prenatal exposure to ethanol. *Journal of Pediatrics,* 1978, *92,* 64–67.

Crepel, F., Mariani, J., and Delhaye-Bouchard, N. Evidence for a multiple innervation of Purkinje cells by climbing fibers in the immature rat cerebellum. *Journal of Neurobiology,* 1976, *7,* 567–578.

Dambska, M., Haddad, R., Kozlowski, P. B., Lee, M. H., and Shek, J. Telencephalic cytoarchitectonics in the brains of rats with graded degrees of microencephaly. *Acta Neuropathologica,* 1982, *58,* 203–209.

Das, G. D. Neural transplantation in mammalian brain: Some conceptual and technical considerations. In R. B. Wallace and G. D. Gas, (Eds.), *Neural Tissue Transplantation Research.*

Davidson, E. H., *Gene Activity in Early Development.* New York: Academic Press, 1976. New York: Springer-Verlag, 1983, pp. 1–64.

Davies, A. M., Bandtlow, C., Heumann, R., Korsching, S., Rohrer, H., and Thonen, H. Timing and site of nerve growth factor synthesis in developing skin in relation to innervation and expression of the receptor. *Nature,* 1987, *326,* 353–358.

Doak, R. L., Hall, A., and Dale, H. E. Longevity of spermatozoa in the reproductive tract of the bitch. *Journal of Reproduction and Fertility,* 1967, *13,* 51–58.

Dobbing, J., and Sands, J. Comparative aspects of the brain growth spurt. *Early Human Development,* 1979, *3,* 79–83.

Dobbing, J., and Smart, J. L. Early undernutrition, brain development and behavior. In S. A. Barnett (Ed.), *Ethology and Development.* London: Heinemann Medical Books, 1973, pp. 16–36.

Dotti, C. G., and Banker, G. A. Experimentally induced alteration in the polarity of developing neurons. *Nature,* 1987, *330,* 254–256.

Eccles, C. U., and Annau, Z. Prenatal methylmercury exposure: II. Alterations in learning and psychotropic drug sensitivity in adult offspring. *Neurobehavioral Toxicology and Teratology,* 1982, *4,* 377–382.

Edelman, G. M. Cell adhesion molecules: A molecular basis for animal form. *Scientific American,* 1984, *250,* 118–129.

Edwards, R. G. Causes of early embryonic loss in human pregnancy. *Human Reproduction,* 1986, *1,* 185–198.

Ehrhardt, A. A., Evers, K., and Money, J. Influence of androgen and some aspects of sexually dimorphic behavior in women with the late-treated androgential syndrome. *Johns Hopkins Medical Journal,* 1968, *123,* 115–122.

Ehrhardt, A. A., Mayer-Bahlburg, H. F., Rosen, L. R., Feldman, J. F., Verdiano, N. P., Zimmerman, I., and McEwen, B. S. Sexual orientation after prenatal exposure to exogenous estrogen. *Archives of Sexual Behavior,* 1985, *14,* 57–75.

Ehrhardt, A. A., and Money, J. Progestin-induced hermaphroditism: I.Q. and psychosexual identity in a study of ten girls. *Journal of Sex Research,* 1967, *3,* 83–100.

Epstein, C. J., Smith, S., Travis, B., and Tucker, G. X chromosomes function before visible X-chromosome inactivation in female mouse embryos. *Nature,* 1978, *274,* 500–503.

Fallon, J. F., and Simandl, B. K. Evidence of a role for cell death in the disappearance of the embryonic human tail. *American Journal of Anatomy,* 1978, *152,* 111–130.

Finlay, B. L., and Slattery, M. Local differences in the amount of early cell death in neocortex predicts adult local specializations. *Science,* 1983, *219,* 1349–1351.

Fowler, R., and Edwards, R. The genetics of early human development. *Progress in Medical Genetics,* 1973, *9,* 49–112.

Freund, G. Neurologic diseases associated with chronic alcohol abuse. In B. Tabakoff, P. B. Sutker, and C. L. Randall (Eds.), *Medical and Social Aspects of Alcohol Abuse.* New York: Plenum Press, 1983, pp. 165–186.

Fuller, G. N., Johnston, D. A., and Wiggins, R. C. The relationship between nutritional adequacy and brain myelin accumulation: A comparison of varying degrees of well fed and undernourished rats. *Brain Research,* 1984, *290,* 195–198.

Fullerton, P. M. Chronic peripheral neuropathy produced by lead poisoning in guinea pigs. *Journal of Neuropathology and Experimental Neurology,* 1966, *25,* 214–236.

Garcia-Bellido, A., Lawrence, P. A., and Morata, G. Compartments in animal development. *Scientific American,* 1979, *241,* 102–110.

Gilbert, S. F. *Developmental Biology.* Sunderland, Mass.: Sinauer Associates, 1985.

Gilles F. H., Shankle, W., and Dooling, E. L. Myelinated tracts and growth patterns. In F. H. Giles, A. Leviton and E. C. Dooling (Eds.), *The Developing Human Brain.* Boston: John Wright, 1983, pp. 117–183.

Globus, A., and Scheibel, A. B. Pattern and field in cortical structure: The rabbit. *Journal of Comparative Neurology,* 1967, *131,* 55–72.

Gurdon, J. B. Embryonic induction—molecular prospects. *Development,* 1987, *99,* 285–306.

Gurdon, J. B., Fairman, S., Mohun, T. J., and Brennan, S. Activation of muscle-specific actin genes in xenopus development by an induction between animal and vegetal cells of a blastula. *Cell,* 1985, *41,* 913–922.

Haddad, R. K., Rabe, A., Lacquer, G. L., Spatz, M., and Valsamis, M. Intellectual deficit associated with transplacentally induced microcephaly in the rat. *Science,* 1969, *163,* 88–90.

Hard, E., Dahlgren, I. L., Engel, J., Larsson, K., Liljequist, S., Lindh, A., and Musia, B. Development of sexual behavior in prenatally ethanol-exposed rats. *Drug and Alcohol Dependence,* 1984, *14,* 51–61.

Haydon, P. G., McCobb, D. P., and Kater, S. B. The regulation of neurite outgrowth, growth cone motility, and electrical synaptogenesis by serotonin. *Journal of Neurobiology,* 1987, *18,* 197–215.

Hecox, K., Cone, B., and Blaw, M. Brainstem auditory evoked response in the diagnosis of pediatric neurologic disease. *Neurology,* 1981, *31,* 832–840.

Hogan, R. N., and Coleman, P. D. Experimental hyperphenylalaninemia: Dendritic alterations in cerebellum of rat. *Experimental Neurology,* 1981, *74,* 234–244.

Hokfelt, T., Johansson, O., and Goldstein, M. Chemical anatomy of the brain. *Science,* 1984, *225,* 1326–1334.

Huttenlocher, P. R. Dendritic development in neocortex of children with mental defect and infantile spasms. *Neurology,* 1974, *24,* 203–210.

Jewett, D. L., Romano, M. N., and Williston, J. S. Human auditory evoked potentials: Possible brainstem components detected on the scalp. *Science,* 1970, *167,* 1517–1518.

Levine, M., Rubin, G. N., and Tjian, R. Human DNA sequences homologous to a protein coding region conserved between homeotic genes of Drosophila. *Cell,* 1984, *38,* 667–673.

Lichtensteiger, W., and Schlumpf, M. Prenatal nicotine affects fetal testosterone and sexual dimorphism of sacchain preference. *Pharmacology, Biochemistry and Behavior,* 1985, *23,* 439–444.

MacLusky, N. J., and Naftolin, F. Sexual differentiation of the central nervous system. *Science,* 1981, *211,* 1294–1303.

McEwen, B. S., Luine, V. N., and Fischette, C. T. Developmental actions of hormones: From receptors to function. In S. S. Easter, K. F. Barald, and B. M. Carlson (Eds.), *From Message to Mind,* Sunderland, Mass.: Sinauer Associates, 1988, pp. 272–287.

McGivern, R. F., Clancy, A. N., Hill, M. A., and Noble, E. P. Prenatal alcohol exposure alters adult expression of sexually dimorphic behavior in the rat. *Science,* 1984, *224,* 896–898.

Miller, M. Effects of alcohol on the generation and migration of cerebral cortical neurons. *Science,* 1986, *233,* 1308–1311.

Morrissey, R. E., and Mottet, N. K. Neural tube defects and brain anomalies: A review of selected teratogens and their possible modes of action. *Neurotoxicology,* 1980, *2,* 125–162.

Nobin, A., and Bjorklund, A. Topography of monoamine neuron systems in the human brain as revealed in fetuses. *Acta Physiologica Scandinavica,* 1973, *388,* 1–40.

Norton, W. T., and Poduslo, S. E. Myelin in rat brain: Changes in myelin composition during brain maturation. *Journal of Neurochemistry,* 1973, *21,* 759–773.

O'Leary, D. D. M., Stanfield, B. B., and Cowan, W. M. Evidence that the early postnatal restriction of the cells or origin of the callosal projection is due to the elimination of axonal collaterals rather than to the death of neurons. *Developmental Brain Research,* 1981, *1,* 607–617.

Patsalos, P. N., and Wiggins, R. C. Brain maturation following administration of phenobarbitol, phenytoin, and sodium valproate to developing rats as to their dams: Effects on synthesis of brain myelin and other subcellular membranes. *Journal of Neurochemistry,* 1982, *39,* 915–923.

Petit, T. L., and Isaacson, R. L. Anatomical and behavior effects of colchicine administration to rats late in utero. *Developmental Psychobiology,* 1976, *9,* 119–129.

Petkovich, M., Brand, N. J., Krust, A., and Chambon, P. A human retinoic acid receptor which belongs to the family of nuclear receptors. *Nature,* 1987, *330,* 444–450.

Phoenix, C. H. Prenatal testosterone in the nonhuman primate and its consequences for behavior. In R. C. Friedman, R. M. Richart, and R. L. VandeWiile (Eds.), *Sex Differences in Behavior.* New York: Wiley, 1974, pp. 19–32.

Phoenix, C. H., Goy, R. W., Gerall, A. A., and Young, W. C. Organizing action of prenatally administered testosterone propionate on the tissues mediating mating behavior in the female guinea-pig. *Endocrinology,* 1959, *65,* 369–382.

Purves, D., and Litchman, J. W. Elimination of synapses in the developing nervous system. *Science,* 1980, *210,* 153–157.

Purves, D., and Lichtman, J. W. *Principles of Neural Development.* Sunderland, Massachusetts: Sinauer Associates, 1985.

Quadagno, D. M., Briscoe, R., and Quadagno, J. S. Effect of perinatal gonadal hormones on selected nonsexual behavior patterns: A critical assessment of the nonhuman and human literature. *Psychological Bulletin,* 1977, *84,* 62–80.

Rabe, A., and Haddad, R. K. Methylazoxymethanol-induced microencephaly in rats: Behavioral studies. *Federation Proceedings,* 1972, *31,* 1536–1539.

Rakic, P. Defects of neuronal migration and the pathogenesis of cortical malformations. *Progress in Brain Research,* 1988, *73,* 15–37.

Reinisch, J. M., and Sanders, S. A. Early barbiturate exposure: The brain, sexually dimorphic behavior and learning. *Neuroscience and Biobehavioral Reviews,* 1982, *6,* 311–319.

Reuhl, K. R., Chang, L. W., and Townsend, J. W. Pathological effects of in utero methylmercury exposure on the cerebellum of the golden hamster: I. Early effects upon the neonatal cerebellar cortex. *Environmental Research,* 1981, *26,* 281–306.

Rodier, P. M. Behavioral effects of antimitotic agents administered during neurogenesis. In E. P. Riley and C. V. Vorhees (Eds.), *Handbook of Behavioral Teratology.* New York: Plenum Press, 1986, pp. 185–209.

Rudeen, P. K., Kappel, C. A., and Lear, K. Postnatal or in utero ethanol exposure reduction of the volume of the sexually dimorphic nucleus of the preoptic area in male rats. *Drug and Alcohol Dependence,* 1986, *18,* 247–252.

Samorajski, T., Lancaster, F., and Wiggins, R. C. Fetal ethanol exposure: A morphometric analysis of myelination in the optic nerve. *International Journal of Developmental Neuroscience,* 1986, *4,* 369–374.

Schlessinger, A. R., Cowan, W. M., and Gottlieb, D. I. An autoradiographic study of the time of origin and pattern of granule cell migration in the dentate gyrus of the rat. *Journal of Comparative Neurology,* 1975, *159,* 149–176.

Seiger, A., and Olson, L. Late prenatal ontogeny of central monoamine neurons in the rat. *Zeitschrift Anatomishe Entwgesh,* 1973, *140,* 281–318.

Sengelaub, D. R., and Finlay, B. L. Early removal of one eye reduces normally occurring cell death in the remaining eye. *Science,* 1981, *213,* 573–574.

Shatz, C. J. Prenatal development of the cat's retino-geniculate pathway. *Journal of Neuroscience,* 1983, *3,* 482–499.

Shiota, K., Uwabe, C., and Nishimura, H. High prevalence of defective human embryos at the early postimplantation period. *Teratology,* 1987, *35,* 309–316.

Siiteri, P., and Wilson, J. D. Testosterone formation and metabolism during male sexual differentiation in the human embryo. *Journal of Clinical and Endocrinological Metabolism,* 1974, *38,* 113–125.

Smart, Y. C., Roberts, T. K., Clancy, R. L., and Cripps, A. W. Early pregnancy factor: Its role in mammalian reproduction—research review. *Fertility and Sterility,* 1981, *35,* 397–402.

Smith, J. C. A mesoderm-inducing factor is produced by a Xenopus line. *Development,* 1987, *99,* 3–14.

Snyder, K. D. Congenital mercury poisoning. *New England Journal of Medicine,* 1971, *284,* 1014–1016.

Spemann, R. H. *Embryonic Development and Induction.* New Haven, Conn.: Yale University Press, 1938.

Spyker, J. M., and Smithberg, M. Effects of methylmercury on prenatal development in mice. *Teratology,* 1972, *5,* 181–190.

Stanfield, B. B., O'Leary, D. D. M., and Fricks, C. Selective collateral elimination in early postnatal development restricts cortical distribution of rat pyramidal tract neurons. *Nature,* 1982, *298,* 371–373.

Swaab, D. F., and Fliers, E. A sexually dimorphic nucleus in the human brain. *Science,* 1985, *228,* 1112–1115.

Thiersch, J. B. Effect of N-desacetyl-thio-colchicine (YC) and N-desacetyl-methyl-colchicine (MC) on rat fetus and litter in utero. *Proceedings of the Society of Experimental Biology and Medicine,* 1958, *98,* 479–483.

Toth, L. E., Slawin, K. L., Pintar, J. E., and Nguyen-Hun, M. C. Region-specific expression of mouse homeobox genes in the embryonic mesoderm and central nervous system. *Proceedings of the National Academy of Sciences,* 1987, *84,* 6790–6794.

Vingan, R. D., Dow-Edwards, D. L., and Riley, E. P. Cerebral metabolic alterations in rats following prenatal exposure: A deoxyglucose study. *Alcoholism: Clinical and Experimental Research,* 1986, *10,* 22–26.

Wallhauser, E., and Scheich, H. Auditory imprinting leads to differential 2-deoxyglucose uptake and dendritic spine loss in the chick rostral forebrain. *Developmental Brain Research,* 1987, *31,* 29–44.

Ward, I. L. Female sexual behavior in male rats treated prenatally with an anti-androgen. *Physiology and Behavior,* 1972, *8,* 53–56.

Wiggins, R. C. Myelin development and nutritional insufficiency. *Brain Research Review,* 1986, *4,* 151–175.

Wiggins, R. C., and Fuller, G. N. Early postnatal starvation causes lasting brain myelination. *Journal of Neurochemistry,* 1978, *30,* 1231–1237.

Wilcox, A. J., Weinberg, C. R., O'Connor, J. F., Baird, D. D., Schlatterer, J. P., Canfield, R. E., Armstrong, E. G., and Nisula, B. C. Incidence of early loss of pregnancy. *New England Journal of Medicine, 1988, 319,* 189–194.

Wolpert, L. Positional information and the spatial pattern of cellular differentiation. *Journal of Theoretical Biology,* 1969, *25,* 1–48.

Yakovlev, P. A., and LeCours, A. R. The myelogenetic cycles of regional maturation of the brain. In A. Minkowski (Ed.), *Regional Development of the Brain in Early Life.* Oxford: Blackwell Scientific Publishers, 1967, pp. 3–70.

Yamamoto, H., Chan-Palag, V., Steinbusch, H. W. N., and Palay, S. L. Hyperinnervation of arrested granule cells produced by the transportation of monoamine-containing neurons into the fourth ventricle of the rat. *Anatomy and Embryology,* 1980, *159,* 1–15.

Zenzes, M. T., Wolf, U., Gunther, E., and Engel, W. Studies on the function of H-Y antigen: Dissociation of reorganization experiments on rat gonadal tissue. *Cytogenetics and Cell Genetics,* 1978, *20,* 365–372.

CHAPTER 5

Behavioral Teratogenesis

The preceding chapter examined how the brain develops and some of the ways such development can be subverted by agents called *teratogens*. The word *teratogen* comes from the Greek *teras*, ''monster.'' Although we no longer refer to children born with malformations as monsters, we still call agents that cause birth defects *teratogens*. Teratology (the study of agents that cause birth defects) is closely related to the general area of *developmental toxicology* which looks at all chemically induced perturbations in development. For some developmental toxicologists, teratogenesis is a toxic effect restricted to the embryo or fetus (Hutchings, 1985).

At one time, teratology concerned itself only with structural anomalies. This focus expanded about 20 years ago to include physiological and functional anomalies as well, resulting in the present discipline of behavioral teratology.

Behavioral Teratology

Behavioral teratology is the study of behavioral abnormalities resulting from prenatal or (in animals) early postnatal influences. Behavior may be abnormal because it is not observed in a species but is typical of it or because it occurs less or more frequently than normal.

One of the first problems with such a definition is the meaning of behavior. Students with a background in psychology are already aware of the confusion and lack of agreement about what behavior is or isn't. This confusion is reflected in the definitions contained in the many general psychology textbooks that have been written. What's even more confusing is that all of these definitions make sense and can be defended by those who have written them. In this book, *behavior* refers to any observable activity or any functional changes in the body that can be inferred by observation. This is a very broad perspective, broad enough to include the migration of nerve cells during development and the aggressive behavior of children in school—the levels of analysis are unfettered.

To a major extent, behavioral teratology is still very much influenced by psychology because evaluation of behavior, especially learning and memory function, is one of psychology's strong points. When psychology started to emerge in the 1950s, behavior continued to be an important variable but it was no longer the only variable worth measuring. And with the development of techniques for measuring what was going on at the cellular level of the brain, it became possible to make inferences about how changes at the level of the neuron might relate to the behavior of the whole animal or person.

Whereas a permanent record can be made of neuronal damage, behavior is ephemeral; it leaves no permanent trace apart from what is observed and recorded at the time of its occurrence (Butcher *et al.*, 1980). What is recorded is also very sensitive to many postnatal variables, most of which are capable of producing similar effects on behavior so that it is difficult to determine causation.

Both psychobiology and behavioral teratology presume behavior has a biological basis and both consider that basis to be just as worthy of investigation as behavior itself. In fact, it was only after behavioral teratologists broadened their focus to include biological measures that the field gained a solid foundation and with it, respectability, says Donald Hutchings (1983), a pioneer in the field. According to Hutchings, behavioral teratology's early years were characterized by measurements of behavior only. Clearly, different agents produced behavioral changes, but there were no studies of biochemical or histological changes in the brain with which to correlate behavioral changes. Biochemical or histological abnormalities underlying behavioral anomalies were undoubtedly lurking in the brain, but without knowing where and what these were and without organizing principles, there was no way of putting behavioral teratology on a sound biological foundation.

The change came in the 1970s when new investigators began studying agents known to interfere with brain development and began relating the effects they observed to the principles and concepts of teratology (Vorhees, 1986). These efforts gave behavioral teratology a sound biological foundation. Behavioral teratologists no longer simply assumed a teratogen caused some behavioral anomaly because of some abnormality in the brain. Now they started searching for such abnormalities. Those interested in more background and perspective on behavioral teratology can consult the new *Handbook of Behavioral Teratology* edited by Riley and Vorhees (1986).

At present, behavioral teratology is an inductive science. There are no basic laws from which to make deductions as there are in other sciences like physics or chemistry. In physics $F = m \times a$ where F is force, m is mass, and a is acceleration. If you know any two of these parameters, you can deduce the third. In behavior such elegant and simplistic laws do not exist. Instead, behavioral teratology now relies on a number of principles derived from teratology for organization and prediction. A summary of some of the principal ways teratogens may interfere with CNS development is presented in Table 5.1.

The most influential contributor to the formulation of these principles was James Wil-

Table 5.1. Possible Changes by which Teratogens Can Interfere with CNS Development

Cell death
Altered differentiation
Decreased proliferation of cells
Decreased migration of cells
Structural damage to tissues
Change in frequency of cell death
Altered synaptogenesis
Altered myelination

son. In *Environment and Birth Defects,* published in 1973, Wilson sorted the available data up to that time into six organizing principles. Although based almost entirely on structural malformations, these principles are still valid and have subsequently been refined and updated by Vorhees (1986) to include behavioral anomalies. This chapter draws upon both of these excellent presentations.

Principles of Teratology

Principle 1. The Principle of Teratogenic Response

Teratogens generally cause a spectrum of birth defects rather than a single defect. Depending on agent, dosage, time of exposure, and species, any of a number of anomalies may occur ranging from death to behavioral deficits in the absence of any physical anomalies.

Birth defect (also called *congenital anomaly* or *congenital defect*) is a general term referring to any structural, functional, or biochemical anomaly originating before birth or shortly afterwards. *Structural defect* (also called *congenital malformation*) refers to a defect observable to the naked eye or by microscope; it can be present on the surface or inside the body. There can be one or many defects. They can be hereditary or nonhereditary. A *malformation syndrome* is a recognizable pattern of structural defects. A *deformity* is an alteration in shape or structure of a previously normally formed part. Clinodactyly (curved 5th finger) is an example of a deformity.

A congenital malformation is a congenital anomaly, but the reverse is not necessarily true—a congenital anomaly is not always a congenital malformation. A person born with phenylketonuria (PKU) has a congenital anomaly but not a congenital malformation. Someone with PKU may have no structural defects and look normal but he has a biochemical disorder.

Depending on the level of analysis, every behavioral anomaly will ultimately be traceable to some biochemical or structural defect in the brain. Linking behavioral damage to such biochemical or structural changes is less than straightforward, because behavior represents an integration of many systems, and such changes will undoubtedly be widespread rather than confined to any single area. If multiple sites are affected, associating a particular change in a particular site with some behavioral anomaly will be a difficult undertaking. However, such biochemical, physiological, and subtle morphological changes in brain tissue are in and of themselves worthy of investigation.

Principle 2. The Principle of Mechanisms

The Principle of Mechanisms states that teratogens produce their effects by interfering with induction or differentiation. Details of such changes have been presented in previous chapters and are summarized in Figure 5.1.

Mutations have previously been discussed in the context of changes in germinal cells. Mutations are alterations in the sequences of DNA. Such alterations can also affect developmental sequences in somatic cells resulting in abnormal development. Chromosomal aberrations have likewise been previously discussed. These involve major losses or additions to chromosomes. Such aberrations are associated with major structural and functional abnormalities.

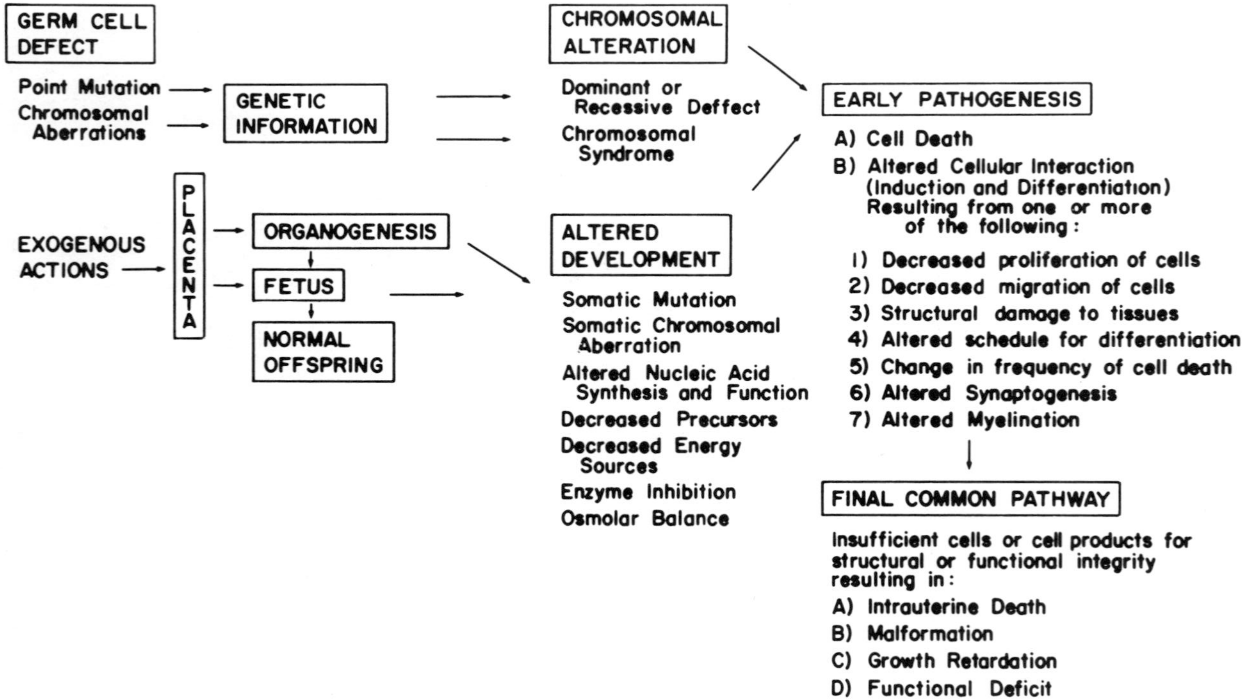

Figure 5.1. Basic mechanisms in prenatal determination of behavioral anomalies. (Adapted from Hutchings, 1985, and Tuchmann-Duplessis, 1975.)

Mitotic disturbances are interferences with cell division either through interference with spindle formation (e.g., colchicine), or separation of chromatids (e.g., x-rays), or impaired DNA biosynthesis (e.g., mitomycin C). Alterations in nucleic synthesis and function refer to changes in DNA or RNA apart from those involving mitotic disturbances. These often involve interference with synthesis or transcription of RNA (e.g., actinomycin C). Decreased availability of precursors, substrates, and coenzymes either because of maternal undernutrition or because of agents that impair absorption of nutrients across membranes (e.g., chloramphenicol or tetracycline) can result in decreased availabilty of substrates needed for fetal growth and development. Apart from the availability of substrates already mentioned, some agents may disrupt anaerobic and aerobic metabolism (e.g., 6-aminonicotinamide) thereby depriving cells of the energy they need for proliferation and migration. Some agents may also interfere with the activity of enzymes like ornithine decarboxylase (e.g., dioxin) resulting in abnormal development. Because the embryo and fetus are primarily composed of fluid, alterations in fluid pressures and composition of different fluids in the embryo could also interfere with developmental processes.

A teratogen may initiate all or any of the pathogenic responses shown in Figure 5.1 but the end result is the same: (1) either too few cells survive or (2) alterations occur in differentiating cells resulting in anomalies in structural or functional integrity.

It is not always clear whether the mechanism(s) for teratogenic effects involve some direct effect on the conceptus or one that arises indirectly as a result of some influence on the mother. For instance, a teratogen may cause the mother to eat less or may alter the transport of nutrients from her gastrointestinal tract to her blood, it may alter her uterine vasculature or its tonus or blood supply, or it may alter her endocrine glands resulting in abnormal hormonal balances. Even in experimental teratology, it is possible to control only for these nonspecific factors superficially.

Principle 3. The Principle of Access

The Principle of Access states that to be a teratogen, an agent must gain access to the developing embryo/fetus in high enough concentrations to impact on developmental processes. Only after a teratogen has combined with the part(s) of the cell with which it interacts to produce its effects can it be a possible teratogen. Factors affecting such access are summarized in Table 5.2.

Absorption. People introduce most teratogens into their body through their mouths and these are eventually absorbed into the blood from the gastrointestinal tract. Teratogens may also enter the body by other routes, such as inhalation (e.g., smoking tobacco or marihuana), sniffing (e.g., cocaine), or by intravenous injection (e.g., narcotics). Route of administration is a major consideration in experimental teratology and is examined in depth in Appendix A.

Transportation of substances across membranes including the placenta occurs mainly by passive diffusion. Like the movement of water from areas of high concentration to areas of lower concentration, the movement of teratogens across membranes is related to their dilution at a particular site. Blood levels produced by a given dose of teratogen may therefore be affected significantly by the volume of the administration. For instance, a dose of 5 mg/kg of drug X may produce a higher blood level if given in a volume of 5 ml/kg than if in 50 ml/kg. However, because of their limited solubility, some substances may be more active when administered in larger rather than small volumes. This is because the process

Table 5.2. Factors Influencing Passage of Teratogens from Mother to Fetus

Maternal factors
Absorption (route of administration, concentration gradient, molecular size, lipid solubility, ionization)
Biological rhythms
Age and parity
Diet
Distribution
Protein binding
Blood levels
Drug interactions
Amount of body fat
Metabolism
Species differences
Tolerance, cross tolerance
Uterine blood flow
Contractions
Hypertension
Placental factors
Placental blood flow
Molecular size of chemical molecule
Lipid solubility of chemical molecule
Plasma protein binding
pH of maternal blood
Thickness and surface area of the placental membrane
Placental membrane permeability

of dissolution has to precede that of absorption. Although volume considerations have been carefully examined in pharmacology, they have received little attention in teratology.

Absorption is also affected by molecular size. Small molecules have little difficulty passing through membrane pores. Substances with a molecular weight around 1000 will not easily cross the placenta. Most drugs have a molecular weight below 500 and therefore are not restricted. Large molecules like Trypan Blue are unable to cross the placenta but are still teratogenic because they become concentrated in the placenta and prevent transport of nutrients from the mother to the fetus.

Another way teratogens can pass into and across cells is by dissolving in cell membranes. Cell membranes, including those of the placenta, are mainly made up of protein and lipid material. The more soluble a teratogen is in lipids and protein, the more easily it can be absorbed across such membranes.

Yet another factor affecting absorption is ionization. A substance that has become ionized is one that has split into electrically charged particles. Because only the un-ionized form of a teratogen is lipid soluble, ionization retards or inhibits movement across cellular membranes. Teratogens that are weak acids or bases have little difficulty crossing the placenta; those that become ionized have considerable difficulty.

Lidocaine, a drug commonly used to reduce pain associated with giving birth, is a weak base that rapidly crosses the placenta in its un-ionized form. Under normal conditions, drug levels in the fetus are similar to levels in the mother. However, if the fetus's blood becomes acidotic, the drug dissociates in the fetus and does not pass back to the mother, and levels in the fetus accumulate (Biehl *et al.*, 1978). In addition, increased acidity reduces the protein binding of lidocaine by blood so that there is more free drug to enter the fetal brain (see below). These increased levels could be behaviorally teratogenic to the fetus.

Some substances potentiate the actions of other teratogens by affecting their ionization in the gastrointestinal tract. For example, alcohol increases acidity in the stomach so that acidic teratogens like diazepam are absorbed more quickly (MacLeod *et al.*, 1977), whereas those that are bases are ionized and absorbed more slowly.

The time of day a substance is taken may also affect its teratogenic action. For example, pentobarbital is most lethal to mice when given early in the morning (7 a.m.) and least when injected at noon (Lindsay & Kullman, 1966).

Both behavior and physiological functioning have endogenous biological rhythms. These rhythms may affect teratogenicity as well as toxicity. Rats and mice are more active at night than during the day, whereas humans are less active during the night. Liver enzyme activity also has a circadian (about 24 hours) rhythm (Rapoport *et al.*, 1966). Because the magnitude of the biological response to drugs is in many cases dependent upon the rate at which they are metabolized, individual differences in teratogenicity may correspond to the time of day in which exposure occurs.

Metabolic activity and endocrine function also exhibit seasonal variability (Haus and Halberg, 1970) and this may account for some reports of seasonal variations in malformation rates. For example, anencephaly in humans occurs much more often among children born during the winter months (Slater *et al.*, 1964). In mice, vitamin A is much more teratogenic when exposure occurs during the summer compared with winter (Kalter & Warkany, 1961), whereas cortisone induces cleft palate in mice more often if they are exposed during the winter months (Kalter, 1959).

A third factor affecting teratogenic response is maternal age. Maternal age affects teratogenicity in many ways (see Principle 8); one way is by affecting absorption of compounds into the blood.

Figure 5.2 shows that in pregnant rats, blood alcohol levels (BALs) are higher and peak later in older animals than in younger animals. Following a second injection of alcohol, older pregnant females had a peak BAL of 421 mg% at 4 hr postinjection, whereas middle-age and young mothers had peak BALs of 361 and 310 mg% at 2 and 1 hr postinjection, respectively. A similar, but less marked trend was observed following the first injection. Figure 5.2 also shows that the BAL curves for the three age groups closely parallel each other during the elimination phase, suggesting similar kinetics in alcohol elimination.

An increase in BAL due to maternal age or any factor would, theoretically, put the conceptus (and mother) at greater risk for an adverse pregnancy outcome. This increased risk is associated with increased maternal mortality (Abel & Dintcheff, 1985) and fetal mortality (see Table 5.3).

Diet may also affect teratogenic response. McClain and Rohrs (1985) placed pregnant mice on either a commercial rodent diet (Purina Rodent Laboratory Chow) or a purified diet (AIN-76) and injected them intraperitoneally with 50 mg/kg diphenylhydantoin (DPH) or saline on gestation days 12–14. About 21% of those on the Purina diet had pups with cleft palate compared with 75% for those consuming the AIN diet (see Figure 5.3). Sub-

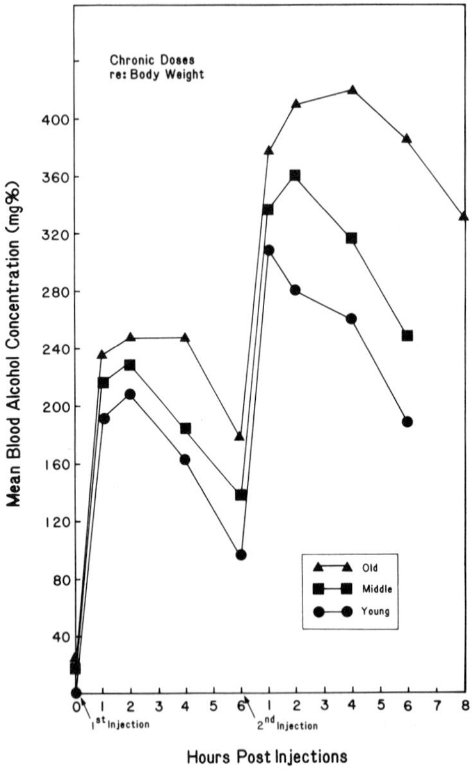

Figure 5.2. Mean (±S.E.) blood alcohol concentrations (BAC) for chronically treated young, middle, and old dams on gestation day 19 with dosages given relative to body weight (3.5 g/kg, twice daily). There was a monotonic increase in BAC with increasing age. (Data from Church *et al.*, submitted.)

stitution of carbohydrates (cornstarch for sucrose, which comprises 50% of the AIN diet) did not change the incidence of DPH-induced cleft palate but substitution of fatty acids (linseed oil for cornstarch, which comprises 5% of the diet) reduced the incidence to that similar to what was observed with Purina Lab Chow.

The reason for these differences between diets was that plasma DPH levels were significantly higher in animals fed the purified diet because of lower drug-metabolizing activity associated with diet consumption (McClain and Rohrs, 1985). When linseed oil was substituted for corn oil, metabolic activity in animals consuming the AIN diet was similar to those in animals consuming the Purina diet.

Table 5.3. Effects of Maternal Age on Effects of Prenatal Exposure to Alcohol (3.5 g/kg, p.o., Twice Daily) (Means ± S.E.)[a]

Variable	Young	Middle	Old
No. pregnant dams	13	13	21
No. implants	12.2±0.7	12.3±0.7	12.0±0.7
No. live fetuses	10.6±0.9	11.3±0.7	7.0±1.3
No. resorptions	1.6±0.6	1.0±0.3	5.0±1.3
Fetal weight	2.3±0.1	2.1±0.1	1.9±0.1[b]

[a] From Abel and Dintcheff, 1985.
[b] $n = 13$ for this mean value, omitting data from 8 litters with 100% fetal resorptions.

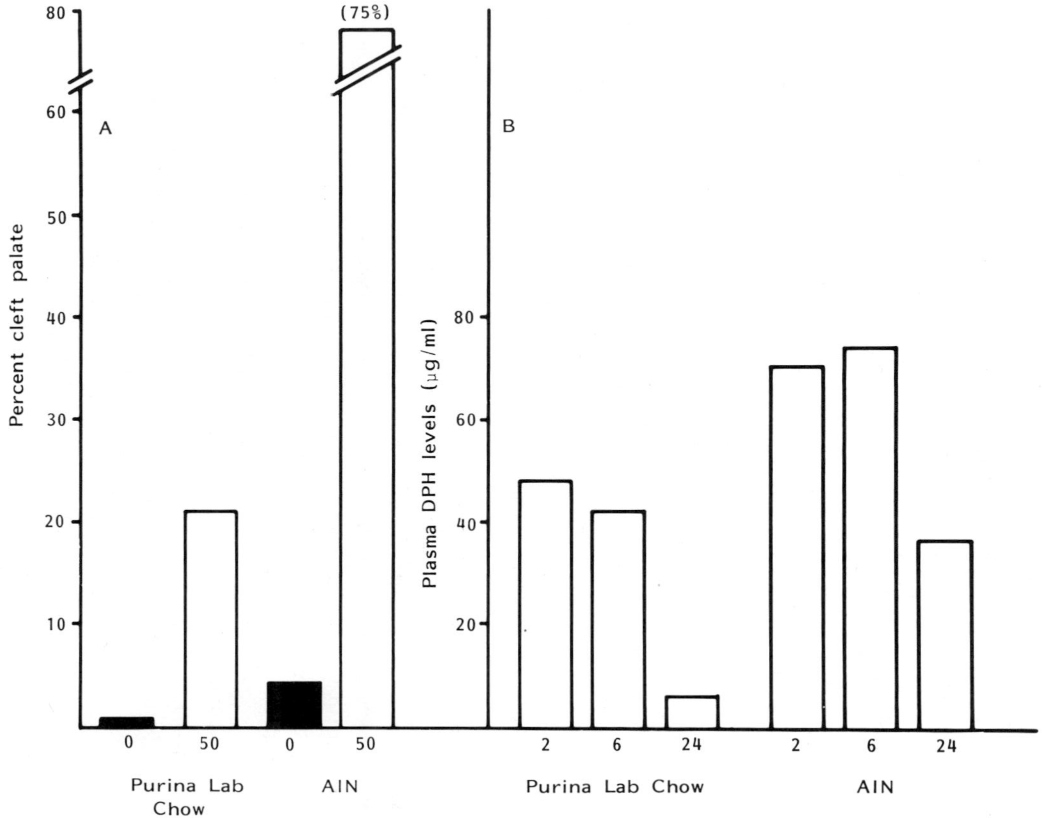

Figure 5.3. (A) Effects of diet (Purina Lab Chow and AIN-76) on diphenylhydantoin (DPH)-induced cleft palate in mice. (B) DPH plasma levels in pregnant mice fed different diets. Drug (50 mg/kg, i.p.) was administered on gestation days 12–14. (Figure adapted from McClain and Rohrs, 1985.)

Distribution. Once teratogens enter the blood, they may become "bound" to plasma proteins like albumin. Because only unbound teratogens can pass through cellular membranes, protein binding affects teratogenicity through its effect on the concentration of free drug available for passage to the fetus. Protein binding can also increase rate of clearance of a teratogen from blood by transporting it to the liver, where it is stripped from its binding sites and metabolized.

Not all substances are bound to plasma proteins with the same strength. Some substances, like delta-9-tetrahydrocannabinol (THC), the active ingredient in marihuana, are almost completely bound to human plasma protein.

In comparing maternal and fetal plasma levels, it is important to compare unbound plasma levels rather than total free levels to get an accurate idea of possible differences between maternal and fetal levels. The higher reported levels of THC in fetal compared with maternal circulation (e.g., Ho *et al.*, 1971) could very well be due to comparisons between total rather than unbound drug levels. When the concentration of a teratogen in the blood decreases, the amount of teratogen previously rendered inactive because of binding now becomes available for entry into cells. Consequently, although protein binding may remove a portion of a teratogen from free circulation, it may also serve to prolong its duration of action because molecules will move back into free circulation from binding sites as blood levels decrease.

Protein binding will affect access of a teratogen to the placenta only when its concentration in the blood is quite low to begin with. When peak blood levels are relatively high, the number of binding sites becomes quickly saturated. The amount of teratogen bound to plasma protein may then be at a maximum, but the actual percentage of bound teratogen is usually negligible compared with the total amount of teratogen that is free to diffuse across cell membranes.

During pregnancy, protein binding of drugs such as phenobarbital and phenytoin is decreased (Chen *et al.*, 1982). This is because there is relatively less albumin in the blood during pregnancy (D'Arcy & McElnay, 1982). Because of decreased protein binding there should be higher levels of unbound drugs in pregnant women compared with nonpregnant women. This means that more drug should be available for transplacental passage to the fetus than would be predicted on the basis of blood levels in the nonpregnant woman.

Some teratogens like diazepam are bound more strongly to protein in fetal than maternal blood. Plasma concentrations of total teratogen (bound and unbound) are higher in the newborn than the mother but the level of unbound teratogen in the mother and fetus are similar (Levy *et al.*, 1981). After birth, the highly protein bound portion of a teratogen is partially displaced by a sharp increase in free fatty acids so that the newborn is exposed to higher levels of free teratogen than the mother (Nau *et al.*, 1983). Because of immature metabolic activity, diazepam also remains in the neonatal body for a longer time than in the adult. The combined factors of displaced protein binding and immature metabolic activity may account for the occurrence of "floppy infant syndrome" associated with prenatal diazepam exposure (e.g., Gilberg, 1977).

Because protein binding is especially reduced in women with preeclampsia (Studd *et al.*, 1970), preeclampsia is a risk factor not only for intrauterine growth retardation, but also for teratogenicity due to effects on protein binding of teratogens. However, there are no studies comparing teratogenic effects in women with or without preeclampsia.

Effects on protein binding may be one explanation for interactions between teratogens. If two teratogens complete for the same binding sites, the teratogen with the higher affinity for plasma protein will displace the other. This results in an increase in the concentration of free teratogen for the displaced teratogen, thus possibly enhancing its teratogenic activity. If teratogen X, which is normally 99% bound to plasma protein, is displaced somewhat by teratogen Y so that X is only 95% bound, the amount of teratogen X now available to cross the placenta is increased fivefold. This change could significantly increase the teratogenicity of a hitherto nonteratogenic chemical.

Potentiation is a formidable problem that can result in very dramatic effects. For example, doses of cyclophosphamide or 5-fluorouracil that produce 26% and 10% malformation rates in rats when given individually produce a 100% malformation rate when combined (Wilson, 1964). A comparable interaction between alcohol and THC on fetotoxicity is illustrated in Figure 5.4.

Interactions between potential teratogens and other compounds can result in one of four possibilities. In addition to potentiation like that between cyclophosphamide and 5-fluorouracil and THC and alcohol, teratogens can have an additive effect such that the end result represents the summation of their additive effects. A third possibility is no significant biological consequence of simultaneous exposure to two or more compounds. The final possibility is that one compound interferes with the teratogenic impact of another. An interesting example of this last possibility is Randall and Anton's (1984) study showing that when aspirin is given together with alcohol, the number of malformed offspring and the prenatal mortality normally caused by alcohol is reduced (see Figure 5.5).

Metabolism. Unless there is a mechanism for inactivating teratogens or eliminating them from the body, their effects could conceivably go on for a long time or at least until the cells responding to a teratogen become exhausted.

One way the activity of a teratogen may be terminated is by altering its physical or chemical properties. Most of these enzymatic transformations take place in the liver but only a portion is transformed during each pass through the liver. The general effect of these transformations is that the polarity of a substance is increased. This makes it less lipid soluble and thereby decreases its activity. In some cases, however, the metabolite becomes more active than the parent compound.

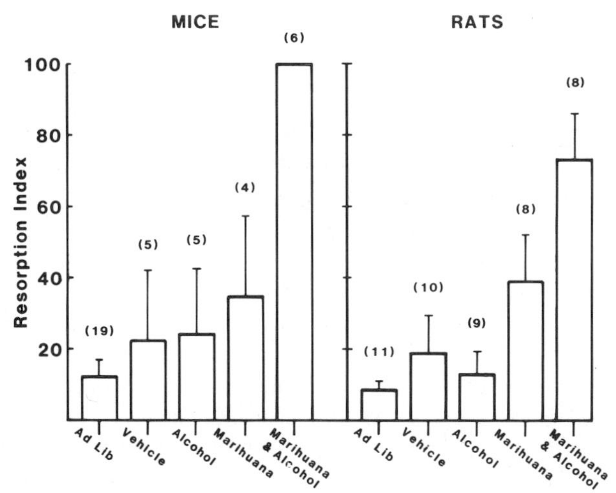

Figure 5.4. Interactive effects of marihuana and alcohol on fetotoxicity in rats and mice.

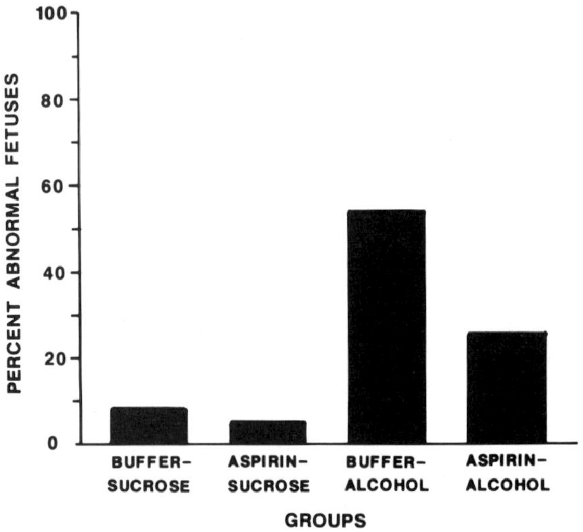

Figure 5.5. Interactive effects of alcohol and aspirin on birth defects in mice. Photo courtesy of Dr. C. Randall.

Some of the problems that make extrapolations of teratogenic effects in animals to humans tentative are that different species have different enzymatic pathways or concentrations of enzymes in various tissues. Thalidomide, for instance, causes birth defects in humans but not in rats or mice because only the human liver converts it to a toxic metabolite (Gordon *et al.*, 1981).

Pregnancy also affects liver enzyme differently among species and this could affect comparisons. For example, human liver drug metabolizing activity increases during pregnancy but decreases by about 50% in the rat. As a result, clearance of drugs such as diphenylhydantoin is increased during pregnancy in humans (Chen *et al.*, 1982).

Just as there may be competition between drugs for binding sites on plasma protein and cellular receptors, there may also be competition between drugs for the sites in the liver where metabolizing enzymes are located. Such competition is the basis for the treatment of alcoholism with the drug Antabuse (Disulfiram). Normally, alcohol is converted by the body into acetaldehyde, which in turn is converted into acetic acid by the enzyme aldehyde dehydrogenase. Antabuse, however, competes with acetaldehyde for the metabolic activity of this enzyme. As a result, acetaldehyde does not become transformed into acetic acid. As the amount of acetaldehyde in the blood increases, unpleasant physiological reactions such as nausea and vomiting begin to occur (Hald & Jacobsen, 1948). The alcoholic who receives Antabuse on a regular basis supposedly is motivated to abstain from alcohol because of anticipation of these unpleasant reactions, but many alcoholics drink anyway. As a result of their higher blood acetaldehyde levels, more acetaldehyde arrives at the placenta and because acetaldehyde is much more toxic than alcohol, there is a greater likelihood of a teratogenic response.

Besides affecting the metabolism of drugs by competition for metabolic enzymes, some compounds act directly on these enzymes to inhibit their activity whereas other drugs are able to stimulate the activity of these same enzymes. Such changes could result in different outcomes in experimental studies in which animals are chronically exposed to teratogens compared with acute exposure or exposure only during pregnancy.

Stimulation of drug-metabolizing activity generally leads to a more rapid inactivation of drugs. As a result, blood peak levels are reduced and there is a more rapid rate of drug disappearance. Consequently, larger and larger doses may have to be administered to produce the original pharmacological or behavioral effect. This condition of decreased responsiveness to the effects of a drug resulting from prior exposure to it is called *tolerance*. However, many of the microsomal enzymes lack substrate specificity and they will act upon a wide number of compounds. This can result in the related phenomenon of cross-tolerance wherein tolerance to one drug confers tolerance to other drugs as well.

Many of the enzymes involved in drug metabolism are either not present or are nonfunctional in the fetus. This is advantageous to the fetus. If a lipophilic drug were metabolized to a more polar and therefore less lipophilic metabolite in the fetus, it might not be able to pass back through the placenta and would remain in the fetus. Similarly, if the fetus were able to metabolize alcohol to acetaldehyde, it would be at much greater risk for developmental anomalies because acetaldehyde is much more toxic. After birth, there is a dramatic increase in metabolic function. This is also advantageous because it allows the newborn to break down any drugs retained during pregnancy.

Tolerance may also occur as a result of adaptation of the nervous system to certain drugs and this adaptation may also extend to other substances via cross-tolerance. Such cross-tolerance may be a factor in newborn behavior because a woman who is tolerant to a compound such as alcohol may be cross-tolerant to a general anesthetic. This means more of the anesthetic will have to be administered and her fetus will receive a greater amount of anesthetic than a fetus born to a woman who is not an alcoholic. Intrapartum anesthesia during delivery may result in several adverse newborn effects, such as behavioral depression, altered EEG activity, poor visual tracking and orientation to novel stimuli, decreased motor maturity, and increased irritability. Some of these disturbances may occur as long as 4 months after obstetric medication (Friedman *et al.*, 1978). Consequently, many of the behavioral effects reportedly associated with alcohol could just as readily be due to increased exposure of sedatives or analgesics during labor. As a result, observations of behavioral anomalies in children born to women who drink or use ilicit drugs may be confounded with effects of intrapartum medication.

Uterine Blood Flow. During pregnancy, uterine blood vessels are maximally dilated so as to maximize the supply of nutrients and oxygen to the fetus. In the rat, the extent of blood flow in late pregnancy is directly correlated with the number of placentae (P). In the virgin rat, blood flow to the uterine horn is about 0.158 ml/min. In the 12-day pregnant rat, it is increased to 0.116 (P) + 0.02 ml/min. At gestation day 20, it increases still further to 0.890 (P) + 0.3 ml/min (Lasuncion *et al.*, 1987). A reduction in blood flow would result in decreased transmission of nutrients and fetal hypoxia because of decreased oxygen transfer from the mother (Skillman *et al.*, 1985).

Because the blood rapidly distributes teratogens through the body, the higher the blood flow in the area of administration, the more rapid its absorption, whereas any substances that restrict uterine blood flow will produce varying degrees of fetal hypoxia. One such drug is cocaine. Maternal administration of cocaine to pregnant sheep produces a significant and dose-dependent decrease in total uterine blood flow. Maximal reductions of 24%, 34%, and 47% occur within 5 minutes after intravenous injection of 0.5, 1.0, and 2.0 mg/kg. This decrease in uterine blood flow is associated with a significant reduction in fetal blood oxygen and a significant increase in fetal heart rate. When cocaine is administered directly to the fetus, there is no significant change in fetal blood oxygen and increases in heart rate are not dose-related and are smaller than what occurs after maternal administration (Woods *et al.*, 1987). Because these effects on fetal oxygenation occur only when cocaine is administered to the mother, the changes must be a consequence of the reduction in uterine blood flow. The mechanism of action for this effect involves cocaine's well-known blockage of catecholamine reuptake at nerve terminals. Because catecholamines cause vasconstriction, inhibition of their reuptake results in vasoconstriction and decreased uterine blood flow.

Although the effects of prenatal exposure to cocaine have not been studied extensively, there are clinical reports of an increased risk of pregnancy complications, such as placental absorption, and spontaneous abortion and poor response to environmental stimuli in newborns associated with such exposure (Chasnoff *et al.*, 1985).

On the other hand, vasoconstriction of uterine blood vessels will also delay transplacental passage of potential teratogens by decreasing blood circulation through the placenta (Borell *et al.*, 1965). Because of this effect on placental blood flow, some clinicians advise that when analgesics are administered intravenously during labor, such administrations coincide with uterine contractions to reduce possible fetal effects (Haram *et al.*, 1978).

Placental Factors. The placenta is part of the conceptus and is the point of contact between it and the mother. It controls nutritional access to the conceptus and has special secretory and regulatory functions of its own that maintain pregnancy.

Teratogens may affect embryonal/fetal development directly or indirectly by affecting placental transport of nutrients from the maternal blood to the conceptus. For example, cadmium is considerably more toxic to the fetus when administered to the pregnant mother than directly to the fetus, even though fetal levels are about 10 times higher after direct fetal injection (Levin & Miller, 1980). Cadmium damages the placenta and also reduces transplacental transport of essential elements like zinc. This lowers zinc-dependent thymidine kinase activity in the embryo, which in turn results in a significant decrease in DNA synthesis.

Movement of teratogens from the blood through the placenta to the fetus is affected by the same factors that affect absorption across membranes throughout the body. However, various species have placentas with different numbers of cellular layers. Rodents and primates have a single chorion and fetal capillary bed separating maternal and fetal blood; other species have several layers. The method of placental transfer in species like the rodent also differs in early pregnancy from later pregnancy. In early pregnancy, rodents have a yolk sac placenta, whereas in late pregnancy the placenta system in rodents is like that in humans.

Principle 4. The Principle of Susceptibility

The Principle of Susceptibility states that just because a compound may gain access to the embryo or fetus—and most do—such access does not guarantee a teratogenic response. Hutchings (1983) states this principle in more colorful terms:

> it is misleading to portray the embryo as having a kind of gossamer fragility that would be silently ravaged by alien invaders. Rather . . . the embryo has multiple lines of defense, is a feisty combatant, and, even if knocked down and out, has enormous powers of recuperation and repair. (p. 8)

Susceptibility goes to the heart of the problem of extrapolating from studies in animals to humans. Just because a substance proves to be a teratogen in some animal model does not mean it will be a teratogenic in other animal models or in humans. However, if a substance is teratogenic in one species and not another, determining the reason for this difference may go a long way toward identifying its mechanism of action and those susceptibility factors in humans that make them more resistant or less resistant to its actions.

A corollary of the principle of susceptibility states that the exact same agent may cause a defect in one species and no effect in another or in one member of a species and not another. In the case of thalidomide, nearly every mother with an abnormal infant had taken the drug between the 3rd and the 8th week after conception. However, several children exposed to the drug during this time had no abnormalities (e.g., Kajii *et al.,* 1973).

Both the genotype of the embryo/fetus and the mother can affect susceptibility to mutagens or teratogens. The genotype of the conceptus affects the response to whatever it is that passes through the placenta; the genotype of the mother determines what will be delivered to the placenta.

Genotype of the Conceptus. Few studies have evaluated behavioral outcomes to teratogens in terms of genetic susceptibility. Therefore, the following examples illustrating genetic susceptibility mainly concern structural damage due to teratogens.

An elegant way of studying the influence of genotypic influences in humans is by comparing dizygotic twins. Because dizygotic twins develop in an almost identical intrauterine environment (there are slight differences such as position) but are genetically different, discordance reflects either different susceptibility or responsiveness. Susceptibility is indicated when only one twin is affected; responsiveness is indicated when both are affected but one is affected more seriously than the other.

An example of differential responsiveness is Christoffel and Salafsky's (1975) report of dizygotic twins born to an alcoholic woman. Although one of the infants had many of the physical features characteristic of the fetal alcohol syndrome at birth, the other was minimally affected and probably would have escaped recognition had his twin not been so severely affected. Differential susceptibility has also been noted in dizygotic twins whose mothers took anticonvulsants or thalidomide during pregnancy (Lenz, 1966; Loughnan *et al.,* 1973). In each case, only one of the twins exhibited teratogenicity.

An example of genotypic contributions affecting teratogenic response is Finnell and Chernoff's (1984) study of diphenylhydantoin-induced malformations in different strains of mice. Eleven common malformations were recorded. When all the strains were combined, creating a heterogenous population, the authors were able to arrange the pattern of malformations into hierarchies with those occurring in at least 25% of prenatally exposed fetuses (e.g., delayed and deficient bone growth, dilated cerebral ventricles, kidney defects) and those occurring in less than 25% of the exposed fetuses (e.g., heart anomalies, cleft palate) (see Table 5.4). This hierarchy was similar to that seen in humans exposed to diphenylhydantoin. When individual strains of mice were examined, the same hierarchy was no longer present. Instead, different strains of mice, each with their own genotype, had their own hierarchies of malformations. Finnell and Chernoff (1984) suggest that multiple malformations (the rule rather than the exception—see Principle 6) due to teratogens reflect the existence of vulnerable genes or vulnerable cells for each organ, which differ among different species.

The implications of this conclusion for behavior are (1) the need to sample different kinds of behavior because only some behaviors may be affected and (2) the need to compare the same behaviors among different species and strains.

One of the few comparisons of strain differences in response to teratogens using behavior as the dependent variable is an evaluation of sodium salicylate by Buelke-Sam and

Table 5.4. Frequency of Malformations in Phenytoin-Treated Mice (60 mg/kg) for Combined and Individual Strains[a]

All	C3H	C57	SWV
Distal phalanges (64%)	Distal phalanges (59%)	Occiput (78%)	Cerebral ventricles (62%)
Occiput (57%)	Occiput (59%)	Distal phalanges (67%)	Distal phalanges (66%)
Cerebral ventricles (44%)	Cerebral ventricles (32%)	Facial (37%)	Kidney (55%)
Kidney (38%)	Kidney (28%)	Hypoplastic digits (34%)	Sternebra (53%)
Hypoplastic digits (36%)	Hypoplastic digits (17%)	Cerebral ventricle (30%)	Hypoplastic digits (52%)
Sternebra (29%)	Cardiac (15%)	Kidney (28%)	Occiput (30%)
Cardiac (25%)	Facial (11%)	Cardiac (28%)	Vertebral centra (31%)
Facial (22%)	Sternebra (7%)	Sternebra (22%)	Cardiac (29%)
Vertebral centra (17%)	Vertebral centra (7%)	Ocular (20%)	Facial (19%)
Ocular (13%)	Palate (2%)	Vertebral centra (11%)	Ocular (17%)
Cleft palate (2%)	Ocular (0%)	Cleft palate (0%)	Cleft palate (3%)

[a]From Finnell and Chernoff, 1984.

her co-workers (1984). These researchers administered this drug to Long Evans and Sprague–Dawley rats on gestation days 8–10 and tested offspring activity at 30 days of age. Long Evans rats exhibited a dose-related decrease in activity, whereas Sprague–Dawley rats exhibited a dose-dependent increase. Although animals were not dosed and tested concurrently, the opposite direction in effects was seen in all replicates suggesting that the effect on activity was not due to some artifact. The implications of this difference in response are far reaching in terms of genotypic contributions and in screening for behavioral teratogens and warrant further and more in-depth analysis with other teratogens.

One way that vulnerable genes or cells may contribute to genotypic differences in responsiveness to teratogens may be through receptor activity. A classic example of individual differences in receptors is the inability of certain people to detect the taste of phenylthiourea (Fischer, 1967). Either the receptors necessary for detecting phenylthiourea are not present or they are present but nonfunctional in many people.

Because chemical substances act on some cells and not others, and biological activity of many substances is changed markedly by very minor alterations in their chemical structures, cells must contain reactive receptor cell sites that are complementary in configuration to those of the substances that act upon them.

Whenever a teratogen and its receptors interact, certain physiological and biochemical changes occur at the site of action that act as a stimulus to produce a response. In pharmacology, drugs that initiate responses by interacting with receptors are termed *agonists;* compounds that prevent such interactions and that produce no responses of their own at the receptor are referred to as *blocking agents* or *antagonists* (anti-agonists).

Diazepam (Valium) is an example of a teratogen whose effects involve an interaction with specific receptors in the brain. By itself, *in utero* diazepam exposure alters the adult response to stress in rats; administration of diazepam along with its antagonist has no effect on the adult stress response (Simmons *et al.*, 1984.).

Because the mechanism of action for many drugs begins with an interaction between those drugs and their receptors, the absence of such receptors makes a teratogen-receptor interaction impossible Such an absence may be due to immature development. Similarly, if receptors are present but affinity or binding are not yet mature, no response or minimal responsiveness is likely.

Altered activity of an appropriate receptor should also affect the actions of a teratogen whose mechanism of action involves some interaction with receptors. Much of the work in this area has been reviewed by Kimmel (1981). Differences in receptors may explain why cortisone produces cleft palate in mice and rabbits but not in rats.

Stating that a teratogen acts on a particular receptor does not itself constitute an explanation for the actions of that teratogen. Instead, it is really only the first in a series of steps that eventually leads to a response.

All teratogens do not exert their effects via receptor mechanisms. Drugs like alcohol depress cellular activity by temporarily altering the properties of cellular and subcellular membranes in a way that cannot be attributed to their molecular configuration. This general effect referred to as structural nonspecificity is shared by sedative-hypnotic and other depressant drugs.

Genotype of the Mother. Differences in responsiveness to teratogens may also be due to differences in the way substances are broken down in the mother's body. Because of qualitative differences in the metabolic pathways by which drugs are inactivated or quantitative differences in the metabolic enzymes common to a number of species, the same dose of drug may be teratogenic in one species but not in another. This is especially possible if a metabolite rather than the parent compound is the active agent. If drug effects are due to metabolites and these metabolites cannot be produced by the fetus, this too could account for differences in maternal–fetal responsiveness.

Measurement of maternal blood levels of teratogens may be especially valuable in assessing genetic susceptibility. If fetal anomalies occur more often in one species or strain than in another, although both receive the same amount of teratogen, this increased incidence may be due to differences in maternal absorption, distribution, metabolism, or elimination kinetics (see Principle 3), resulting in different maternal and fetal teratogen levels and duration of fetal exposure.

Dosages of various compounds, when expressed in terms of milligrams per kilogram body weight, may bear little relationship to the actual blood levels of a drug in different species of animals because each species differs not only in the kinds of metabolic enzyme systems it possesses but also in the rates of absorption into the blood, binding to plasma proteins, kidney excretion, and other factors. (Brodie & Reid, 1971). To compare different species there must be a standard from which to make comparisons. The best standard is plasma levels of free drug in each species (Brodie & Reid, 1971). For example, delta-9-tetrahydrocannabinol (THC), the principal psychoactive ingredient in marihuana, is not water soluble and has to be dissolved in oily vehicles when given to animals. These oily solvents sometimes form depots at their sites of administration and absorption of drugs is very poor. To achieve some response it is therefore necessary to administer very high doses of these drugs or very little will get into the blood. With THC, the doses are sometimes as much as 500 times greater than what people expose themselves to when they smoke marihuana. Smoking allows the drug to enter the blood through the lungs, and this is a very efficient method. The blood levels achieved after orally administering a dose of 50 mg/kg of THC to a rat are roughly comparable to the blood levels achieved after smoking two marihuana cigarettes. The fact that the administered dose is about 500 times higher is irrelevant. What is relevant is the blood level.

Comparisons of blood levels of a teratogen should be based on peak levels. To identify peak levels, familiarity with the pharmacokinetics of each teratogen is necessary. Without such information, blood levels might be assayed too soon or too long after peak levels occur.

Blood levels of teratogens should be sampled in pregnant animals because levels in nonpregnant animals may be higher or lower than those in pregnant animals, depending on drug dosage. This is illustrated in Figure 5.6. When pregnant animals were injected with 2

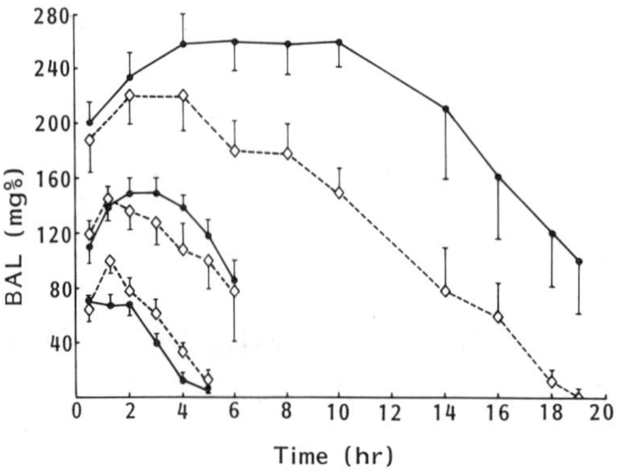

Figure 5.6. Blood alcohol levels (BALs) (mean ± S.E.) in pregnant (0—0) and nonpregnant (◇----◇) rats following oral administration of alcohol. Lower curves: 2 g/kg; intermediate curves: 4 g/kg; upper curves: 6 g/kg (not all points are included). BALs are lower in pregnant compared with nonpregnant rats with administration of a dose of 2 g/kg, are similar after a dose of 4 g/kg, and are markedly higher after a dose of 6 g/kg. A minimum of seven animals are represented by each curve. (From Abel, 1979.)

g/kg of alcohol, their blood alcohol levels were slightly lower than levels of nonpregnant animals. At a dose of 4 g/kg, pregnant animals began to achieve slightly higher levels than nonpregnant animals, and at 6 g/kg, levels in pregnant animals were considerably higher. Using blood alcohol levels in nonpregnant animals to deduce levels in pregnant animals given the same dosage may therefore be misleading (see Principle 3).

The influence of genotype on blood levels of a teratogen and the effects of that influence on occurrence of malformations is illustrated by Chernoff's (1980) study of blood alcohol levels in pregnant mice. Chernoff placed CBA, C3H, and C57 mice on liquid diets containing either 20% or 0% ethanol-derived calories (EDC) from gestation day 0–18. Fetal and maternal blood alcohol levels were highest for the CBA strain followed by the C3H and C57 strains. The percentage of abnormal fetuses mirrored maternal blood alcohol levels, which in turn was related to maternal alcohol dehydrogenase (ADH) activity (the enzyme that metabolizes alcohol). CBA mice had the highest blood alcohol levels, the highest percentage of anomalies, and the lowest ADH activity; C57 mice had the lowest blood alcohol levels and percentage of anomalies and the highest ADH activity.

These results indicate that strain differences in susceptibility to alcohol's teratogenic effects are affected by maternal genotype. Alcohol is the teratogen, but the mother's genotype, which determines her metabolic enzyme activity, affects how much alcohol is present in her blood and how fast it gets there and therefore, how much is delivered to her embryo/fetus.

Principle 5. The Principle of Critical Periods

The Principle of Critical Periods states that susceptibility to a potential teratogen depends on when exposure occurs. The concept of the critical period is a cornerstone of both embryology (see Chapter 4) and teratology. Of all the principles in teratology, this principle implies a unique response to a foreign substance in the developing organism compared with an adult. Whereas all the other principles in teratology could also be applied to other related disciplines like pharmacology or toxicology, this principle applies only to teratology.

The critical period for malformations is organogenesis, the time when cell groups and tissues are forming into organs. In humans, this occurs between the 3rd and 13th week of gestation. In the case of thalidomide, malformations occurred only when mothers took the drug after being 34 days pregnant and before they were 50 days pregnant. Children exposed on gestation days 34–37, were born with duplicated thumbs; those exposed on gestation days 38–43 had abnormal ears, kidney defects, shortened arms, and urogenital and respiratory defects. Children exposed after organogenesis had no abnormal limbs or organs.

However, human embryos all develop at different rates (Nishimura *et al.*, 1968; 1974) and therefore some embryos develop faster than others, some slower. There is no prize for being the first embryo on the block to develop a cerebellum and no penalty for being the last. Therefore, the critical periods concept is not rigid as to timetable. This may be one reason dizygotic twins or litter mates are not equally affected by a teratogen.

A difference in genetically programmed cell cycle durations would result in exposure to the same amount of teratogens at slightly different times. On gestation days 11 and 12 in the rat, the duration of the cell cycle is about 8 hours and there are only about 3 generations of cells each day (Kohler, 1970). Some individuals may go through 3 generations of cell divisions a day, others only 2.5 or 3.5. Some teratogens are "phase dependent"; that is they are only teratogenic if they encounter cells in a particular phase of their cell cycle of DNA replication (see Chapter 2). For example, hydroxyurea acts on cells only when they are in their "S" phase (when DNA duplication is occurring), whereas colchicine acts only during mitosis when spindles are forming. Because dizygotic twins and litter mates are not genetically identical, they may be undergoing all cycles at slightly different times and this may affect their susceptibility.

The major factor affecting teratogenesis is the extent of differentiation cells have undergone when they are exposed to a teratogen. Prior to implantation, the developing embryo is not very susceptible to teratological influences because its cells are not differentiated. A teratogen may be embryotoxic, however, and may kill so many cells that the embryo dies. Alternatively, a teratogen may kill only a few cells and those remaining may multiply to replace those that are lost.

Although the preimplantation embryo is less susceptible to teratogens, it is not totally refractory; morphological anomalies have been reported prior to implantation for ethylene oxide (Generoso *et al.*, 1987), x-rays (Russel and Montgomery, 1966), lead (Jacquet *et al.*, 1976), and cyclophosphamide (Gottschewski, 1964). Cyclophosphamide also retards development by about 24 hours when exposure occurs prior to implantation (Spielmann *et al.*, 1979).

Although these agents seem to violate the principle of critical periods, alternative explanations are possible. Because of slow absorption, metabolism, or elimination from the body (see Principle 3), some teratogens may be stored in maternal tissues and then slowly released back into the blood so that the time of embryological exposure is later than maternal exposure. Phenobarbital, for example, has a half life of about 5 days (Waddell & Butler, 1957). This means 50% of the administered dose or its metabolite will still be present in the body 5 days after it is first taken. Thalidomide remains in the rabbit blastocyst for more than 58 hours (Fabro *et al.*, 1965). In such cases, initial exposure would have begun some time before the critical period but effects do not begin to occur until the organism enters its critical period.

Other agents may exert their effects as a result of a gradual impact that begins even before conception but that makes its impact felt during the critical period. For example, withholding Vitamin B from the diet is not teratogenic if begun during organogenesis but it is teratogenic if begun before (Warkany & Nelson, 1940). This is because the mother's body has reserves of Vitamin B. If the depletion begins only when the critical period begins, organogenesis may occur before the reserves of this vitamin have been lowered to a critical level.

Other teratogens, like some cancer chemotherapeutic agents, require activation over a period of time before they become teratogenic (Chaube & Murphy, 1968). In some cases, this activation period may involve stimulation of the enzymes that metabolize the teratogen. As a result of more effective metabolism, a greater amount of metabolite is produced. If the metabolite is teratogenic, the increased metabolic efficiency would transform an otherwise nonteratogenic compound into one that is teratogenic.

Effects on the nervous system can also occur after the critical period. Although birth is an important landmark for certain functions like respiration, it is relatively insignificant for brain development (see Chapter 5). Because different types of neurons are formed at different times of development, neural mechanisms underlying different types of behavior may be subject to injury all through development, especially after the critical period for organogenesis.

In one of the few behavioral studies in which a teratogen has been administered during and after organogenesis, Driscoll and her co-workers (1982) found that alcohol exposure during gestation days 7–20 or 14–20 in the rat adversely affected passive avoidance performance of offspring, whereas administration during days 7–13 had no such effect. Furthermore, the deficits in animals exposed during days 14–20 were as severe as those exposed for the longer period. Similarly, prenatal exposure to phenobarbital after organogenesis had the same impact as during organogenesis on neuronal deficits (Bergman *et al.*, 1980). In all probability, exposure to teratogens during the last few days prior to birth when histogenesis is occurring is more important for behavior in animals than exposure during organogenesis.

Principle 6. The Principles of Dose–Response Relationships

The Principle of Dose–Response Relationships states that the extent of abnormal development is related to the amount of exposure. At low doses, a teratogen kills or otherwise adversely affects a percentage of developing cells resulting in a decreased population of cells. This may be expressed as abnormal behavior or a decrease in body weight. At a higher dose, more cells are affected resulting in malformations and continued growth retardation. At higher doses still, so many cells are affected that the whole organism dies (see Figure 5.7). Extent of damage can be expressed as either greater damage to a specific organ or an increase in the number of organs affected as dosage increases.

For those teratogens whose effects can be arranged along such a continuum, behavior represents the most sensitive index of teratogenic insult, provided, of course, the teratogen is capable of producing a behavioral effect (Vorhees, 1986).

Dose–response relations like those shown in Figure 5.7 are not always obtained in behavioral teratology; in many instances, "negative" dose–response relations have been

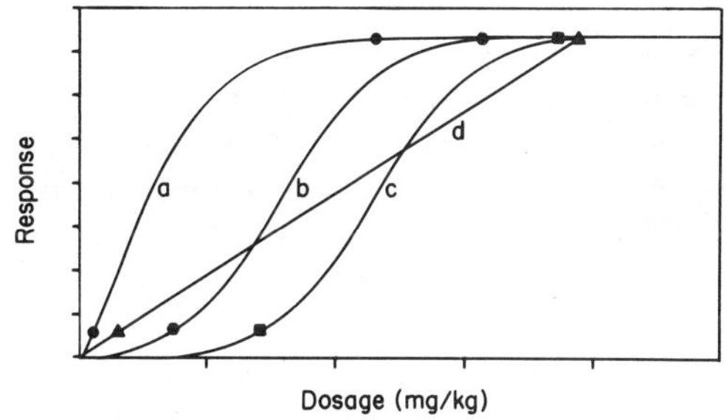

Figure 5.7. Hypothetical dose–response relations between a teratogen and different effects. Curve a depicts a *graded* function in which the dependent variable (e.g., motor activity) takes on progressively increasing values as the dose is increased. Because behavior is a very sensitive indicator of teratogenic effect, the curve is on the far left, indicating a lower dose is required to produce an effect compared with effects represented by curves b and c. The function depicted in curve a indicates not only that behavior is the most sensitive of the four measures but that very little increments in dose result in very large effects.

Curves b and c depict *quantal* relationships and represent all-or-none responses for structural malformations (curve b) and fetal death (curve c). Parallel curves like b and c suggest a teratogen is acting on the same population of receptors. Curves b and c are different from a in that the latter represent the number of subjects at each dose with a particular response, whereas curve a represents a quantitative impact (e.g., hyperactivity). Graded functions would be more common in malformation studies. Because malformations require exposure to a higher dose of teratogen, curve b is to the right of curve a, and because death requires more teratogen exposure than b, curve c is further to the right than b. Function d contrasts with a, b, and c in indicating direct proportionality between dose and response such as might occur in intrauterine growth retardation. Note that in this hypothetical situation, a teratogen acting on both behavior and intrauterine growth always has a greater impact on behavior. When comparing malformations and growth, however, intrauterine growth retardation is initially the more sensitive response but after a point more malformations begin to occur with smaller increments in dose compared with growth retardation.

observed; that is, higher doses produce less, if any, change from control levels compared with lower doses (e.g., Nelson, 1981; Tachibana, 1982). One reason dose–response relations are sometimes elusive in behavioral teratoglogy is that death and abnormal development may not be similar expressions of response to the same teratogen. At high doses, a teratogen may produce increased fetal mortality and survivors may be more resistant to any behavioral teratogenic effects of the agent (Nelson, 1981).

Alternatively, the dose–response relationship may be so steep that a small increase in dosage causes malformations or death, whereas a small decrease is below threshold (Vorhees, 1986). If only a few dosages are studied, as is typically the case in behavioral teratology because of the large numbers of animals that must be maintained, the threshold may be exceeded by only one dose.

When agents are first tested on animals, several dosage levels are chosen including a relatively high dose that will probably cause toxic effects. Fractions of this dose are then chosen for further study. One measure of acute toxicity is the LD_{50}, the dose that produces 50 % lethality in a group of animals. Dosages are then administered at fractions of the LD_{50} dose. The advantage of this top-down evaluation is efficiency compared with the bottom-up approach because the former establishes a range to work in.

A different strategy for choosing dosage levels is to pick doses that have some resemblance to the amounts used by humans. In this regard, the effective dose (ED_{50}) may be more valuable than the LD_{50}. This is the dose that has pharmacologically therapeutic effects in 50 % of those tested. In animal studies, Vorhees (1986) recommends matching doses to human ED_{50} values. For example, if anticonvulsant drugs were being studied, a reasonable dose around which to structure a dose–response study would be the dose that inhibits seizures in 50% of the animals tested. If antianxiety drugs were being studied, a starting point might be the dose that overcomes the suppressive effects of conditioned shock suppression in 50% of the animals tested. The disadvantage of this approach, as pointed out by Vorhees (1986), is that what is ''relevant'' or what is an ED_{50} in one species may not be so in another. Yet another approach is to evaluate the behavioral effects of a particular compound in the mother and then use the ''behaviorally effective'' (BE_{50}) dose as a focus (Hutchings *et al.*, 1984). Very often toxic effects in the mother are indicated by decreased food and water consumption or decreased weight gain. Doses would then be administered across a range above and below this value. This would have the advantage of identifying dosages with marked maternal toxicity.

The relation between the maternal ED_{50} dose and the dose that produces no embryological/fetal deaths but does cause teratogenic effects is shown in Figure 5.8.

In Figure 5.8, the ''no observable effect level'' (NOEL) in the embryo has the same dosage range as in the mother. However, the embryotoxic dose range is narrower than the pharmacological range for the mother. Doses at the upper pharmacological range in the mother may be toxic to the embryo causing structural malformations. This is very like what occurred in connection with thalidomide—doses that produced sedation in mothers produced teratogenic responses in the embryo. Another dose–response relation encountered in teratology is shown in profile C. In this situation, there is no embryotoxic effect at pharmacological doses in the mother. It is only with maternally toxic doses that embryotoxic effects occur.

Many compounds are more teratogenic if given as a single dose than chronically, even if chronic exposure involves even higher doses. For example, a 200 ug/kg dose of mitomycin D given to rats on gestation day 9 produces a 28% malformation rate in survivors, whereas a dose of 250 ug/kg/day given on gestation days 0 through 9 produces a 9% malformation rate (Wilson, 1964). Conceivably, this may be due to some maternal adaptation to the agent prior to the critical period.

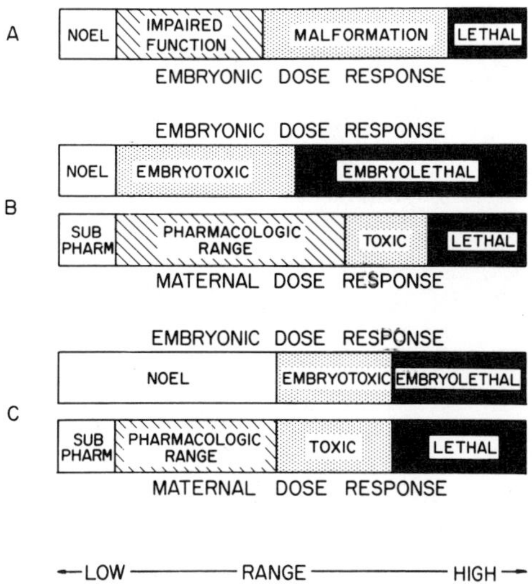

Figure 5.8. Hypothetical dose–response profiles for teratogen such as thalidomide with different toxic-effect ranges in the mother and conceptus. Relationship A depicts the range of possible effects on the embryo from no observable effect level (NOEL) to death. Below the threshold level for death, there are two embryonic responses each with their own threshold. Impaired function is shown to the left of malformations because its threshold is lower. Possible relation between maternal and embryonic responses are shown in B and C. In B, the threshold dose for no response is the same for mother and embryo. A dose that produces a pharmacological effect on the mother also causes malformations in embryos. As the dose increases from the subpharmacological through the pharmacological range, the threshold for toxicity is eventually reached. However, before the mother experiences such toxic effects, the threshold for embryolethality is reached.

In Profile C, subpharmacological and pharmacological doses do not produce malformations in the embryo. It is only when the mother is exposed to toxic doses (e.g., decreased eating, loss of body weight) that malformations begin to occur in the embryo. Because the dose range is similar for both, it is possible that the teratogen is acting indirectly on the embryo via the mother. (From Hutchings, 1985.)

Principle 7. The Principle of Developmental Delay

Unlike missing arms or legs, behavioral damage may be only transient, reflecting developmental delay rather than permanent damage. An example of such developmental delay is Abel's (1982) study of the prenatal effects of alcohol on the behavior of rats. As shown in Figure 5.9, animals prenatally exposed to alcohol performed worse than control animals at 17 and 48 days of age but by 114 days of age, differences were no longer significant.

Other teratogens may have a "sleeper" effect. Animals prenatally exposed to the antimitotic agent azacytidine, for instance, are not hyperactive prior to weaning but are hyperactive when they reach puberty (Rodier *et al.*, 1975), and animals prenatally exposed to diazepam do not have norepinephrine levels below normal until after puberty (Simmons *et al.*, 1984b).

"Sleeper" effects have been attributed to late maturation of CNS mechanisms controlling various functions (Rodier, 1986). It is only when these mechanisms are mature that deficits can be seen because prior to maturation the functions they serve do not occur frequently enough for differences to be noted.

Principle 8. The Physiological, Pathological, and Psychological Status of the Mother Can Affect Susceptibility to Teratogens or Can Induce Teratogenesis

Consumption of high levels of alcohol during pregnancy is associated with anomalies in children. If this was all that was known about the relation between alcohol and anomalies, it would not be enough to identify alcohol conclusively as a teratogen. The reason the relation would still be equivocal is that alcohol is associated with many other factors that could also cause anomalies. Some of these are listed in Table 5.5.

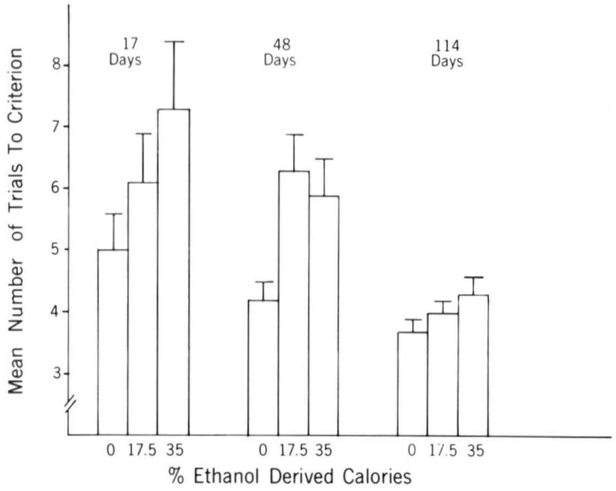

Figure 5.9. Effects of prenatal alcohol administration on passive avoidance learning in rats tested at different ages. Pregnant rats consumed liquid alcohol diets containing either 0%, 17.5%, or 35% ethanol-derived calories (EDC) from gestation days 5–20. Animals given the 0 and 17.5% diets were pair fed to 35% EDC animals. Vertical lines indicate standard errors. (From Abel, 1982.)

Alcoholic women are often underweight and sickly because of their drinking and this can affect the health of their unborn children. Such risk factors are circumstances that increase an individual's likelihood for some adverse effect. Conceivably, some behavior consequences of *in utero* exposure to teratogens may be secondarily related to teratogen-induced maternal undernutrition, which itself is associated with cognitive impairment in offspring. In addition to reducing the nutrient pool available to the embryo/fetus, undernourishment can reduce blood flow to the uterus (Lasuncion *et al.*, 1987) resulting in an even greater decrease of nutrients to the fetus and in fetal hypoxia (Skillman *et al.*, 1985).

Poor maternal nutrition can also contribute to the teratogenicity of other agents. Vitamin B compounds, for example, attenuate cortisone-induced cleft palate in mice (Stean & Peer, 1956) and defects due to hypervitaminosis A (Miller & Woollam, 1958). Dietary absence of vitamin B compounds removes this attenuating influence. Likewise, hypocaloric diets have a synergistic effect on the teratogenic effect of drugs such as cortisone (Kalter, 1960). Maternal health can thus modify the actions of potential teratogens or can cause behavioral effects that could mistakenly be attributed to a teratogen.

Maternal age is a major factor affecting teratogenicity. Influences of maternal age on absorption were examined in connection with Principle 3. In general, reproductive efficiency declines after 35 years of age. The incidence of Down syndrome, for instance, increases with maternal age (see Chapter 2), as does the incidence of congenital anomalies such as cardiac malformations and hydrocephalus (Kiely *et al.*, 1986).

Among the reasons for age-related effects on reproductive efficiency are decreased fetal oxygenation due to factors such as increased hypertension (Grimes & Gross, 1981) and reduced uterine blood flow to the placenta (Naeye, 1983).

The effects of maternal age are confounded with parity. The latter is one of the main risk factors contributing to fetal alcohol syndrome (Sokol *et al.*, 1986). However, teasing apart the separate contributions of age and parity is very difficult in human teratological studies. Conceivably, the reported contribution of parity to this syndrome is the result of advanced maternal age.

Table 5.5. Maternal and Embryological/ Fetal Factors Associated with Alcoholism

Maternal risk factors
Age (under 15 or over 35)
Increased parity (more than five children)
Short birth intervals (less than 24 months)
Malnutrition
Chronic disease (hypertension, kidney disease)
Previous stillbirths
Low socioeconomic status
No prenatal care
Risk factors during pregnancy
Infection
Drug use
Smoking
Malnutrition
Low weight gain
Premature rupture of membranes
Prematurity
Anesthesia
Prolonged labor
Excessive bleeding during delivery
Breech birth

Anemia is the most common hematological abnormality during pregnancy, affecting up to 5% of all pregnancies (Paulone *et al.,* 1987). A common cause of anemia during pregnancy is iron deficiency and a common result is preeclampsia and eclampsia, which can affect protein binding of teratogens (see Principle 2). Maternal anemia is associated with increased risk of preterm birth, stillbirth, and neonatal death (Arias, 1984). However, anemia itself becomes a risk factor to the fetus only when oxygen delivery is reduced below 50% of normal. This is because there is a compensatory fetal increase in oxygen extraction from the blood up to a 50% decrease. Beyond that, fetal oxygen consumption decreases (Paulone *et al.,* 1987).

Maternal liver cirrhosis, gastrointestinal disorders, anemia, hypoglycemia, infection, and many other conditions are commonly associated with drug use and chronic drinking. Infection of the fetus can result in hypoxia, fever, premature delivery, and intrauterine death if exposure occurs after organogenesis or malformation if exposure occurs during organogenesis. How these conditions affect the subsequent adult behavior of the developing fetus has not been studied. There is evidence, however, that women who suffer from ill health during pregnancy have children who are characterized by "poor health, physical defects, developmental lags, behavioral disturbance and habit disorders" (Stott & Latchford, 1976) more often than mothers who experience no health problem during pregnancy.

Although there is no recognized single cause of drug abuse or excessive drinking, these behaviors are often symptomatic of a stressful life situation. The relationship between maternal stress during pregnancy and subsequent effects on offspring behavior in animals has been amply documented (see review by Archer & Blackman, 1971). In humans, maternal stress (marital discord, overcrowding, attitudes about pregnancy, etc.) also contributes to obstetric complications (Laukaran & VandenBerg, 1980) and can lead to childhood disorders such as bed wetting, distractibility, sleep disturbances, disorders in reading ability, and cognitive impairment (e.g., Pasamanick & Lilienfield, 1955).

Social class is a factor in many birth defects. Anencephaly occurs almost twice as often among women in the lower socioeconomic classes (Blomberg, 1980). Low socioeconomic class is confounded with many factors, all of which could be included in the general category of stress.

The fact that many teratogens render the thermoregulatory mechanisms controlling body temperature inoperative makes environmental temperature an especially critical variable in teratogenicity. When control over heat-regulating mechanisms is lost, poikilothermia results, and body temperature will rise or fall according to whatever the critical ambient room temperature happens to be (Shemano & Nickerson, 1958). Above this temperature, body temperature will rise; below it, it will fall. Because decreases in temperature tend to slow down enzymatic reactions, a lowering of body temperature will prolong the action of any compound whose inactivation occurs by means of enzymatic degradation, as with barbiturates.

Appendix A. Routes of Administration

The route of administration can have a profound effect on the rate of absorption and the transformations to which a substance may be subjected before it finally comes into contact with its site of action. There are two main routes of administration—enteral and parenteral. The enteral route usually means oral administration, abbreviated "p.o.," from the Latin *per os,* "by mouth," and generally involves absorption from the gastrointestinal tract. Parenteral routes involve subcutaneous (s.c.), intramuscular (i.m.), intraperitoneal (i.p.), or intravenous (i.v.) injection.

The route of administration has a direct bearing on the speed of onset and the intensity of a compound's action. Compounds do not exert effects until they are present at their sites of action at some concentration equal to or greater than necessary to trigger a response. Because an equilibrium or steady-state condition tends to become established between the

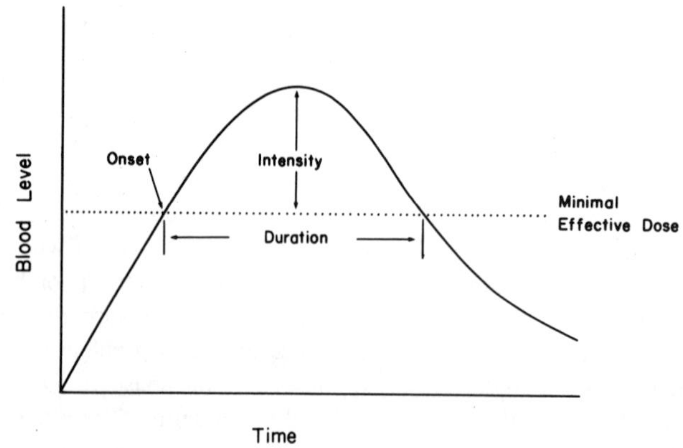

Figure 5.10. Hypothetical relation between blood level of a compound and its onset, intensity, and duration of effects.

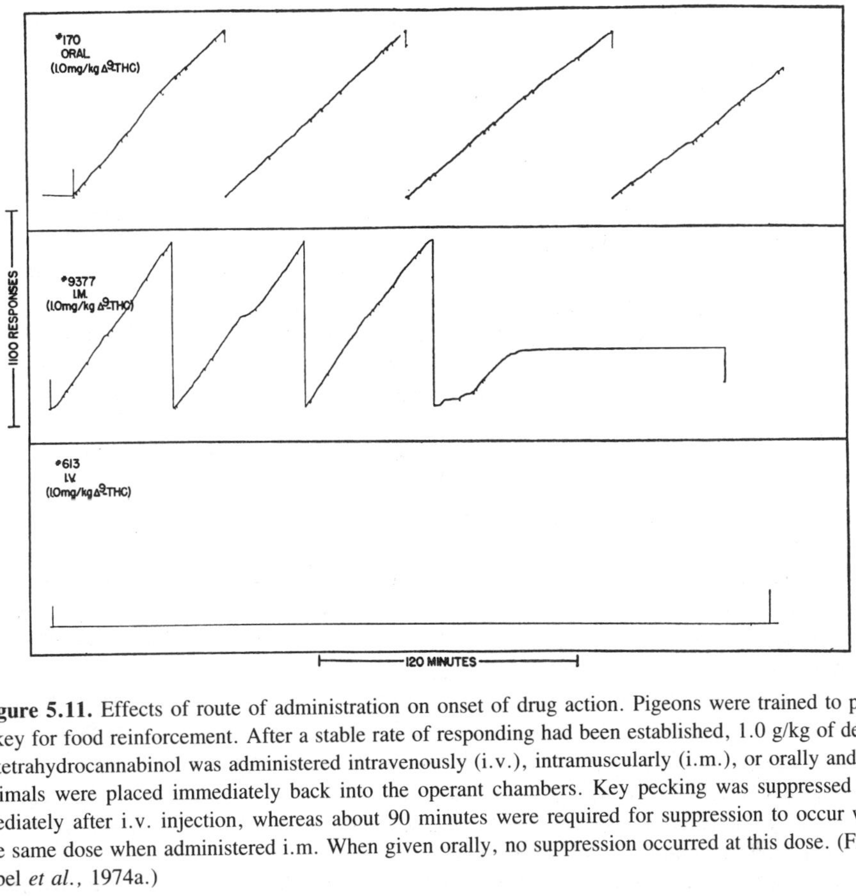

Figure 5.11. Effects of route of administration on onset of drug action. Pigeons were trained to peck a key for food reinforcement. After a stable rate of responding had been established, 1.0 g/kg of delta-9-tetrahydrocannabinol was administered intravenously (i.v.), intramuscularly (i.m.), or orally and the animals were placed immediately back into the operant chambers. Key pecking was suppressed immediately after i.v. injection, whereas about 90 minutes were required for suppression to occur with the same dose when administered i.m. When given orally, no suppression occurred at this dose. (From Abel *et al.*, 1974a.)

concentration of a compound at its site of action and the concentration of that drug in the blood, there will also be a threshold blood level that corresponds to the threshold level of a compound at its site of action. Figure 5.10 illustrates the relationship between the hypothetical blood levels and onset, intensity, and duration of action.

An example of the behavioral consequences of route of administration on behavior is shown in Figure 5.11.

Each route of administration obviously carries with it attendant biological considerations. In animal studies, where the purpose is to determine the potential of a substance to produce teratogenic effects, intravenous or intraperitoneal routes will result in maximum blood concentrations (for exceptions, see below). If the purpose of a study is to examine the teratogenic potential for humans, then the route of administration should be similar to that used by humans.

Some compounds are teratogenic when given by one route and nonteratogenic when given by another. For example, marihuana and dexamphetamine are both teratogenic in animals when given intraperitoneally but not when given orally (Abel, 1980; Nora *et al.*, 1965). Thalidomide, on the other hand, is teratogenic when given orally but not when administered intraperitoneally (Cahen, 1966). Route of administration is thus a major consideration in teratological studies.

Oral Administration

A drug such as aspirin has little difficulty in passing out of the gastrointestinal tract. Insulin however, will not even pass the first membrane barrier of the gastrointestinal tract because insulin is acted upon and destroyed by enzymes present in the stomach. Therefore, insulin has no physiological effect if taken by mouth. The examples of aspirin and insulin point out the value of knowing about factors that can affect the passage of substances from the moment they are taken into the body. Without such information, it is hardly possible to understand how potential teratogens are able or not able to produce certain effects.

There are three main ways of administering compounds orally to animals. These are by intubation, by inclusion in the diet, and through drinking fluid.

For intubation, compounds are typically injected directly into the oral cavity by means of a needle with a rounded tip. Since rats and mice cannot regurgitate, there is no loss of compound due to vomiting. For animals that can vomit, compounds can be introduced into the gastrointestinal tract through a tube placed into the mouth and then worked gently down the throat almost to the stomach. Administration of drugs by the oral route is often used to obtain parametric information concerning the extent of absorption, the onset and duration of activity, the possible irritant effects of a new drug, and the like, all of which are very important factors if a drug is being considered for use in humans.

Intubation is a very labor-intensive method and considerable planning has to go into decisions such as concentration of solution and fractions in which the total dose is delivered because delivery of a large volume of fluid into the stomach is an artificial situation. Some of the disadvantages associated with this technique include the stress associated with intubation and possible perforation of the esophagus. Other advantages and disadvantages of this procedure are summarized in Table 5.5.

Inclusion of test substances in the diet is another widely used procedure in animal studies. For test materials such as diphenylhydantoin, which are solid, the lab chow is ground into a fine powder and the test material can be added to the desired concentration (e.g., Hansen & Billings, 1986).

Liquid diet administration is an alternative for water-soluble test substances such as alcohol. In such cases, the compound is added to commercially available liquid diets such as Sustacal. These diets can be supplemented with vitamins and minerals for animals. There are also liquid diets specially formulated for animal nutritional requirements in conformity with recommendations by the National Academy of Sciences and the American Institute of Nutrition.

In the liquid diet procedure, compounds are mixed with the diet and placed in the animal's cage. Because this liquid diet is the only source of food or fluid, animals will ingest the liquid diet and in so doing will self-administer the compound. However, the amount of ingestion and the time of administration are left to the animal's discretion. Because rodents tend to eat and drink through the day and night, intersubject variability in intake and blood levels are increased. Also, there may be precipitation or nonhomogeneous dispersion of drugs or supplements in the diet. Table 5.5 lists other advantages and disadvantages associated with this method.

The easiest method for oral administration is placement of compounds in an animal's drinking water. However, as in the case of liquid diets, this is not possible if compounds

Table 5.6. Advantages and Disadvantages of Methods of Administration

Method	Advantages	Disadvantages
Intubation	1. Each animal receives same dose at same time 2. Sustained BAL high doses 3. Dosage level can be maintained all through pregnancy 4. Can determine dose–response reactions 5. Large number of animals can be treated 6. Inexpensive	1. Stressful 2. Possible gastric irritation 3. Death due to injection in lung 4. High mortality rate with high doses 5. Labor intensive 6. Decreased food and fluid intake
Placement in drinking fluid	1. Nonstressful 2. Ease of administration 3. Large numbers of animals 4. Inexpensive	1. Taste of compound may be aversive so animals will not voluntarily drink 2. Wide variability in ingestion 3. Weight loss and dehydration 4. Decreased consumption just before parturition 5. Bottles may leak
Liquid diet	1. Nonstressful 2. Balanced diet possible 3. Ease of administration 4. Large numbers of animals can be treated	1. Labor intensive—diets must be prepared fresh 2. Diarrhea 3. Poor dose–response considerations 4. Decreased consumption just before parturition 5. Expensive 6. Bottles may leak
Schedule-induced polydipsic	1. High intake	1. Expensive 2. Not all animals dipsic 3. Not suitable for large numbers of animals 4. Labor intensive 5. Weight reduction required

are insoluble in water (e.g., cannabinoids). Another problem is that animals may reject the fluid because of its taste. For example, animals will begin to reject water that contains 5% solutions of alcohol or higher and therefore may become dehydrated. To permit ingestion of higher concentrations, it may be necessary to begin with concentrations below 5% and to gradually increase the concentration. However, if administration is to occur only during pregnancy, this may not be possible because of the short gestational age of the rat and mouse. Other advantages and disadvantages of this method are shown in Table 5.6.

Operant procedures using schedule-induced polydipsia have also been used to administer compounds to pregnant animals (Samson, 1981). This technique is based on the tendency of food-deprived animals to drink large amounts of fluid quickly if food is offered in small amounts every few minutes. Compounds are placed in the animal's fluid and are ingested during these brief and intense drinking periods. The main disadvantage of this procedure is that it requires programming equipment to deliver the fluid.

Automated gastric infusion procedures have also been used to administer compounds to animals. One of the interesting developments in this regard has been the "pup in the cup" procedure, which allows administration of substances to rat pups removed from their mothers (e.g., Diaz *et al.*, 1977).

In this procedure, rat pups are lightly anesthetized and a small tube is placed through the mouth and into the stomach and then out through the stomach and skin. The tube is held in place in the stomach by a flange and the other end is connected to an infusion pump that delivers a liquid diet containing the compounds to be studied.

Another form of oral administration in early postnatal studies involves ingestion of a compound during nursing. One such example is ingestion of alcohol by neonates suckling on dams that have consumed alcohol (Abel, 1974). The difficulty with this procedure is that quantification of dose in milk requires analysis of milk samples. Also, administration of a compound to the dam may affect her ability to produce milk or may affect the composition of her milk.

Absorption of compounds from the gastrointestinal tract is affected by many factors. As noted in conjunction with Principle 3 (see above), chemical considerations such as the acidity of a drug and the pH of the stomach and intestine will determine if the drug will be absorbed at all.

When test materials are placed in the diet, the total amount administered per day may be high but the peak blood level will be much lower than if the test material is administered in one total exposure. If a threshold level is not achieved by the former, despite prolonged exposure, no effect may occur.

Compounds absorbed from the gastrointestinal tract go directly to the liver. Because the liver is an active site of metabolism, blood levels of parent compounds given in this way may be a great deal lower than if given by one of the other routes of administration.

The presence of food in the stomach is yet another consideration in oral administration because food may interfere with the access of a compound to the stomach wall. Food particles may also form an insoluble complex with a compound so that it cannot pass out of the gatrointestinal tract. In addition, a compound may be degraded by enzymes secreted by the stomach in the process of breaking down the food.

Psychological factors are another important source of variability affecting administration from gastrointestinal sites because the "emotional state" of a subject will influence the secretion of gastrointestinal fluids that could alter the pH of the stomach, thus altering the degree of a drug's ionization and thereby affecting its absorption. Emotional factors could also affect blood flow through this area and this could also affect rate of drug absorption.

Subcutaneous Injection (s.c.)

This method of subcutaneous injection involves injecting a compound directly under the skin. Relatively constant rates of absorption can be achieved with this route but the actual speed of absorption varies as a function of the vehicle in which the compound is dissolved, the area over which the solution spreads, and the rate of blood flow in the vicinity of the drug. To obtain a slow, reliable rate of absorption from subcutaneous sites, compounds are sometimes implanted under the skin in the form of compressed pellets or sustained release capsules that allow the compounds within to drain out slowly. Inserting such pellets is a relatively simple procedure with animals such as the mouse. The animal is lightly anesthetized (chloroform should be avoided because it produces kidney damage in mice), a fold is made in the skin above the neck, and an incision is made. The pellet or capsule is then inserted in the incision and moved toward the back of the animal. If the mouse is kept in isolation and the incision is small enough so that there is not much bleeding, the wound need not be stitched closed. The ideal shape for such pellets is a flat disc, because the exposed area will diminish only slightly as the disc becomes thinner, thus ensuring a relatively constant level of the drug in the bloodstream.

One disadvantage of this route of administration worth considering is that some drugs such as alcohol and cocaine are highly irritating if given subcutaneously and those that are dissolved in an oily solvent may be absorbed poorly. As a result, peak blood levels may not become high enough to produce any observable response unless a very large dosage of drug is injected.

Inhalation

Studies of exposure to environmental pollutants and volatile substances often expose animals in specially built inhalation chambers. The animal is simply placed into the chamber, the substance is introduced, and the animal inhales it. To simulate occupational exposure, animals may be placed in such chambers for 6 hours a day for 5 days per week for many weeks.

There are several technical difficulties associated with inhalation studies, however, such as uneven distribution of pollutants throughout the chamber. Many chambers, for instance, contain "dead spaces" in which there is low atmospheric distribution.

Intramuscular Injection (i.m.)

In administering drugs by intramuscular injection, the needle is placed within the skeletal muscles. Before the injection is actually given, however, the plunger of the syringe should be pulled back to make sure the needle has not entered a blood vessel.

As in the case of subcutaneous injections, the anatomical region into which a drug is directed via the muscles can also affect the rate of its absorption. In general, absorption from muscle tissue tends to be fairly rapid if the drug is dissolved in an aqueous vehicle. Drugs dissolved in oily vehicles, however, may form depots at the injection site. This will have the effect of slowing the rate of entry of the drug into the circulation. In addition, drugs such as alcohol may not be suitable for intramuscular injection because of their highly irritant effects.

Because animals such as the rat may react rather violently to being handled for the first time, they should be "tamed" prior to drug administration by frequent handling to habituate them to such treatment (Abel, 1971).

Intraperitoneal Injection (i.p.)

Although rarely done with humans, the injection of drugs into the peritoneal cavity is among the most commonly used routes in animal experiments. Drugs are usually taken up very rapidly from the peritoneal cavity into the bloodstream, but drugs dissolved in oily vehicles are absorbed much more slowly than those in aqueous solutions. A comparison of blood levels of THC dissolved in sesame oil and following three different routes of admin-

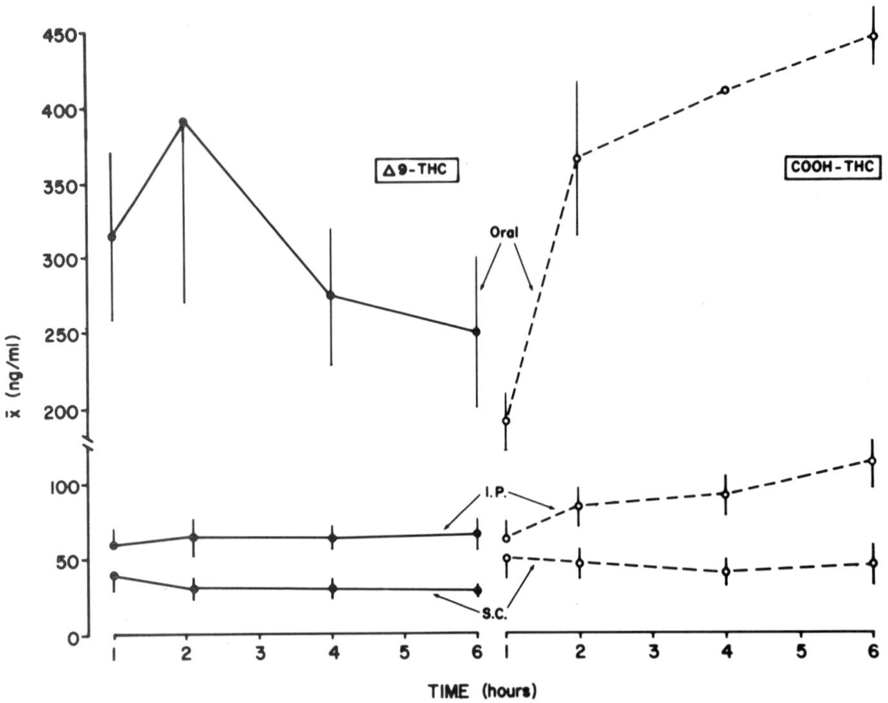

Figure 5.12. Plasma levels (X ± S.E.) of THC and its metabolite, COOH-THC, in female rats injected orally, intraperitoneally (i.p.), or subcutaneously (s.c.) with 50 mg/kg THC dissolved in sesame oil. (From Abel & Subramanian, 1987.)

istration is shown in Figure 5.12. Although oral administration typically results in much lower blood levels than i.p. administration with water-soluble drugs, this experiment showed the opposite relation when drugs have to be administered in oil vehicles.

Interestingly, THC is considerably more teratogenic when administered i.p. than orally in animals (Abel, 1980). This suggests the increased teratogenicity associated with the i.p. route is due to artifacts. For example, despite its relative ease, i.p. injections have the potential for puncturing of the uterus of fetuses or causing local irritation. Another problem associated with intraperitoneal injections, particularly if given to the same animal over a long period of time, is the possibility of peritonitis.

Intravenous Injection (i.v.)

By placing a compound directly into the bloodstream a very rapid onset of action is produced and the concentration initially in the blood is known. Some drugs are self-administered this way because of the sudden "rush" that comes with placing a large concentration into the blood. Compounds that are poorly absorbed from other sites can also be administered in this way.

If given too quickly, however, the injection may produce undesirable cardiovascular and respiratory effects (e.g., a sharp increase in blood pressure or shallow and irregular breathing). Special dangers are also associated with the possible entry of air into the blood. If compounds are to be given over a long period of time, chronic intravenous injection may cause damage to the veins resulting in leakage of the drug into surrounding subcutaneous and muscle tissue.

Solvent Considerations

The solvent in which a compound is dissolved or suspended is termed the *vehicle.* A solvent should have no intrinsic activity of its own. Unfortunately, this is not always the case for compounds that are not water soluble. Some water-insoluble compounds (e.g., THC) must be dissolved in organic solvents that may affect their absorption and distribution. Disulfiran, for example, is not teratogenic when carboxymethyl cellulose is the vehicle but is very teratogenic when dissolved in dimethyl sulfoxide (DMSO) (Robens, 1969). Likewise, pyrimethamine, an antimalarial drug, is much more teratogenic when DMSO or alcohol are used as vehicles (Anderson & Morse, 1966).

For some teratogens, there is an interplay between vehicle and route of administration. Hypervitaminosis A, for example, is teratogenic when given orally only if dissolved in an oily vehicle. If given intraperitonially, it is only teratogenic if given as an aqueous solution (Kalter, 1968).

References

Abel, E. L. Habituation as a factor in early handling. *Journal of Comparative and Physiological Psychology,* 1971, *74,* 219–222.

Abel, E. L. Alcohol ingestion in lactating rats: Effects on mothers and offspring. *Archives Internationales de Pharmacodynamie et de Therapie,* 1974, *210,* 121–127.

Abel, E. L. Prenatal effects of alcohol on adult learning in rats. *Pharmacology Biochemistry and Behavior,* 1979, *10,* 239–243.

Abel, E. L. Prenatal effects of cannabis. *Behavioral and Neural Biology,* 1980, *29,* 137–156.

Abel, E. L. In utero alcohol exposure and developmental delay of response inhibition. *Alcoholism: Clinical and Experimental Research,* 1982, *6,* 369–376.

Abel, E. L. Alcohol enhancement of marihuana-induced fetotoxicity. *Teratology,* 1985, *31,* 35–40.

Abel, E. L., and Dintcheff, B. A. Factors affecting the outcome of maternal alcohol exposure: II. Maternal age. *Neurobehavioral Toxicology and Teratology,* 1985, *7,* 263–266.

Abel, E. L., McMillan, D. E., and Harris, L. S. Delta-9-tetrahydrocannabinol: Effects of route of administration on onset and duration of activity and tolerance development. *Psychopharmacology,* 1974, *35,* 29–34.

Abel, E. L., and Subramanian, M. G. Effects of route of administration and method of analysis on plasma cannabinoid levels in rats. Unpublished manuscript,

Anderson, I., and Morse, L. M. The influence of solvent on the teratogenic effect of folic acid antagonist in the rat. *Experimental Molecular Pathology,* 1966, *5,* 134–145.

Archer, J. E., and Blackman D. E. Prenatal psychological stress and offspring behavior in rats and mice. *Developmental Psychobiology,* 1971, *4,* 193–248.

Arias, F. *High Risk Pregnancy and Delivery.* St. Louis: Mosby, 1984.

Bergman, A., Rosselli-Austin, L., Yedwab, G., and Yanai, J. Neuronal deficits in mice following phenobarbital exposure during various periods in fetal development. *Acta Anatomica,* 1980, *108,* 370–373.

Biehl, D., Shnider, S. M., Levinson, G., and Callender, K. Placental transfer of lidocaine: Effects of fetal acidosis. *Anesthesiology,* 1978, *48,* 409–412.

Blomberg, S. Influence of maternal distress during pregnancy on fetal malformations. *Acta Psychiatrica Scandinavica,* 1980, *62,* 315–330.

Borell, U., Fernstrom, I., Ohlson, L., and Wiquist, N. Effect of uterine contractions on the uteroplacental blood flow at term. *American Journal of Obstetrics and Gynecology,* 1965, *93,* 44–57.

Brodie, B. B., and Reid, W. D. The value of determining the plasma concentration of drugs in animals and man. In B. N. LaDu, H. G. Mandel, and E. L. Way, (Eds.), *Fundamentals of Drug Metabolism and Drug Disposition.* Baltimore: Williams & Wilkins, 1971.

Buelke-Sam, J., Kimmel, C. A., Nelson, C. J., and Sullivan, P. A. Sex and strain differences in the developmental activity profile of rats prenatally exposed to sodium salicylate. *Neurobehavioral Toxicology and Teratology,* 1984, *6,* 171–175.

Butcher, R. E., Wootten, V., and Voorhees, C. V. Standards in behavioral teratology testing: Test variability and sensitivity. *Teratogenesis, Carcinogenesis, and Mutagenesis,* 1980, *1,* 49–61.

Cahen, R. L. Experimental and clinical chemoteratogenesis. *Advances in Pharmacology,* 1966, *4,* 263–349.

Chasnoff, I. J., Burns, W. J., and Schnoll, S. H. Cocaine use in pregnancy. *New England Journal of Medicine,* 1985, *313,* 666–669.

Chaube, S., and Murphy, M. L. The teratogenic effects of the recent drugs active in cancer chemotherapy. *Advances in Teratology,* 1968, *3,* 181–237.

Chen, S. S., Perucca, E., Lee, J. N., and Richens, A. Serum protein binding and free concentrations of phenytoin and phenobarbitone in pregnancy. *British Journal of Clinical Pharmacology,* 1982, *3,* 547–554.

Chernoff, G. F. The fetal alcohol syndrome in mice: Maternal variables. *Teratology,* 1980, *22,* 71–75.

Christoffel, K. K., and Salafsky, I. Fetal alcohol syndrome in dizygotic twins. *Journal of Pediatrics,* 1975, *87,* 963–967.

Church, M. W., Abel, E. L., Dintcheff, B. A., and Matyjasik, C. Maternal age, blood alcohol concentration, and pregnancy outcome in the rat. Unpublished manuscript,

D'Arcy, P. F., and McElnay, J. C. Drug interactions involving the displacement of drugs from plasma protein and tissue binding-sites. *Pharmacology and Therapeutics,* 1982, *17,* 211–220.

Diaz, J., Schain, R. J., and Bailey, B. G. Phenobarbital-induced brain growth in artificially reared rat pups. *Biology of the Neonate,* 1977, *32,* 77–82.

Driscoll, C. D., Chen, J. S., and Riley, E. P. Passive avoidance performance in rats prenatally exposed to alcohol during various periods of gestation. *Neurobehavioral Toxicology and Teratology,* 1982, *4,* 99–104.

Fabro, S., Smith, R. L., and Williams, R. T. The persistence of maternally administered 14C-thalidomide in the rabbit embryo. *Biochemical Journal,* 1965, *97,* 14.

Finnell, R. H., and Chernoff, G. F. Variable patterns of malformation in the mouse fetal hydantoin syndrome. *American Journal of Medical Genetics,* 1984, *19,* 463–471.

Fischer, R. Genetics and gustatory chemoreception in man and other primates. In M. R. Kare and O. Maller (eds.), *The Chemical Senses and Nutrition.* Baltimore: Johns Hopkins Press, 1967, 61–81.

Friedman, S. L., Brackbill, Y., Caron, A. J., and Caron, R. F. Obstetric medication and visual processing in 4- and 5-month old infants. *Merrill-Palmer Quarterly,* 1978, *24,* 111–128.

Generoso, W. M., Rutledge, J. C., Cain, K. T., Hughes, L. A., and Braden, P. W. Exposure of female mice to ethylene oxide within hours after mating leads to fetal malformation and death. *Mutation Research,* 1987, *176,* 269–274.

Gillberg, C. "Floppy infant syndrome" and maternal diazepam. *Lancet,* 1977, *2,* 244.

Gordon, G. B., Spielberg, S. P., Blake, D. A., and Balasubramanian, V. Thalidomide teratogenesis: Evidence for a toxic arene oxide metabolite. *Proceedings of the National Academy of Sciences,* 1981, *78,* 2545–2548.

Gottschewski, G. H. M. Mammalian blastopathies due to drugs. *Nature,* 1964, *201,* 1232–1233.

Grimes, D. A., and Gross, G. K. Pregnancy outcomes in black women aged 35 and older. *Obstetrics and Gynecology,* 1981, *58,* 614–620.

Hald, J. E., and Jacobsen, E. The formation of acetaldehyde in the organism after injection of antabuse (tetraethyliuramdisulphide) and alcohol. *Acta Pharmacologia et Toxicologia,* 1948, *4,* 305–310.

Hansen, D. K., and Billings, R. E. Effect of route of administration of phenytoin teratogenicity in A/J mice. *Journal of Cranofacial Genetics and Developmental Biology,* 1986, *6,* 131–138.

Haram, K., Bakke, O. M., Johannessen, K. H., and Lund, T. Transplacental passage of diazepam during labor: Influence of uterine contractions. *Clinical Pharmacology and Therapeutics,* 1978, *24,* 590–599.

Haus, E., and Halberg, F. Circadian rhythm in level and timing of serum corticosterone in standardized inbred mature mice. *Environmental Research,* 1970, *3,* 81–106.

Ho, B. T., Fritchie, G. E., Englert, L. F., McIsaac, W. M., and Idanpaan-Heikkila, J. E. Marihuana: Importance of the route of administration. *Journal of Pharmacy and Pharmacology,* 1971, *23,* 309–310.

Hutchings, D. E. Behavioral teratology: A new frontier in neurobehavioral research. In E. M. Johnson and D. M. Kochar (Eds.), *Handbook of Experimental Pharmacology,* New York: Springer-Verlag, 1983, 207–235.

Hutchings, D. E. Prenatal opioid exposure and the problem of causal inference. In T. M. Pinkert (Ed.), *Current Research on the Consequences of Maternal Drug Abuse.* Rockville, Md.: National Institute on Drug Abuse, 1985, pp. 6–19.

Jaquet, P., Leonard, A., and Gerber, G. B. Action of lead on early divisions of the mouse embryo. *Toxicology,* 1976, *6,* 129–132.

Kajii, T., Kida, M., and Takahashi, K. The effect of thalidomide intake during 113 human pregnancies. *Teratology,* 1973, *8,* 163–166.

Kalter, H. Seasonal variation in frequency of corticosterone-induced cleft palate in mice. *Genetics,* 1959, *44,* 518–523.

Kalter, H. Teratogenic action of a hypocaloric diet and small doses of cortisone. *Proceedings of the Society for Experimental Biology and Medicine,* 1960, *104,* 518–520.

Kalter, H. *Teratology of the Central Nervous System.* Chicago: University of Chicago Press, 1968.

Kalter, H., and Warkany, J. Experimental productions of congenital malformations in strains of inbred mice by maternal treatment with hypervitaminosis. *American Journal of Pathology,* 1961, *38,* 1–21.

Kiely, J. L., Paneth, N., and Susser, M. An assessment of the effects of maternal age and parity in different components of perinatal mortality. *American Journal of Epidemiology,* 1986, *123,* 444–454.

Kimmel, G. L. Developmental aspects of clinical interaction with cellular responses. In C. A. Kimmel and J. Buelke-Sam (Eds.), *Developmental Toxicology.* New York: Raven Press, 1981, pp. 115–130.

Kohler, E. Growth kinetics of mammalian embryos during organogenesis. In R. Bass (ed.), *Metabolic Pathways in Mammalian Embryos during Organogenesis and its Modifications by Drugs.* Berlin: Freie Universitat, 1970, pp. 17–27.

Lasuncion, M. A., Lorenzo, J., Palacin, M., and Herra, E. Maternal factors modulating nutrient transfer to fetus. *Biology of the Neonate,* 1987, 51, 86–93.

Laukaran, V. H., and VandenBerg, B. J. The relationship of maternal attitude toward pregnancy outcomes and obstetric complications. *American Journal of Obstetrics and Gynecology,* 1980, *136,* 374–479.

Lenz, W. Malformations caused by drugs in pregnancy. *American Journal of Diseases of Children,* 1966, *112,* 99–106.

Levin, A. A. and Miller, R. K. Fetal toxicity of cadmium in the rat: Maternal vs fetal injections. *Teratology,* 1980, *22,* 1–5.

Levy, G. Pharmacokinetics of fetal and neonatal exposure to drugs. *Obstetrics and Gynecology,* 1981, *58,*(5), 9–16.

Lindsay, H. A., and Kullman, V. S. Pentobarbital sodium: Variation in toxicity. *Science*, 1966, *151*, 576–577.

Loughnan, P. M., Gold, H., and Vance, J. C. Phenytoin teratogenicity in man. *Lancet*, 1973, *1*, 71–80.

MacLeod, S. M., Giles, H. G., and Patzalek, G. Diazepam actions and plasma concentrations following ethanol ingestion. *European Journal of Clinical Pharmacology*, 1977, *11*, 345–349.

McClain, P. M., and Rohrs, J. M. Potentiation of the teratogenic effects and altered disposition of diphenylhydantoin in mice fed a purified diet. *Toxicology and Applied Pharmacology*, 1985, *77*, 86–93.

Miller, J. W., and Woollam, D. H. M. Effect of vitamin B complex on the teratogenic activity of hypervitaminosis A. *Nature*, 1958, *182*, 940.

Naeye, R. L. Maternal age, obstetric complications, and the outcome of pregnancy. *Obstetrics and Gynecology*, 1983, *61*, 210–216.

Nau, H., Luck, W., Kuhnz, W., and Wegener, S. Serum protein binding of diazepam, desmethyldiazepam, furosemide, indomethacin, warfarin, and phenobarbital in human fetus, mother, and newborn infant. *Pediatric Pharmacology*, 1983, *3*, 219–227.

Nelson, B. K. Dose/effect relationships in developmental neurotoxicology. *Neurobehavioral Toxicology and Teratology*, 1981, *3*, 255.

Nishimura, H., Tahimura, T., Semba, R., and Uwabe, C. Normal development of early human embryos: Observation of 90 specimens at Carnegie Stages 7 to 13. *Teratology*, 1974, *10*, 1–5.

Nishmura, H., Takano, K., Tanimura, T., and Yasuda, M. Normal and abnormal development of human embryos: First report of the analysis of 1,213 intact embryos. *Teratology*, 1968, *1*, 281–290.

Nora, J., Trasler, D. G., and Fraser, F. C. Malformations in mice induced by dexamphetamine sulfate. *Lancet*, 1965, *2*, 1021–1022.

Pasamanick, B., and Lilienfield, A. M. Association of maternal and fetal factors with development of mental deficience: I. Abnormalities in the prenatal and perinatal periods. *Journal of the American Medical Association*, 1955, *159*, 155–160.

Paulone, M. E., Edelstone, D. I., and Sheld, A. Effects of maternal anemia on utero-placental and fetal oxidative metabolism in sheep. *American Journal of Obstetrics and Gynecology*, 1987, *156*, 230–236.

Randall, C. L., and Anton, R. F. Aspirin reduces alcohol-induced prenatal mortality and malformations in mice. *Alcoholism: Clinical and Experimental Research*, 1984, *8*, 513–515.

Rapoport, M. I., Feigin, R. D., Bruton, R. D., and Beisil, W. R. Circadian rhythm for trytophan pyrolase activity and its circulating substrate. *Science*, 1966, *153*, 1642–1644.

Riley, E. P. and Vorhees, C. V. (eds.) *Handbook of Behavioral Teratology*. New York: Plenum Press, 1986.

Robens, J. F. Teratogenic studies of carboxymethyl cellulose, disulfiram, and thiram in small laboratory animals. *Toxicology and Applied Pharmacology*, 1969, *15*, 152–163.

Rodier, P. M. Time of exposure and time of testing in developmental toxicology. *Neurotoxicology*, 1986, *7*, 69–76.

Rodier, P. M., Webster, W. S., and Langman, J. Morphological and behavioral consequences of chemically-induced lesions of the CNS. In N. Ellis (Ed.), *Aberrant Development of Human Infancy: Human and Animal Studies*. Hillsdale, N.J.: Erlbaum, 1975. pp. 169–176.

Russell, L. B., and Montgomery, C. S. Radiation-sensitivity differences within cell-division cycles during mouse cleavage. *International Journal of Radiation Biology*, 1966, *10*, 151–164.

Samson, H. H. Maternal ethanol consumption and fetal development in the rat: A comparison of ethanol exposure techniques. *Alcoholism: Clinical and Experimental Research*, 1981, *5*, 67–74.

Shemano, I., and Nickerson, M. Effect of ambient temperature on thermal response to drugs. *Canadian Journal of Physiology*, 1958, *36*, 1243–1249.

Simmons, R. D., Kellogg, C. K., and Miller, R. K. Prenatal diazepam exposure in rats: Long-lasting receptor-mediated effects on hypothalamic norepinephrine-containing neurons. *Brain Research*, 1984, *293*, 73–83.

Simmons, R. D., Miller, R. K., and Kellogg, C. K. Prenatal exposure to diazepam alters central and peripheral responses to stress in adult rat offspring. *Brain Research,* 1984, *307,* 39–46.

Skillman, C. A., Plessinger, M. A., Woods, J. R., and Clark, K. E. Effect of graded reductions in utero-placental blood flow on the fetal lamb. *American Journal of Physiology,* 1985, *249,* H1098–H1105.

Slater, B. C. S., Watson, G. I., and McDonald, J. C. Seasonal variation in congenital abnormalities. *British Journal of Preventive and Social Medicine,* 1964, *18,* 1–7.

Sokol, R. J., Ager, J., Martier, S., Debanne, S., Ernhart, C., Kuzma, J., and Miller, S. I. Significant determinants of susceptibility to alcohol teratogenicity. *New York Academy of Sciences,* 1986, *477,* 87–102.

Spielmann, H., Eibs, H. G., and Jacob-Muller, U. Cyclophosphamide treatment prior to implantation: The effects on embryonic development. In T. V. N. Persaud (Ed.), *Teratology Testing.* Baltimore: University Park Press, 1979, pp. 95–112.

Strean, L. P., and Peer, L. A. Stress and an etiologic factor in the development of cleft palate. *Plastic Reconstructive Surgery,* 1956, *18,* 1–8.

Studd, J. W. W., Blainey, J. D., and Bailey, D. E. Serum protein changes in the pre-eclampsia–eclampsia syndrome. *Journal of Obstetrics of the British Commonwealth,* 1970, *77,* 796–801.

Tachibana, T. Instability of dose-response results in small sample studies in behavioral teratology. *Neurobehavioral Teratology and Toxicology,* 1982, *4,* 117–118.

Vorhees, C. V. Principles of behavioral teratology. In E. P. Riley and C. V. Vorhees (Eds.), *Handbook of Behavioral Teratology.* New York: Plenum Press, 1986, pp. 23–48.

Waddell, W. J., and Butler, T. C. The distribution and excretion of phenobarbital. *Journal of Clinical Investigation,* 1957, *36,* 1217–1226.

Warkany, J., and Nelson R. C. Appearance of skeletal abnormalities in the offspring of rats raised on a deficient diet. *Science,* 1940, *92,* 383–384.

Wilson, J. G. Teratogenic interaction of chemical agents in the rat. *Journal of Pharmacology and Experimental Therapeutics,* 1964, *44,* 429–436.

Wilson, J. G. *Environment and Birth Defects,* New York: Academic Press, 1973.

Woods, J. R., Plessinger, M. A., and Clark, K. E. Effect of cocaine on uterine blood flow and fetal oxygenation. *Journal of the American Medical Association,* 1987, *257,* 957–961.

CHAPTER 6

Identifying Human Behavioral Teratogens

Designs and Methods

Although over 800 agents have now been found to produce some form of teratogenicity in animals, only a few (30) are considered teratogenic in humans (Shepard, 1986) and none of these has been so identified solely on the basis of behavioral anomalies. Although behavioral damage, especially mental retardation, represents a far greater and more intractable problem than structural damage, behavioral teratogens *per se* are not included in any list of agents causing anomalies in humans. This is because immediate identification of anomalies is still society's primary concern and "silent teratogens" do not make their influence felt until long after birth. Nevertheless, narcotics, marihuana, PCBs, and other compounds clearly warrant more than passing attention because of their behavioral teratogenicity. In terms of cost alone, care and treatment of people with behavioral dysfunctions due to prenatal CNS damage represents a far greater economic burden to society than caring for individuals with structural defects (e.g., Abel & Sokol, 1987).

Identification of human teratogens originates when (1) an unusual anomaly, a pattern of anomalies, or a common anomaly occurs more often than usual, (2) the suspected agent is consistently associated with an anomaly, (3) exposure precedes outcome and the relationship is precise enough so that the presence of the teratogen predicts outcome, and (4) exposure occurs during a period in development when anomalies are likely to occur.

The frequency of occurrence of a particular anomaly is described in terms of incidence and prevalence. *Incidence* refers to the number of new cases of an anomaly entering a population at a particular time; *prevalence* refers to the number of existing cases at a particular time regardless of when they entered. For structural birth defects, the prevalence is always lower than the incidence because many malformed embryos and fetuses abort spontaneously, or are stillborn, or in the case of malformations like anencephaly, die shortly after birth. By contrast, the prevalence of congenital behavioral anomalies is always higher than the incidence because they are usually not recognized until several years after birth.

Before the incidence or prevalence of any birth defect can be determined, there are three basic issues that have to be considered: (1) the criteria for the birth defect, (2) its ascertainment—the accuracy and thoroughness with which instances of the birth defect are recorded, and (3) the population being evaluated.

Definition of Birth Defects

Some birth defects are much more easily recognized than others. No one would argue about whether an infant has a missing limb or not but there could be disagreement about whether his or her eyes are smaller than normal (and therefore possibly indicative of microcepaly because the eyes are part of the brain).

If a teratogen is to be identified without any doubt as to its role as a causal agent, criteria for labeling a feature as a birth defect must be agreed upon. Several years ago, a criterion of 4% was chosen as a basis for labeling structural features as birth defects (Marden *et al.*, 1964). Any feature that occurs in 4% or less of infants of the same race was considered an anomaly; if the feature occurred in over 4% of infants of the same race, it was considered normal variation.

Criteria for behavioral anomalies are less easily established for many reasons, one of which is a lack of consensus about what is normal. This is particularly difficult in the case of two of the most common behavioral anomalies—mental retardation and attention deficit disorder.

Mental retardation implies subnormal intelligence. But how subnormal does intelligence have to be before it qualifies for the label "mental retardation"?

Instruments designed to measure mental retardation are considerably less precise than those used to measure blood pressure or other biological parameters. Although some organic disorder is assumed to be responsible for mental retardation, structural changes in the brain are not always detectable. Nor do organically recognizable disorders always result in behavioral disorders. Hydrocephalus, for example, is not always associated with behavioral dysfunction, nor is microcephaly, phenylketonuria, or other organic impairments.

The instrument for measuring mental retardation is the intelligence test. The two main intelligence tests in current use are the Stanford Binet and the Wechsler (see below). These tests evaluate the number of correct answers to test items in terms of standards for a particular age and express these in terms of an "Intelligence Quotient" or IQ with a score of 100 being average "normal" intelligence.

In 1959, the American Association on Mental Deficiency (AAMD) said anyone with an IQ score of less than 85 (1 standard deviation below the average) was mentally retarded. In 1973, the AAMD revised its criterion and said that anyone with an IQ score of less than 70 (2 standard deviations below normal) was mentally retarded (Grossman, 1973). This shift in criterion reduced the number of mentally retarded Americans from about 35 million (16% of the population) to about 2.2 million (2.3% of the population) (Zigler & Cascione, 1984).

Same people. Same scores. Different standard—millions fewer retarded Americans. People hitherto considered to have borderline or dull mental retardation (IQ 70–85) were now considered to fall into the limits for normal intelligence (although their performance on many tasks would still be poor). A shift in criteria for normality can obviously mean a tremendous difference in prevalence.

Most (75% to 80%) people who are mentally retarded are mildly retarded (IQ 55–69). They are usually not identified until they enter school when they encounter new cognitive demands (Accardo & Caputi, 1979). Moderately and severely retarded children are usually identified prior to entering school because their handicaps are more evident. Estimating the prevalence of mental retardation is therefore affected by the age of the people studied. Any study that looks for an association between exposure to a teratogen and mental retardation would be more likely to find one if children of school age rather than younger children were examined, all other factors being equal.

Table 6.1. Criteria for Mental Retardation according to American Association on Mental Deficiency[a]

1. IQ < 70
2. Below expected standards of personal independence
3. Diminished social responsibility
4. Problems occurring before 18 years of age

[a]From Grossman, 1973.

Whereas an IQ below 70 is a cardinal criterion of mental retardation, AAMD defines mental retardation in terms of three additional criteria (see Table 6.1):

> (mental retardation refers to) . . . significantly subaverage general intellectual functioning existing concurrently with deficits in adaptive behavior, and manifested during the developmental period. (Grossman, 1973, p. 11)

Even though an individual has an IQ below 70, if he or she is personally independent and socially responsible according to expectations for his or her age group, that person would not be considered mentally retarded by the AAMD. For school-age children, expectations usually mean age-related academic skills and ability to follow classroom regulations. The fourth criterion for mental retardation is recognition during the "developmental period," that is, up to 18 years of age. People who develop subaverage intelligence after 18 because of disease or brain damage are not considered mentally retarded.

If mental retardation is defined in terms of the AAMD's criteria, no infant can be diagnosed as mentally retarded and no teratogen can be identified as causing mental retardation. A child may be diagnosed as having brain damage but this does not necessarily mean mental retardation.

Attention deficit disorder (ADD) is another behavioral disorder without a consensus about criteria. Prior to the 1980s, it was called by a variety of terms such as "Strauss syndrome," "hyperactivity," "hyperkinetic syndrome," "minimal brain dysfunction," "minimal cerebral dysfunction," "organic behavior syndrome," and "minimal brain damage." In North America, about 2% to 4% of all children are estimated to suffer from ADD (Weiss, 1975). In England and Europe, estimates are far lower suggesting that either the diagnostic criteria are vague and therefore subjective, testing is different in these two areas, or children in the United States are actually different from those in England and Europe.

When clinicians first began characterizing this disorder following an outbreak of encephalitis in the 1920s, they described children now recognized as having ADD as hyperactive, impulsive, emotionally unstable, antisocial, disruptive in school, and nonresponsive to discipline. Further studies during the 1940s and 1950s by A. A. Strauss and his coworkers (e.g., Strauss & Lehtinen, 1947; Strauss & Kephart, 1955) led to the belief that the disorder was due to brain injury and it became known as the "Strauss syndrome."

In the 1950s and 1960s, the disorder was considered the result of perinatal brain damage similar to mental retardation but differing in degree: mental retardation resulted from major brain damage, hyperactivity from "minimal brain damage."

In the 1970s, a lack of consensus about both the causes of the disorder and its symptoms resulted in adoption of the present term, "attention deficit disorder" (ADD) and criteria for its diagnosis. These criteria include (1) an inability to sustain attention combined with impulsivity, (2) occurrence before 7 years of age, and (3) existence for at least 6 months. A more complete list of the diagnostic criteria for ADD are presented in Appendix A.

The attentional and impulse control problems are readily apparent in the classroom. ADD children seem unable to focus on academic tasks like mathematics and reading (Porrino *et al.*, 1983a) and have trouble finishing assignments. They seem like they are not listening. Their work is generally sloppy and full of errors. During play at school or during physical education, these children may switch from game to game without being able to stay with one long enough for completion, but their activity level is not significantly different from non-ADD children (Porrino *et al.*, 1983b). Some researchers have argued hyperactive children are not really overactive, it is just that their activity occurs at inappropriate times, such as during school, and therefore gets labeled hyperactive. However, when hyperactive and nonhyperactive children wore monitors that recorded their activity for 24 hours for 7 days, they were found to be more active than controls not only on week days (i.e., school days) but also on weekends (i.e., nonschool days) and at all hours, including sleep (Porrino *et al.*, 1983b). These observations clearly indicate hyperactivity is not simply an inappropriate response in a structured setting like a classroom.

Although hyperactivity has been emphasized as a cardinal feature of this syndrome, it is not always present; even when present, it tends to decrease in adolescence. Attention problems, on the other hand, are always present and usually continue beyond adolescence.

Ascertainment

Unless every case of an anomaly is identified and recorded, estimates of incidence or prevalence will be inaccurate. Even with the best intentions, efforts can be subverted for unexpected reasons.

In Los Angeles, for example, many children with anencephaly are coded as "stillbirth" in some hospitals (Sever, 1983). Although illegal, this is done for religious reasons because some priests will not baptize anencephalics (Hook, 1982).

In studies that rely on birth certificates, incomplete reporting may have occurred because a defect wasn't recognized in the short time before the certificate was completed, the information may not have been transferred from hospital records to the birth certificate, or only the most severe anomalies may have been recorded.

Population Being Evaluated

Study site is a major factor in incidence and prevalence. For example, most identified cases of fetal alcohol syndrome (FAS) in the United States have come from study sites where the race of a majority of the mothers was black or Indian and socioeconomic status (SES) was low. The estimated incidence for FAS at these sites is 2.6 per 1,000 compared with 0.6 per 1,000 for sites where the mothers were mainly white and of middle socioeconomic status (Abel & Sokol, 1987). The highest incidence of FAS occurs among women of the Apache and Ute tribes of the American Southwest (19.5 per 1,000).

As noted in Chapter 5 (Principle 8), differences in maternal age, race, socioeconomic status, and other factors may influence teratogenic risk. Among the reasons for the higher incidence of FAS among women with low SES may be a greater degree of problem drinking; a bias in suspecting heavy drinking among low-SES women and therefore looking harder for FAS in their children; an interaction between alcohol and certain, as yet unknown, risk factors more prevalent among lower-SES women; or higher conception rates among low-SES alcoholics than among high-SES alcoholics, to name just a few possibilities. The characteristics of a population can thus greatly influence incidence and prevalence estimates.

Procedures for Identifying Teratogens

The designs and analytical methods used to identify possible human teratogens can be divided into three basic types. The first is descriptive and depends on basic observations and survey techniques. These are often the first indications of abnormalities. Once a teratogen is suspected, a second, more sophisticated analytic procedure can be chosen that relies on unbiased assignment of subjects to specific treatment groups. The final and most sophisticated approach would be an experimental evaluation in which subjects are examined in the relatively pristine environment of the laboratory.

Descriptive Studies

Medical Case Reports

Most teratogens are first identified by physicians. This is the way thalidomide was identified. Simultaneously, two physicians, one in Germany, the other in Australia, reported the birth of children without limbs to mothers who had taken thalidomide during pregnancy. This was a rare malformation so the physicians were quickly aware of something out of the ordinary.

However, case reports only suggest possible relations; they do not prove causation. They also tend to be based on very few individuals, sometimes only one or two. Case studies are also limited by the number of patients a physician sees and by his or her astuteness in recognizing something out of the ordinary. Because case data are based on so few instances, the relation between a structural birth defect and maternal exposure may be only coincidental. It is merely a starting point for identifying teratogens that cause structural anomalies or for additional studies of behavioral anomalies.

An example of the latter is the case study of Shaywitz *et al.* (1981) involving two preschool-age children with a history of prenatal alcohol exposure. In addition to behavioral anomalies previously associated with the syndrome, the clinicians found that both were hypervigilent to auditory stimuli, both had poor sound discrimination, and both had developmentally delayed language skills. These auditory and language problems were subsequently investigated by Church and Gerkin (1988) and Larsson and her co-workers (1985).

Surveillance Programs

Some cities, states, and countries have established birth defects surveillance programs for early detection of congenital anomalies. By gathering data over long periods of time, these programs can establish background rates. It can then be determined if a teratogen or mutagen has been introduced into a population by looking for fluctuations in these rates.

The largest birth defects surveillance program in the United States is the Birth Defects

Monitoring Program (BDMP) started in 1974 and directed by the Centers for Disease Control (CDC) in Atlanta, Georgia. This nationwide program for monitoring malformations in the United States collects data from about 1,200 hospitals with obstetrical services. Every case includes information about such things as sex, race, date of birth, birthweight, and date of discharge.

The most crucial ingredient in any surveillance program is ascertainment. Although the BDMP is the largest surveillance system in the country, recording about 1 million births per year, the data are somewhat biased because hospitals voluntarily participate and therefore the data are not randomly collected; physicians do not use the same criteria for diagnosis and differ in their diagnostic skills; there is no geographically well-defined population; and there is variability in how well and how thorough diagnoses are made and recorded in rural versus urban hospitals, in different parts of the country, and in hospitals serving different socioeconomic groups.

A more modest surveillance program also sponsored by the CDC is the Metropolitan Atlanta Congenital Defects Program (MACDP) initiated in 1967. This program collects data from 20 area hospitals in 5 counties around Atlanta involving about 24,000 births per year. This is a much more thorough program than the BDMP and has multiple procedures for ascertainment that include sources for gathering information about infants with structural, chromosomal, and biochemical anomalies diagnosed before one year of age. These cases are identified through records of local cytogenetics laboratories and genetic counseling clinics in addition to hospital records and there is follow up of children with certain defects. Data about anomalies are routinely sent from these sources to a center where they are tabulated and analyzed. Such data can be used to determine occurrence of specific anomalies in a geographically defined population at specific times and are especially valuable for detecting new teratogens. Such data can also be used to conduct case-control studies (see below) and to assess the impact of prenatal diagnosis and genetic counseling.

Behavioral anomalies, however, cannot be detected by most surveillance programs. This is because behavioral anomalies are almost impossible to detect at birth. The only surveillance system in the world presently collecting data about behavior is the Health Surveillance Registry in British Columbia, Canada (Lowry *et al.*, 1975). This registry was begun in 1952 and has been collecting information on the incidence of mental retardation (classified according to IQ levels) in addition to structural anomalies. The registry obtains information from more than 80 sources including special schools and organizations for retarded individuals, hospital discharge diagnoses for children under 7, genetics clinics, and the like, and individuals may be registered at any time. Data files are continuously updated with new and relevant diagnostic information. This registry is unique in being the only source providing reliable estimates of the incidence and prevalence of mental retardation in a large population. However, a similar data base is now under consideration for ascertainment of mental retardation and cerebral palsy in California (Grether, 1987).

Analytical Evaluations

Retrospective studies involve scouring pregnancy and birth records or birth defect registry records for similar anomalies and are outcome-to-antecedent in strategy. Records with the same anomalies are then evaluated to see if there is any common element. One such example is Flint's (1983) study of childhood deafness in Glasgow, Scotland. Children with hearing disorders were identified from clinical records and their parents were then contacted and asked to complete a questionnaire. On the basis of the answers to these questionnaires, causes of deafness were identified in 208 out of 250 cases. Genetic disorders, such as Down syndrome, were responsible for 32% of the cases, infections acquired in childhood such as meningitis, accounted for 22%, perinatally acquired factors such as anoxia accounted for 17%, and prenatally acquired factors such as rubella accounted for another 14%.

Hagberg *et al.* (1981) identified 91 children with mild mental retardation (IQ 50–70) among 24,498 children born in Gothenburg, Sweden, between 1966 and 1970. Information concerning pregnancy, birth, and the perinatal period was obtained from obstetric and birth records and from medical records. Prenatal causes accounted for 23% of the cases (21 out of 91). Out of these prenatal causes, 33% (7 out of 21) were attributed to prenatal alcohol exposure.

Retrospective studies such as these, however, are only suggestive of possible causal relationships because of inherent biases. For instance, parents may selectively or incorrectly recall exposures that occurred prior to or during pregnancy, they may not want to answer questions about their children, or hospitals may not allow their medical records to be examined by researchers because they want to preserve patient confidentiality. If hospitals will not make their records available unless a patient has signed a consent form and if some patients refuse, they will not be included in the study. This could result in a bias because those that do not give their consent may differ from those that do on a variable of significance.

Case Control

A refinement of the retrospective study is the case control evaluation. In these studies, a particular anomaly is identified and children or their mothers (the "cases") are compared with unaffected subjects (the "controls") for exposure to a suspected teratogen. If exposure is greater among cases, it could suggest a causal influence.

In case control studies, there is no random assignment to exposed and nonexposed groups. Instead, cases are identified, controls are chosen, and exposure or nonexposure is determined for each group.

An example is the case control study of Edmonds *et al.* (1975), which evaluated CNS malformations among women married to men who worked with vinyl chloride. On the basis of BDMP records, there appeared to be an increased incidence of such malformations in Ohio. One possibility was that the increase was due to paternal exposure to vinyl chloride in factories located in the state. To evaluate this possibility, hospital records were examined to identify and confirm specific cases. Controls were then chosen from infants born immediately before and after the cases and were matched on several factors. The occupations of the parents of cases and controls were then determined and no significant difference was found.

Case control studies are relatively inexpensive and easy to conduct. The major problem is choosing the appropriate controls. If only birth records are inspected, there may be no way of evaluating length of exposure, which could be a critical factor contributing to teratogenicity. Case control studies involving interviews of subjects can also be biased by the same factors as retrospective studies. Parents of a child with a birth defect may be so distraught they may scrutinize their past behavior more carefully for possible exposures than would controls. If the interviewers are aware of the nature of the study or the sought-for relationships, they may probe subjects for background information more diligently than they would controls. Trained interviewers are therefore an important component in any such study.

Prospective Studies

Prospective epidemiological studies are among the best ways to evaluate teratogens. In contrast with the outcome-to-antecedent strategy of retrospective studies, prospective studies proceed along an antecedent-to-outcome direction. Such studies may begin with a woman's prenatal care visits and follow her through pregnancy. At some time during these visits, preferably during the first visit and early in the pregnancy, a thorough history of her health, family background, eating and drinking habits, and the like is obtained by a standardized questionnaire. Data are collected thoroughly and dependably and there is little deliberate bias in recording. There is also less bias due to memory problems because background information is collected prior to birth. The amount of risk attached to a particular teratogen can then be estimated by determining how many children with a specific defect are born and how many of them had mothers with a specific exposure history. Comparisons can be made between women who were exposed or were not exposed.

Although subject to less bias than retrospective studies, prospective studies are only as good as the cooperation of those answering the questionnaires. Some patients may not wish to complete such questionnaires; others may deliberately underestimate use of some agent such as alcohol. When underestimation is suspected, more accurate self-report data may be obtained by a deception called the "bogus pipeline." With this technique, the patient is told that her verbal or written response may be independently checked by laboratory tests of her blood and urine. In one such study 14% of the pregnant patients said they drank when responding to questionnaire only, compared to 27% when told that lab tests also would be conducted to verify their responses (Lowe *et al.*, 1986).

Prospective studies may also be biased if those women who are most at risk are uncooperative and refuse to participate or do not attend prenatal care clinics at all. For example, the incidence of fetal alcohol syndrome (FAS) is 2.2 per 1,000 if data are collected retrospectively but only 1.8 if collected prospectively (Abel & Sokol, 1987). The reason it is higher in retrospective studies is that women most at risk for FAS do not seek prenatal care. Therefore they aren't included in any prospective studies. Women who do seek prenatal care may be more concerned with their health and the health of their unborn children and may drink less than women who do not seek such care.

There are currently several prospective studies of teratogens that include behavioral assessments of children. Teratogens that have been evaluated include alcohol in Seattle (e.g., Streissguth *et al.*, 1986), Cleveland (e.g., Golden *et al.*, 1982), and Sweden (e.g., Larsson *et al.*, 1985); marihuana in Ottawa, Canada (e.g., Fried, 1982); medicinal agents by the Collaborative Perinatal Project (e.g., Naeye & Peters, 1984), PCBs in Michigan (e.g., Jacobson *et al.*, 1985) and North Carolina (Roger *et al.*, 1986); heroin in Copenhagen, Denmark (e.g., Olofsson *et al.*, 1983), and Houston (e.g., Wilson *et al.*, 1979); for methadone in New York (e.g., Johnson & Rosen, 1982); and amphetamines in Stockholm, Sweden (e.g., Billing *et al.*, 1985).

Although longitudinal prospective survey studies often provide valuable insights into relations between substances and their possible behavioral teratogenic influences, because of the various methodological problems associated with these procedures, epidemiologists have developed more sophisticated prospective designs like the cohort study.

Cohort Studies are a special kind of prospective study in which individuals are followed over a long period of time. As for prospective studies in general, groups are identified in terms of their exposure or nonexposure to a given agent before the occurrence of any anomaly, which reduces any possibilty of recall bias. This can be an either-or dichotomy or a difference between groups in amount of exposure. Groups are then followed over time and pregnancy outcome is recorded for each member of the cohort. Because cohort studies often involve large numbers of individuals, incidence and prevalence rates can be determined for exposed and nonexposed individuals and multiple observations can be made. The drawbacks of cohort studies are that they are very expensive and very long term and many of those who initially agree to participate either lose interest or move away. The critical factor is to control for confounding to overcome lack of random assignment.

The largest example of this kind of study is the Collaborative Perinatal Project conducted between 1959 and 1966 involving about 55,000 pregnancies. As part of this study, children at 7 years of age were evaluated for spelling, reading, arithmetic skills, attention span, and motor activity as related to maternal smoking during pregnancy (Naeye & Peters, 1984). Women were categorized as nonsmokers, light smokers (1–19 cigarettes per day), or heavy smokers (20 or more cigarettes per day). After controlling for many confounding variables, it was found that children of heavy smokers had slightly lower spelling and reading scores, had more frequent short attention spans, and were more often hyperactive than children of nonsmokers. Because this was a very large data base, the authors were also able to compare siblings whose mothers smoked during only one of their two pregnancies. This enabled them to control for both genetic and environmental factors. This comparison also showed that siblings prenatally exposed to smoking had lower test scores, had shorter attention spans, and were more hyperactive than siblings not exposed to smoking *in utero* (Naeye & Peters, 1984).

Randomized Controls

The most rigorous of the analytical designs is the randomized control. In this design, patients are allocated randomly to either the exposed or nonexposed control group. In this way, bias in assigning subjects to treatment because of clinical judgment or specific prognostic factor (e.g., previous history of giving birth to children with birth defects) is avoided.

However, it is unconscionable to prescribe a drug for which there is a suspicion of teratogenicity. The only instances of randomized clinical studies using a behavioral endpoint involve tests of newborns whose mothers received obstetrical analgesia. In one such study (Writer *et al.*, 1981), women in labor were given either morphine or bupivacaine and both the clinician and the patient were blind to what was administered. The Scanlon test battery (see below) was administered at 2–6 and 20–26 hours after birth. Several of the infants exposed to morphine were scored as borderline status at 2 hours (abnormal alertness, tone, reflex activity), whereas none of the infants exposed to bupivacaine received such scores. Both groups were normal at 20–26 hours.

Randomized clinical trials are difficult to perform because not all women will consent to a particular procedure. If sample size is too small, it may be difficult to detect differences, especially for uncommon defects. For example, if a drug caused a tenfold increase in a defect that normally occurs in only 1 in a 1,000 births, it would probably not be detected unless several thousand births were studied.

A number of methods are now available for assessing infant developmental status that can and in some cases have been adapted to human behavioral teratology. Although each provides a valuable insight into developmental status at the time of testing, none reflects the total complexity of behavior or is able to completely predict an individual's future intelligence or abilities. Because it is not possible to discuss all the available procedures available for assessing behavior, I have restricted the following survey to those that have been used to examine possible human behavioral teratogens.

Apgar Test

By far the most commonly used and most simple technique for assessing newborn status is the Apgar test devised by Virginia Apgar in 1953. This test is routinely administered in nearly all hospitals in the United States. Newborns are rated at 1 and 5 and sometimes 10 minutes after birth. The infant is given a score of 0, 1, or 2 points each for heart rate, respiratory effort, muscle tone, reflex response, and skin color. A heart rate of 100 to 140 beats per minutes is given a 2; below 100 beats per minute is scored a 1, and no detectable heart rate is given a 0. Respiratory effort is scored as 2 for regular breathing and lusty crying; no visible breathing is scored as 0; in between is a 1. Muscle tone is scored as 2 if the infant spontaneously flexes his or her arms and resists extension of his or her legs. A completely flaccid infant is given a 0; in between, a 1. Reflex response is scored as 2 if the infant cries when the soles of his or her feet are stimulated, 1 if he or she grimaces, and 0 if he or she makes no response. Skin color is scored as a 2 if the infant is pink at 1 minute, 0 if purple. Black children are obviously difficult to score on this measure. The composite score reflects the infant's ability to adapt to its environment. The optimal score is 10.

Several studies have examined Apgar scores in newborns prenatally exposed to drugs (e.g., Streissguth *et al.*, 1982). The significance of lower scores for future behavior, however, is dubious. The Apgar test is a good predictor of neonatal mortality—infants with the lowest scores have higher infant mortalities—and it has some predictive value for cognitive function for very young children. For instance, infants rated between 7 and 9 on the Apgar were less visually attentive at 3 and 9–13 months of age than those scoring 10 (Lewis *et al.*, 1967); those with low Apgar scores (0 to 3) also had lower mental and motor scores in the Bayley scales (see below) than those with scores of 7 to 10 (Serunian & Broman, 1975). When older children are tested, however, the predictive value is poor. In the Collaborative Perinatal Project, for example, no correlation was found between Apgar scores and IQ at 4 years of age when infants were matched for race, sex, and socioeconomic status (Broman *et al.*, 1975). Similarly, in a seven-year followup study involving over 49,000 infants, 55% of those who subsequently developed cerebral palsy had Apgar scores of 7 and 10 at 1 minute and 73% scored 7 and 10 at 5 minutes (Nelson & Ellenberg, 1981).

Prechtl Exam

The Prechtl neurological exam (Prechtl & Beintema, 1977) involves observation of reflex behaviors of the infant while he or she is in a supine and prone position, evaluation of muscle tone, and evaluation of arousal in stages from deep sleep to robust crying. The complete Prechtl exam takes about 30 minutes. There is also a "quick" screening version of the test. "Suspect" infants identified in the short version should then be given the full test (Prechtl, 1970).

After information about feeding behavior and drug exposure are recorded, the infant is observed in his or her crib. Six states are scored ranging from eyes closed, regular respiration, and no movement (State I) to eyes open, gross movements, irregular breathing, crying, and other conditions (Stage VI). The infant's blanket is then removed and the infant's resting posture, spontaneous movements, respiration rate and regularity, and skin color are noted.

The infant is next placed in a supine position on a table and his or her face and skull are examined. Facial expressions (bland, alert, fussing, frowning, crying) are observed and lip and glabella (inhibition of blinking to repeated taps on the forehead) reflexes are elicited.

Next, the infant is undressed and spontaneous motor activity, posture, respiration, skin color, and various other reflex responses are tested. Eyes are then examined for strabismus and nystagmus, shape and size of pupils, reaction to light, reflex blinking to light and sound, corneal reflex, and doll's eye reflex. After these tests, the infant's range of arm and leg movements is observed along with muscle tone. The infant is returned to the supine position and tested for knee jerk, palmar grasp, sucking, and other reflexes. He or she is then put in a prone position and examined for head control and spontaneous head movement; if he or she crawls, coordination is assessed. The infant is then placed upright and placing and stepping movements are observed. He or she is then put in a supine position again and reexamined for spontaneous motor activity and his or her hands are examined to see if the thumb has remained buried in the palm throughout testing. Crying is then provoked and scored for lustiness. A final overall appraisal is made of reactivity (comatose, apathetic, excitability) and reflex and motor function.

The Prechtl exam has been used by Fried (1982; Staisey & Fried, 1983) in a longitudinal prospective evaluation of prenatal marihuana and alcohol exposure. Marihuana use (Fried, 1982) was assessed prior to giving birth and women were divided into nonusers, light users (1 joint or less per week), moderate users (1–5 joints per week), and heavy users (more than 5 joints per week). Infants were examined at 9 and 30 days of age. Infants whose mothers were heavy users were less responsive to visual stimuli and more of them had tremors at 9 days but were no longer different at 30 days.

A similar protocol was used to assess the effects of alcohol (Staisey & Fried, 1983). At 9 days of age, infants exposed to alcohol *in utero* had decreased muscle tone, increased spontaneous startle reactions, decreased neck tone on the pull-to-sit test, decreased glabella and knee jerk reflexes, and decreased extension when legs were supine. These responses were interpreted as reflecting decreased muscle tone. When infants were retested at 30 days of age, prenatally exposed infants differed from controls only in the extension of their legs when supine.

Like the Apgar, the Prechtl is also ineffective in predicting long-term cognitive function except for severely impaired infants (Prechtl, 1965) and has a high false-positive rate. In one study, six times more infants were scored as neurologically abnormal than were found to be so at 18 months (Bierman-Van Eedenburg *et al.*, 1981).

The Brazelton Test

Most behavioral studies in human teratology have used the Brazelton Neonatal Behavioral Assessment Scale (Brazelton, 1973). This is a half-hour test consisting of 17 reflex items (see Table 6.2) and 27 behavioral items (see Table 6.3) administered on the second

Table 6.2. Reflex and Elicited Behaviors Evaluated on Brazelton Test[a]

	X	O	L	M	H	A
Plantar grasp		0	1	2	3	
Hand grasp		0	1	2	3	
Ankle clonus		0	1	2	3	
Babinski		0	1	2	3	
Standing		0	1	2	3	
Automatic walking		0	1	2	3	
Placing		0	1	2	3	
Incurvation		0	1	2	3	
Crawling		0	1	2	3	
Glabella		0	1	2	3	
Tonic deviation of head and eyes		0	1	2	3	
Nystagmus		0	1	2	3	
Tonic neck reflex		0	1	2	3	
Moro		0	1	2	3	
Rooting (intensity)		0	1	2	3	
Sucking (intensity)		0	1	2	3	
Passive movements of right arm, left arm, right leg, left leg		0	1	2	3	

[a]X, response omitted; O, response not elicited; A, asymmetry of response; L, low; M, medium; H, high.

or third day of life. The premise behind the Brazelton is that the newborn is able to adapt to his or her environment so as to be able to elicit responses necessary for his or her growth and development such as getting the attention of his or her caregiver and protecting himself or herself from negative stimuli.

The aim of the Brazelton is to assess the infant's optimal performance. This means the examiner has to be flexible so that each item is administered when the infant is exhibiting the appropriate level of alertness from deep sleep to robust crying.

The test contains three major components and there is no single index summarizing the scores from the individual or component tests. Although rarely used, the first component examines elicited reflexes, which are rated on a 4-point scale from 0 (no elicited response) to 3 (maximal response); the second rates the infant for "attractiveness" (how much effort the examiner must use to get the infant to respond), "interfering variables" (the influence of the environment on the examination), "need for stimulation" (how the infant will react to or use stimulation to become organized), and activities the infant uses to quiet himself or herself such as "hand to mouth," "sucking with nothing in mouth," and "locking onto visual or auditory stimuli."

The third component is the behavioral evaluation. This is the major component of the test and scores 27 items on a 9-point scale. Among these are several measures of habitation such as response decrements to a flashlight shone in the eyes, to a rattle and a bell, and to a pinprick; and five measures of orientation such as response to inanimate visual and auditory stimuli alone, and response to a combination of animate visual and auditory stimuli. Other behavioral items include alertness, activity, irritability, and startle.

The Brazelton scale has been used in several behavioral teratological investigations. Coles and her co-workers (1985) administered the Brazelton to 3-day-old infants whose

Table 6.3. Behavioral Measures Evaluated by Brazelton Neonatal Behavior Assessment Scale[a]

1. Response decrement to light (2,3)
2. Response decrement to rattle (2,3)
3. Response decrement to bell (2,3)
4. Response decrement to pinprick (1,2,3)
5. Orientation inanimate visual (4 only)
6. Orientation inanimate auditory (4,5)
7. Orientation animate visual (4 only)
8. Orientation animate auditory (4,5)
9. Orientation animate visual and auditory (4 only)
10. Alertness (4 only)
11. General tonus (4,5)
12. Motor maturity (4,5)
13. Pull-to-sit (3,5)
14. Cuddliness (4,5)
15. Defensive movements (4)
16. Consolability (6 to 5,4,3,2)
17. Peak of excitement (6)
18. Rapidity of buildup (from 1,2 to 6)
19. Irritability (3,4,5)
20. Activity (alert states)
21. Tremulousness (all states)
22. Startle (3,4,5,6)
23. Lability of skin color (from 1 to 6)
24. Lability of states (all states)
25. Self-quieting activity (6,5 to 4,3,2,1)
26. Hand–mouth facility (all states)

[a]Numbers in parentheses indicate optimal score.

mothers prospectively stated that they had consumed 24 to 28 drinks per week all through pregnancy, or only during the first trimester, or did not drink at all. Evaluations were done by examiners unaware of maternal drinking histories.

Infants whose mothers did not drink and those whose mothers stopped drinking by the second trimester did not differ in arousal and both were awake less often, were less restless, were more alert and attentive to their environments, and showed a more positive reaction to the examiners than infants whose mothers drank throughout pregnancy.

Infants whose mothers drank only during the first two trimesters of pregnancy were more active than nonexposed controls and had more immature motor function.

The Brazelton test has also been used to assess the impact of prenatal exposure to marihuana (e.g., Fried, 1980), narcotics (e.g., Strauss *et al.*, 1975), obstetrical analgesics (e.g., Standley *et al.*, 1974), cocaine (e.g., Chasnoff *et al.*, 1985), and PCBs (e.g., Jacobson *et al.*, 1984). However, it has a high false-positive rate for identifying children with future anomalies. Tronick and Brazelton (1975) rated 53 infants at 3 days of age on the Brazelton and neurologically at 7 years of age. The neurological part of the Brazelton correctly identified 13 out of 15 (87%) later found to be neurologically damaged and the behavioral assessment correctly identified 12 out of 15 (80%). However, 30 out of 50 infants considered as abnormal/suspect on the Brazelton neurological test and 9 out of 50 likewise considered on the behavioral assessment were later found to be normal.

Scanlon

The Scanlon Early Neonatal Neurobehavioral Scale (Scanlon, 1974) takes about 6 minutes to administer and contains select items from the Brazelton and Prechtl. It involves 15 observations of reflexes, muscle tone, and habituation to light and sound. In contrast to the Brazelton, the Scanlon does not try to arouse the infant or obtain his or her maximal response and the order of testing is standardized. The Scanlon test is mainly used to examine neurobehavioral effects associated with the cerebral effects of maternal obstetrical analgesia on the infant (see above).

Bayley Scales

The Bayley Scales of Infant Development (Bayley, 1969) is one of the most commonly used tests for evaluating mental and motor development during the neonatal period and has been used to assess the influence of prenatal exposure to PCBs (Jacobson *et al.*, 1983), narcotics (Strauss *et al.*, 1979), and smoking (Gusella & Fried, 1984).

The Bayley test is subdivided into mental and motor test components. The Mental Development Index (MDI) contains tests of language skills, abstract reasoning, social interaction, initiation and sensory activity, and fine motor coordination. The Psychomotor Development Index (PDI) evaluates gross body coordination.

Golden *et al.* (1982) used the Bayley tests in their prospective examination of the effects of prenatal alcohol exposure. Twelve 6- to 12-month-old infants whose mothers had been alcohol abusers during pregnancy were matched with 12 control infants for gestational age at birth, sex, and race and for maternal parity, substance abuse, and antepartum and intrapartum risk scores. Each exposed infant and his or her matched control was examined at the same age by a rater who was unaware of maternal drinking history. Alcohol-exposed infants had an average mental development of 86 compared with 105 for controls and an average motor development index of 90 compared with 110 for controls.

Although the Brazelton and Bayley scales may provide accurate assessments of an infant's developmental status at time of testing, performance on these tests has not been found to be a very good predictor of cognitive abilities in school-age or older children (Bayley, 1970; McCall *et al.*, 1977). This limitation in predictability had led to the development of new techniques such as the Fagan visual recognition memory test (Fagan & Singer, 1983), which differ from global procedures like the Brazelton and Bayley and use cognitive responses not dependent on infant motor skills.

Visual Recognition Memory

The Fagan test is based on early studies by Fantz (1965) that showed that infants are able to recognize complex and visual stimuli and have a preference for novel stimuli. The infant is seated on his or her mother's lap. A picture is placed before the infant and then

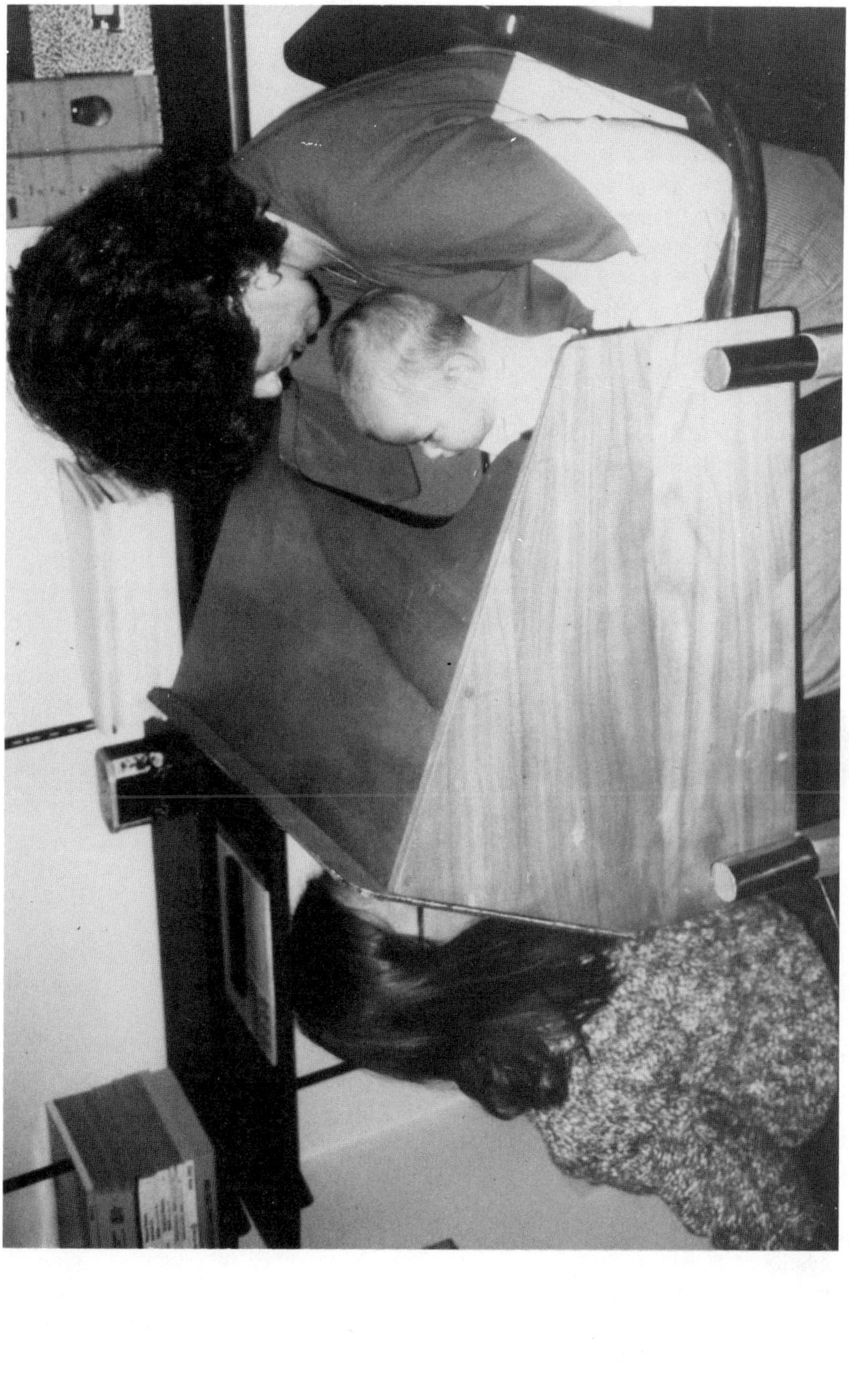

Figure 6.1. Seating of infant in visual recognition task. Photo courtesy of Drs. S. and J. Jacobson.

removed. The infant is then shown the familiar picture along with a novel one. The amount of time the infant looks at the unfamiliar picture is determined by corneal reflections of the picture observed through a peephole located halfway between the two pictures (see Figure 6.1). The time watching the new pictures indicates how much the infant has recalled the original stimulus and has discriminated it from the new one. The test is able to detect effects at lower levels of prenatal exposure to toxins than those that cause decreased birthweight or decreased performance on the Bayley test (Jacobson *et al.*, 1983, 1985).

In contrast with the poor predictability of the Brazelton and Bayley tests, the Fagan test has been found to have very good predictive value for later vocabulary ability and general IQ at 4–7 years of age with correlations ranging from .36 to .66 in nine groups of children (Fagan & McGrath, 1981). Visual recognition memory performance has also been found to predict IQ scores among a high-risk group of preterm children for 34 months to 6 years with correlations of .45 to .66 (Rose & Wallace, 1985).

Thus far, the Fagan test has not been used to evaluate behavioral teratogens except by S. Jacobson and her co-workers in their studies of prenatal exposure to alcohol (Jacobson *et al.*, 1983) and PCBs (Jacobson *et al.*, 1985). The protocol for the PCB study involved recruitment of women on the first or second day after they had given birth. These women were screened for their consumption of PCB-contaminated fish from Lake Michigan and were then divided into those who had not eaten such fish and those who had eaten moderate to large amounts. Exposure was assessed on the basis of self-report, cord serum PCB levels, and breast milk samples of PCBs.

Infants were tested at 7 months of age by an examiner unaware of maternal PCB exposure. The higher the cord serum PCB levels, the poorer the infant's recognition memory. Maternal self-report of fish consumption was also positively related to recognition memory. However, milk level of PCBs was not significantly related to infant recognition.

Because infants can see and hear at birth, visual and auditory processes themselves can also be evaluated to assess neurological integrity using EEG, evoked potentials, and other procedures discussed in Chapter 4. Thus far, these sophisticated techniques, with the exception of EEG, have not been used to any important extent in human behavioral teratology.

Reaction Time

Another measure being used in behavioral teratology is reaction time. Several variations in apparatus and protocols have been developed for reaction time testing in preschool and school-age children (e.g., Rosvold *et al.*, 1956). These tests are able to identify attentional deficits in children with phenylketonuria (Anderson *et al.*, 1969) as well as in children with clinically diagnosed brain damage (Rosvold *et al.*, 1956) and other problems. Performance on these reaction time tests is correlated with other tests of hyperactivity (Sykes, 1973), learning disability (Doyle *et al.*, 1976), and reading level (Noland & Schuldt, 1971).

Streissguth and her co-workers (1986) used a reaction time test to evaluate effects of prenatal alcohol exposure. Maternal drinking habits were assessed prior to birth. Children were evaluated at 6½ to 8½ years of age. The test involved two separate eight-minute presentations of 360 single letters, 60 of which were the "critical stimuli." In the "X task," the children were told to push a button each time the letter X appeared. In the "AX-test" they were told to press the button only when the letter X was preceded by the letter

A. Each letter appeared for about 50 msec at intervals of 1.4 seconds. The test also contained a "distraction" condition in which a click noise was presented to alter vigilance. The data were examined not only for reaction time but also for errors of omission and false alarms (errors of commission).

Reaction time was significantly decreased and errors of omission and false alarms were significantly increased in children whose mothers had consumed alcohol during pregnancy. For errors of omission, there was a linear relationship between amount of drinking and performance for children whose mothers drank a minimum of three drinks per day. Reaction time scores were also dose-related to prenatal alcohol exposure.

Intelligence Tests

There are several tests for measuring intelligence at different ages. Some emphasize certain abilities more than others and performance on these subtests counts for more than performance on other subtests.

Intelligence tests assess problem-solving ability. They rely heavily on language and abstract thinking and minimize motivation and creativity. Despite the many criticisms of such tests, they allow for identification of children whose abilities lie within the average range of intelligence as well as for those that function below (mentally retarded) and above (mentally gifted).

Some tests require auditory, visual, or motor skills more than others. Poor performance due to impaired hearing is not the same as poor performance due to lower intelligence.

Tests also differ in the age range they are designed to assess. Tests that measure infant responses do not have good predictive value for normal children but can reveal developmental delays. Correlations between performance on infant IQ tests and performance in tests at 5 years of age are low. The older a child is tested, the better the correlation with subsequent tests.

The most widely used IQ test for 5- to 15-year-old children is the Wechsler Intelligence Scale for Children–Revised (WISC–R). It has 12 subtests; half assess language-related tasks (vocabulary, similarities, comprehension, information, arithmetic, and digit span) and the other half assess perceptual-motor skills (picture arrangement, picture completion, object assembly, block design, coding, and mazes). Test problems are presented in increasing steps of difficulty. Subtests are scored separately to reveal strengths and weaknesses in particular abilities. A verbal and performance IQ are determined as well as an overall IQ score. Another version, the Wechsler Primary and Preschool Test of Intelligence (WPPSI), has been adapted for testing children 4–6½ years old.

Probably the most familiar of all intelligence tests is the Stanford–Binet. This is the oldest test for preschoolers and allows for testing from age 2 to adulthood. There are six items presented for each 6 months of age until 5 years of age and then six items for every 12 months of age. The test relies considerably on language skills and emphasizes conceptual and abstract thinking. Test results are expressed in terms of mental age (MA), which are then converted into IQ by dividing MA by chronological age (CA).

Streissguth and co-workers (1978a) administered the Wechsler and Stanford–Binet IQ tests to patients aged 9 months to 21 years who had been referred to a dysmorphology clinic. The patients were categorized as having severe, moderately severe, moderate, mild, or very mild fetal alcohol syndrome. IQ scores ranged from 15 to 105 with an average of

65. When these patients were given a second IQ test 1 to 4 years later, 77% of the patients had retest IQs within one standard deviation of initial IQ (Streissguth *et al.*, 1978b).

Summary and Conclusions

Although sophisticated designs and behavioral test procedures are now available for assessing human behavior as a function of prenatal exposure to teratogens, these designs and procedures have not been used extensively. In part, this is because human teratological evaluations are very expensive and take a very long time to perform. There is also the problem of predictability of test data from one age to another and from one test to another. Whereas results from early infant assessment tests do not reflect later cognitive abilities, they do reflect developmental status at time of testing and may possibly reflect developmental delay. With continued refinement of present procedures and introduction of new procedures, the ability of infant, neonatal, and childhood tests to predict later behavior should improve.

Human behavior is the most complex of all life activities and it should be neither surprising nor discouraging that more progress has not been made in identifying human behavioral teratogens. Considerable progress has already been made. There will be more. Enough progress has already been made to for us to know that there are many agents like narcotics, marihuana, PCBs, etc., whose major prenatal effects are behavioral, not structural. A structural handicap can be overcome. Having crippled legs did not prevent Franklin Roosevelt from being President of the United States. Brain damage cannot be treated or cured.

Appendix A. Diagnostic Criteria for Attention Deficit Disorder with Hyperactivity*

A. Inattention, characterized by at least 3 of the following:
   1. Does not complete projects
   2. Does not appear to be listening
   3. Easily distracted, needs quiet place to work
   4. Difficulty performing tasks requiring sustained attention, often asks to have instructions repeated, confuses details
   5. Constantly switching play activities before completion

B. Impulsivity, characterized by at least 3 of the following:
   1. Often acts before thinking
   2. Frequently switches activities
   3. Difficulty organizing work
   4. Requires considerable supervision, needs individual attention to help concentrate
   5. Disrupts class activities (e.g., talks excessively, calls out, or makes noise)
   6. Impatience in awaiting turn

C. Hyperactivity, characterized by at least 2 of the following:
   1. Moves about excessively when awake, climbs onto furniture or cabinets

*Adapted from American Psychiatric Association, *Diagnostic and Statistical Manual of Mental Disorders* (DSM-III), 1980.

2. Sleeps restlessly
3. Unable to sit still or fidgets excessively
4. Does not stay seated for long
5. Always "on the go," seems to run instead of walk, is loud and noisy

D. Onset before seven years of age
E. Characteristics last longer than 6 weeks
F. Problem not due to mental retardation, schizophrenia, or affective disorders

References

Abel, E. L., and Sokol R. J. Incidence of fetal alcohol syndrome and economic impact of FAS-related anomalies. *Drug and Alcohol Dependence,* 1987, *19,* 51–79.

Accardo, P. J., and Capute, A. J. *The Pediatrician and the Developmentally Delayed Child: A Clinical Textbook on Mental Retardation.* Baltimore: University Park Press, 1979.

American Psychiatric Association. *Diagnostic and Statistical Manual of Mental Disorders* (DSM III). Washington, D.C., APA, 1980.

Anderson, V. E., Siegel, F. S., Fisch, R. O., and Wirt, R. D. Responses of phenylketonuric children on a continuous performance test. *Journal of Abnormal Psychology,* 1969, *74,* 358–362.

Apgar, V. A proposal for a new method of evaluation of the newborn infant. *Current Researches in Anesthesia and Analgesia,* 1983, *32,* 260–267.

Bayley, N. *Manual for the Bayley Scales of Infant Development.* New York: Psychological Corporation, 1969.

Bayley, N. Development of mental abilities. In P. H. Mussen (Ed.), *Carmichael's Manual of Child Psychology.* New York: J. W. Ley, 1970.

Bierman-Van Eedenburg, M., Jurgens-Van Der Zee, A. D., and Olinga, A., Huisjes, H. H., and Touwen, B. C. Predictive value of neonatal neurological examination: A follow-up study at 18 months. *Developmental Medicine and Child Neurology,* 1981, *23,* 296–305.

Billing, L., Eriksson, M., Stineroth, G., and Zetterstrom, R. Pre-school children of amphetamine-addicted mothers. *Acta Paediatrica Scandinavica,* 1985, *74,* 179–184.

Brazelton, T. B. *Neonatal Behavioral Assessment Scale.* Philadelphia: Lippincott, 1973.

Broman, S., Nichols, P., and Kennedy, W. A. *Preschool I.Q.: Prenatal and Early Development Correlates.* Hillsdale, N.J.: Erlbaum, 1975.

Chasnoff, I. J., Burns, W. J., Schnoll, S. H., and Burns, K. A. Cocaine use in pregnancy. *New England Journal of Medicine,* 1985, *313,* 666–669.

Church, M. W., and Gerkin, K. P. Hearing disorders in children with fetal alcohol syndrome: Findings from case reports. *Pediatrics,* 1988, *82,* 147–154.

Coles, C. D., Smith, I., Fernhoff, P., and Falek, A. Neonatal neurobehavioral characteristics as correlates of maternal alcohol use during gestation. *Alcoholism: Clinical and Experimental Research,* 1985, *9,* 454–460.

Doyle, R. B., Anderson, R. P., and Halcomb, C. G. Attention deficits and the effects of visual distraction. *Journal of Learning Disabilities,* 1976, *9,* 48–54.

Edmonds, L. D., Falk, H., and Nissim, J. E. Congenital malformations and vinyl chloride. *Lancet,* 1975, *2,* 1098.

Fagan, J. F., and McGrath S. K. Infant recognition memory and later intelligence. *Intelligence,* 1981, *5,* 121–130.

Fagan, J. F., and Singer, L. T. Infant recognition memory as a measure of intelligence. *Advances in Infancy Research,* 1983, *2,* 31–78.

Fantz, R. L. Visual perception from birth as shown by pattern selectivity. *Annals of the New York Academy of Science,* 1965, *118,* 793–814.

Flint, E. F. Severe childhood deafness in Glasgow 1965–1979. *Journal of Laryngology and Otology,* 1983, *97,* 421–425.

Fried, P. A. Marihuana use by pregnant women: Neurobehavioral effects in neonates. *Drug and Alcohol Dependence,* 1980, *6,* 415–424.

Fried, P. A. Marihuana use by pregnant women and effects on offspring: An update. *Neurobehavioral Toxicology and Teratology,* 1982, *4,* 451–454.

Golden, N. L., Sokol, R. J., Kuhnert, B. R., and Bottoms, S. Maternal alcohol use and infant development. *Pediatrics,* 1982, *70,* 931–934.

Grether, J. K. California case file on cerebral palsy and mental retardation. *Journal of Perinatology,* 1987, *8,* 100–104.

Grossman, H. J. (Ed.). *Manual on Terminology and Classification in Mental Retardation.* Washington, D.C.: American Association on Mental Deficiency, 1973.

Gusella, J. L., and Fried, P. A. Effects of maternal social drinking and smoking on offspring at 13 months. *Neurobehavioral Toxicology and Teratology,* 1984, *6,* 13–17.

Hagberg, B., Hagberg, G., Lewerth, A., and Lindberg, U. Mild mental retardation in Swedish school children. *Acta Paediatrica Scandinavica,* 1981, *70,* 445–452.

Hook, E. B. Incidence and prevalence a measures of the frequency of birth defects. *American Journal of Epidemiology,* 1982, *116,* 743–747.

Jacobson, J. L., Jacobson, S. W., Fein, G. G., Schwartz, P. M., and Dowler, J. K. Prenatal exposure to an environmental toxin: A test of the multiple effects model. *Developmental Psychology,* 1984, *20,* 523–532.

Jacobson, S. W., Fein, G. G., Jacobson, J. L., Schwartz, P. M., and Dowler, J. K. The effect of intrauterine PCB exposure on visual recognition memory. *Child Development,* 1985, *56,* 853–860.

Jacobson, S. W., Jacobson, J. L., Dowler, J. K., Fein, G. G., and Schwartz, P. M. Sensitivity of Fagan's recognition memory test to subtle intrauterine risk. Paper presented at the annual meeting of the American Psychological Association, Anaheim, Calif., Sept., 1983.

Johnson, H. L., and Rosen, T. S. Prenatal methadone exposure: Effects on behavior in early infancy. *Pediatric Pharmacology,* 1982, *2,* 113–120.

Larsson, G., Bohlin, A. B., and Tunell, R. Prospective study of children exposed to variable amounts of alcohol in utero. *Archives of Disease in Childhood,* 1985, *60,* 316–321.

Lewis, M., Bartels, B., Campbell, H., and Goldberg, S. Individual differences in attention. *American Journal of Diseases of Children,* 1967, *113,* 461–465.

Lowe, J., Windsor, R. A., Adams, B., Morris, J., and Reese, Y. Use of a bogus pipeline method to increase accuracy of self-reported alcohol consumption among pregnant women. *Journal of Studies on Alcohol,* 1986, *47,* 173–175.

Lowry, R. B., Miller, J. R., Scott, A. E., and Renwick, D. H. G. The British Columbia Registry for Handicapped Children and Adults: Evolutionary changes over twenty years. *Canadian Journal of Public Health,* 1975, *66,* 322–326.

Marden, P. M., Smith, D. W., and McDonald, M. J. Congenital anomalies in the newborn infant, including minor variations. *Journal of Pediatrics,* 1984, *64,* 357–371.

McCall, R. B., Eichorn, D. H., and Hogarty, P. S. Transitions in early mental development. *Monographs of the Society for Research in Child Development,* 1977, *42,* 1–108.

Naeye, R. L., and Peters, E. C. Mental development of children whose mothers smoked during pregnancy. *Obstetrics and Gynecology,* 1984, *64,* 601–607.

Nelson, K. B., and Ellenberg, J. H. Apgar scores as predictors of chronic neurologic disability. *Pediatrics,* 1981, *68,* 36–44.

Noland, E. C., and Schuldt, W. J. Sustained attention and reading retardation. *Journal of Experimental Education,* 1971, *40,* 73–76.

Olofsson, M., Buckley, W., Andersen, G. E., and Friis-Hansen, B. Investigation of 89 children born by drug-dependent mothers. *Acta Paediatrica Scandinavica,* 1983, *72,* 407–410.

Porrino, L. J., Rapoport, J. L., Behar, D., Sceery, W., Ismond, D. R., and Bunney, W. E. A naturalistic assessment of the motor activity of hyperactive boys: I. Comparison with normal controls. *Archives of General Psychiatry,* 1983a, *40,* 681–682.

Porrino, L. J., Rapoport, J. L., Behar, D., Sceery, W., Ismond, D. R., and Bunney, W. E. A naturalistic assessment of the motor activity of hyperactive boys. *Archives of General Psychiatry,* 1983b, *40,* 683–687.

Prechtl, H. F. R. Prognostic value of neurological signs in the newborn infant. *Proceedings of the Royal Society of Medicine,* 1965, *58,* 3–4.

Prechtl, H. F. R. Hazards of over-simplification. *Developmental Medicine and Child Neurology,* 1970, *12,* 522–524.

Prechtl, H. F. R., and Beintema, D. J. The Neurological Examination of the Full-Term Newborn Infant. London: Spastics International Medical Publications and Heinemann Medical Books, 1977.

Roger, W. J., Glader, B. C., McKivvery, J. D., Cavveras, N., Hady, P., Thullen, S., Tinglestad, S., and Tully, M. Neonatal effects of transplacental exposure to PCBs and DDE. *Journal of Pediatrics,* 1986, *109,* 335–341.

Rose, S. A., and Wallace, I. F. Visual recognition memory: A predictor of later cognitive functioning in preterms. *Child Development,* 1985, *56,* 843–852.

Rosvold, H. E., Mirsky, A. F., Sarason, I., Bransome, E. D., and Beck, L. N. A continuous performance test of brain damage. *Journal of Consulting Psychiatry,* 1956, *20,* 343–350.

Scanlon, J. W., Brown, W. U., and Weiss, J. B. Neurobehavioral responses of newborn infants after maternal epidural anesthesia. *Anesthesiology,* 1974, *40,* 121–128.

Serunian, S. A., and Broman, S. H. Relationship of Apgar scores and Bayley mental and motor scores. *Child Development,* 1975, *46,* 696–700.

Sever, L. E. "Incidence and prevalence as measures of the frequency of birth defects". *American Journal of Epidemiology,* 1983, *118,* 608–609.

Shaywitz, S. E., Caparulo, B. K., and Hodgson, E. S. Developmental language disability as a consequence of prenatal exposure to ethanol. *Pediatrics,* 1981, *68,* 850–855.

Shepard, T. H. Human teratogens: How can we sort them out? *New York Academy of Sciences,* 1986, *477,* 105–115.

Staisey, N. L., and Fried, P. A. Relationships between moderate maternal alcohol consumption during pregnancy and infant neurological development. *Journal of Studies on Alcohol,* 1983, *44,* 262–270.

Standley, K., Soule, A. B., Copans, S. A., and Duchowny, M. S. Local-regional anesthesia during childhood: Effect on newborn behaviors. *Science,* 1974, *186,* 186–187.

Strauss, A. A., and Kephart, N. C. *Psychopathology and Education of the Brain-Injured Child* (Vol. 2). New York: Grune & Stratton, 1955.

Strauss, A. A., and Lehtinen, V. *Psychopathology and Education of the Brain-Injured Child* (Vol. 1). New York: Grune & Stratton, 1947.

Strauss, M. E., Lessen-Firestone, J. K., Starr, R. H., and Ostrea, E. M. Behavior of narcotics-addicted newborns. *Child Development,* 1975, *46,* 887–893.

Strauss, M. E., Lessen-Firestone, J. K., Chavez, C. J., and Stryker, J. C. Children of methadone-treated women at five years of age. *Pharmacology Biochemistry and Behavior,* 1979, *2,* 3–6.

Streissguth, A. P., Barr, H. M., and Martin, D. C. Offspring effects and complications of labor and delivery related to self-reported maternal alcohol use during pregnancy. *Developmental Pharmacology and Therapy,* 1982, *5,* 21–32.

Streissguth, A. P., Barr, H. M., Sampson, P. D., Parrish-Johnson, J. C., Kirchner, G. L., and Martin, D. C. Attention, distraction and reaction time at age 7 years and prenatal alcohol exposure. *Neurobehavioral Toxicology and Teratology,* 1986, *8,* 717–725.

Streissguth, A. P., Herman, C. S., and Smith, D. W. Intelligence, behavior, and dysmorphogenesis in the fetal alcohol syndrome: A report on 20 patients. *Journal of Pediatrics,* 1978a, *92,* 363–367.

Streissguth, A. P., Herman, C. S., and Smith, D. W. Stability of intelligence in the fetal alcohol syndrome: A preliminary report. *Alcoholism: Clinical and Experimental Research,* 1978b, *2,* 165–170.

Sykes, D. H., Douglas, V. I., and Morgenstern, G. Sustained attention in hyperactive children. *Journal of Child Psychology and Psychiatry,* 1973, *14,* 213–220.

Weiss, G. The natural history of hyperactivity in childhood and treatment with stimulant medication at different ages: A summary of research findings. *International Journal of Mental Health,* 1975, *4,* 213–226.

Wilson, G. S., McCreary, R., Kean, J., and Baxter, J. C. The development of preschool children of heroin-addicted mothers: A controlled study. *Pediatrics,* 1979, *63,* 135–141.

Writer, W. D. R., James, F. M., and Wheeler, A. S. Double blind comparison of morphine and bupivacaine for continuous epidural analgesia in labor. *Anesthesiology,* 1981, *54,* 215–219.

Zigler, E., and Cascione, R. Mental retardation: An overview. In E. G. Gollin (Ed.), *Malformations of Development.* New York: Academic Press, 1984. Pp. 69–94.

CHAPTER 7

Testing Animal Behavioral Teratogens

Designs and Methods

Mark Twain observed that "man is the only animal that blushes. Or needs to." Trying to develop an animal model for blushing would therefore be pointless from both a physiological and a psychological perspective. Fortunately animals resemble humans in other ways, and these resemblances can be used to make inferences about the human condition. While such inferences are always tenuous, the more closely the results of animal studies resemble observations in humans, the more confident we can be in drawing other inferences from animals to humans. The bottom lines in animal research are (1) how similar animals are to humans in their reaction to the same conditions and, (2) how similar are the methods used in animal and human research.

Earlier chapters of this book dealt at length with the first issue. The second has been examined with considerable expertise by Jane Adams (1986a) in her survey of the similarities and differences between the methods used for testing humans (many of which were described in Chapter 6) and the methods used for testing animals. Since this chapter is devoted to animal research, I felt it worthwhile to begin with a summary of Adams' observations.

Some of the similarities and differences between human and animal studies identified by Adams are summarized in Table 7.1.

Considerable attention is generally paid to growth measures in both human and animal studies. Apart from the fact that birthweight is a reasonably good predictor of future development in both humans and animals (Lochry *et al.*, 1985), it is also the easiest parameter to measure, and therefore it is not surprising to see birthweight data so commonly cited. Motor development is also commonly evaluated on the Brazelton, Prechtl, and Bayley tests and in various reflex assessments in animals. Reliance on such motor tests is largely dictated by the limited behavioral repertoire of young humans and animals. The predictability of such tests for later cognitive effects in humans is poor (see Chapter 6). Comparable examinations of predictability of reflex tests in animals for later behavior have not been made. In light of the heavy reliance on reflex testing in animals, it might be worthwhile to see how well performance on such tests does in fact correlate with later behavioral endpoints.

Both human and animal studies generally rely on a battery of tests. The Apgar test has five subcomponents, the Brazelton many more. Developmental tests in rodents also

Table 7.1. Similarities and Differences in General Approach to Studying Behavioral Teratological Effects on Humans and Animals[a]

Similarities	Differences
1. Considerable attention to birth weight, growth, and motor development	1. More attention to developmental delay in human studies
2. Battery assessment to explore several behavioral endpoints	2. a. Battery of tests usually administered at one age in human studies and several ages in animals b. Narrower age range in examination of human subjects
3. Matching of maternal and environmental characteristics	3. Matching of subjects is on statistical basis in humans relying on very few parameters. Animal studies match in terms of controlled variables. Greater effort during testing to match infants on behavioral state in human studies.
4. Functional deficits rather than physiological changes underlying deficits are described.	4. Human studies generally report incidence of adverse outcome; animal data focus on group differences.

[a]From Adams, 1986.

include several different reflexes and developmental landmarks such as incisor eruption, eye opening, and descent of testes. Human and animal studies differ, however, in how these data are examined. In human studies, performance on each subtest is typically integrated into a single value; in animal studies each subtest is analyzed and evaluated independently. Statistical procedures (e.g., factor analysis, multivariate analysis of variance), are available for an integration of animal data comparable to human studies but are rarely used.

Human studies also put greater emphasis on developmental delay as an important endpoint. Animal studies have generally ignored this issue but are now beginning to include longitudinal evaluations of behavioral changes (e.g., Abel, 1982) and this difference in emphasis is diminishing.

Although both human and animal studies match experimental and control groups, the matching procedures are very different. In human experiments, information is generally collected with respect to age, parity, race, socioeconomic status, smoking, and other data in addition to subjects' use or exposure to the factor of interest. An attempt is then made to match users and nonusers as closely as possible on many of these variables. However, the more variables for matching, the more difficult the matching. Because we cannot be certain that users and nonusers have been matched on all or even many of the factors that may also contribute to the outcome of interest, confounding is always a possibility. In animal studies, the environment and history of the subjects is almost completely in the hands of the experimenter and he or she evaluates the effects of various factors under relatively pristine conditions.

Whereas there is better matching of environmental and maternal factors in animal studies, Adams notes that human studies have the edge in the attention period to the behavioral state of infants at the time of testing. Behavioral state is especially important in the Brazelton exam (see Chapter 7)). In animal studies, many researchers attempt to control for behavioral state by testing animals during only part of the day. Though this may represent the limits of feasibility, Adams suggests that it may not be optimal for assessing some responses such as gross motor effects, which are influenced by sensory stimulation during testing.

A feature shared by both human and animal experiments is their common focus on functional deficits rather than the underlying physiological changes responsible for these deficits. However, considerable progress is being made in animal experimentation to identify changes in the brain responsible for observed behaviors. Although CAT and PET scans (see Chapter 6) are possible in human studies, they are not practical as yet and animal experimentation is still the only viable alternative for looking at underlying mechanisms, especially in the brain.

Also absent from both kinds of investigations are attempts to determine intervention strategies that could affect outcome. However, some initial attempts are being made in this area in animal research (see below).

A final difference identified by Adams is the way data from studies of humans and animals are expressed. In human studies, effects are often reported in terms of how many exposed and control subjects exhibit a particular behavior, whereas animal studies rarely cite the incidence of a particular behavior in such groups. Instead, data are reported in terms of group averages and differences between these averages.

Rationale for Studies in Animals

Because of the vast number of factors associated with exposure to teratogens that could also independently contribute to adverse pregnancy outcomes (cf. Abel, 1981; Riley & Meyer, 1984), clinical and epidemiological studies are often unable to shed light on causal relationships or mechanisms of action. For example, the chronic alcoholic woman is frequently undernourished and is a heavy smoker (Abel, 1984). Factoring out the different contributions of nutrition, smoking, and alcohol to some outcome variable is possible statistically but is extremely difficult in terms of assignment to different groups; the more variables for matching, the more onerous the assignment strategy. Similarly, studies on the mechanisms by which teratogens exert their effect would be difficult, if not impossible, to study in humans for practical as well as ethical reasons.

One viable alternative is to conduct relevant studies in animals and to extrapolate the findings from such studies to humans. The usefulness of such extrapolation will depend on the rigor with which these studies are conducted.

Among the main advantages of animal experimentation is the ''control'' that the experimenter can impose in the laboratory to isolate and study the impact of variables. Studies with animals also permit more intensive and methodological evaluation of important parameters, such as dose–response relations, critical periods of exposure, blood levels of a compound and its metabolites, and genetic factors, than is feasible in human studies. Also, animal studies usually permit evaluation of factors in a much shorter time period than is possible in humans. This is especially important in determining behavioral teratogenic effects where it might take a decade or more to collect relevant human data. Data from animals still need to be corroborated, but animal studies clear away much of the haystack so that there is a better chance of finding the needle.

An important shortcoming of animal studies is their relevance to humans. Although

Table 7.2. Advantages of Using Rodents in Behavioral Teratology Studies

1. Inexpensive to purchase and house
2. Minimal attention required
3. High fertility
4. Well-known reproductive physiology
5. Previous use in teratological studies
6. Extensive battery of information on behavior
7. Short gestation
8. Rapid maturation
9. Ease of timing of pregnancy
10. Ease of obtaining pregnancy
11. Ease of housing and maintaining

this issue will never be answered with total satisfaction, there is no alternative in many cases to animal experimentation.

Using primates like the ape or Old World monkeys might seem to be a possible compromise between the advantages and disadvantages of animal research because they have gestational periods and patterns of development like humans, but they are hard to get and expensive to buy and maintain. Therefore, primates are seldom used in behavioral teratology except to corroborate effects in other species or to examine variables that cannot be assessed in other animal species. Like all creatures, monkeys have their own unique characteristics, which have to be considered in extrapolating results to humans. For instance, brain development in monkeys occurs much earlier than in humans (see Chapter 4) compromising extrapolation somewhat. Another problem with monkeys is that nearly all carry herpes virus simiae (Miller, 1986), which can have adverse effects of its own on embryological and fetal development (Hanshaw *et al.*, 1985). Because monkeys differ in the titers of herpes in their blood, and because relatively few monkeys are typically tested per group in behavioral studies, some effects attributed to a teratogen could be due to either the virus or to its interaction with the test compound. The latter is an especially important consideration in evaluating teratogens that also suppress the immune system (e.g., alcohol, cyclophosphamide) because this will result in the mother becoming more susceptible to herpes and other viral infections (as well as those who work with these animals).

Rodents are still the most popular species for studying behavioral effects. Some of the advantages of using rodents are summarized in Table 7.2. (For a more complete discussion of the advantages and disadvantages of various species in teratology, see Riley and Meyer, 1984.)

Methodological Issues

Methodological factors relating to prenatal exposure to teratogens (e.g., route and method of administration, dose response studies, nutrition, and genetic factors) have been discussed at length in previous chapters and will not be repeated here except as they impact on behavioral testing and inferences.

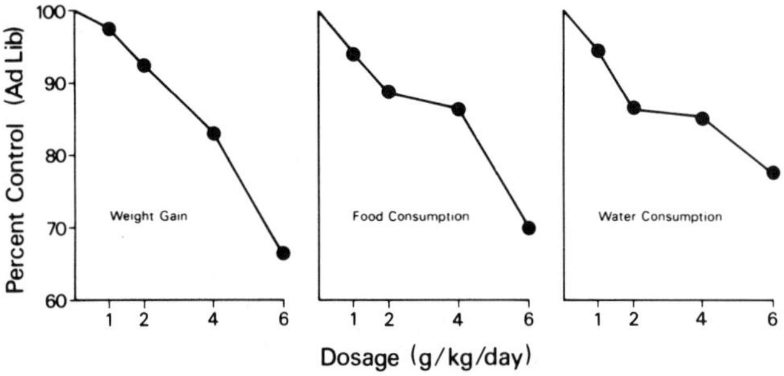

Figure 7.1. Weight gain, food and water consumption in pregnant rats intubated with different amounts of alcohol each day of pregnancy relative to nontreated females.

Pair Feeding

In many cases, substances that are teratogens also cause pregnant animals to reduce their food intake (see Figure 7.1). In such instances, it is problematic whether adverse outcomes in offspring are directly due to the teratogen or indirectly due to secondary effects on maternal nutrition (see Chapter 5, Principle 4).

One approach to this problem is to use the pair-feeding or yoked-feeding technique whereby one group of animals receives only the amount of food and water consumed on the previous day by teratogen-treated animals. In this way, animals are equated for food and fluid intake; the only difference is exposure to teratogens.

A second control would consist of an *ad lib* food and water group to assess the role of reduced food and water intake itself. Comparisons could then be made between treated animals and both pair-fed and *ad-lib*-fed animals. If treated animals differ from *ad-lib*-fed animals but not pair-fed animals, the result would then be attributed to nutritional rather than the direct pharmacological effects of teratogens. However, if treated animals also differed from pair-fed animals, then the result could be attributed to its direct effects.

Table 7.3 illustrates the importance of these controls for interpretation of data. In this study, animals were prenatally treated with 4 or 6 g/kg/day of alcohol throughout gestation. A second group was permitted *ad-lib* access to food and water. During adulthood, animals were food deprived and a mouse was introduced into each cage. As indicated in the table, muricide was significantly decreased for alcohol-treated females compared to *ad-lib*-treated controls. If only an ad lib treated group had been included in this study, the data might have been interpreted as evidence that prenatal alcohol exposure decreased predatory be-

Table 7.3. Prenatal Effects of Ethanol Muricide[a]

	Males					Females				
Group:	C	4	4P	6	6P	C	4	4P	6	6P
N:	10	11	23	13	12	12	10	10	12	11
Percent muricide:	40	27	36	36	42	50	50	36	8	9

[a]C, nontreated, *ad-lib*-fed controls; 4, intubated with 4 g/kg/day; 4P, pair-fed to group 4 animals; 6, intubated with 6 g/kg/day; 6P, pair-fed with group 6 animals.

havior. Inclusion of the pair-fed control, however, indicates that this effect was due to alcohol's reduction on maternal food intake rather than the pharmacological effects of alcohol on the developing nervous system.

In addition to reducing food consumption, some agents like alcohol can affect absorption of other nutrients. A single oral administration of as little as 1 g/kg of alcohol, for instance, can produce lesions in the intestine (Gottfried *et al.*, 1976), the severity of which is related to the concentration of alcohol used (Barone *et al.*, 1974). As a consequence of such damage, absorption of nutrients into the blood could be impaired, thus reducing the nutrient pool potentially available for transmission across the placenta. Alcohol may also impair transmission of nutrients across the placenta so that even if nutrient absorption into blood is not affected, nutrient availability to the fetus may still be compromised.

Thus, even though a study includes pair feeding, it is not possible, given the present state of the art, to conclude positively that a teratogen has not affected fetal development through its effect on nutrient availability.

Associated with the issue of availability is that of utilization. Several studies have shown that many alcoholics are underweight even though their caloric intake may be well over that required to maintain body weight (Smith, 1979). This inability to maintain body weight even when caloric intake is adequate is known as "energy wastage" and is commonly observed in alcoholics and in animal studies. Lieber (1979) compared body weight changes associated with the same caloric intake (2,000 calories) in the form of alcohol or chocolate. With alcohol as the caloric source, body weight did not change over a 44-day period. With chocolate providing calories, body weight increased about 3 kg.

A related consideration is the possibility that agents like alcohol may prompt vomiting and diarrhea in humans with concomitant loss of fluids and impaired absorption of water (up to four times lower) (Smith, 1979). Oily vehicles such as sesame oil, which are used to dissolve drugs like THC, likewise cause diarrhea in animals. These effects on body water may have secondary effects on electyrolyte balance and could indirectly affect fetal body composition (Abel and Greizerstein, 1979; Greizerstein and Abel, 1979).

Age of Testing

The age of testing issue was examined previously in connection with principles of teratology (see Chapter 5, Principle 8). Although behavioral effects observable at one particular age may not be observable at other ages, these age-related differences should not be taken to mean that all of the behavioral effects of prenatal exposure to teratogens are transient. A large number of behavioral differences have also been reported in older animals (e.g., Abel & Dintcheff, 1986). One explanation for the occurrence of transient effects is that prenatal teratogen exposure produces a delay in development (Riley *et al.*, 1979a) and that as the animal matures the effects become less apparent. Another possibility is that with maturation, teratogen-exposed progeny develop compensatory mechanisms for dealing with certain behavioral dysfunctions. Tasks in which age-related differences have been reported are relatively simple ones. They may not challenge a more mature animal with a more developed nervous system but they are capable of uncovering differences between teratogen-exposed and control progency when the nervous system is less mature (Meyer & Riley, 1986). If animals are challenged, these latent functional deficits may become apparent. For example, Means *et al.* (1984) reported that adult animals exposed to alcohol prenatally did not differ in activity except when challenged by administration of the stimulant Ritalin.

Sample Size

The number of animals per group is an often-neglected consideration in behavioral teratology. This factor has been discussed by Palmer (1977) and Riley and Meyer (1984). Group sizes of less than 15 animals have often been found to produce inconsistent results as far as statistical significance is concerned (trends are almost always consistent when experiments are repeated). To achieve a statistical probability of $p < .05$ with group sizes of 10 when none of the control animals exhibit any abnormality, at least 50% of the experimental animals would have to be affected by teratological insult. If 2 out of 10 control animals exhibited some abnormality, then 8 out of 10 experimental animals would have to be similarly affected for statistical significance at this level (Palmer, 1977).

Unit of Analysis

As many as 16 pups may be born in a rat litter. Each pup can be considered a single independent unit or the whole litter could be considered as the unit of analysis. Each alternative has its own implications (Abbey & Howard, 1973; Teicher *et al.*, 1981). If the individual pup is the unit of analysis, a statistical bias may arise because intralitter variability is typically much smaller than interlitter variability. This is because all pups from a single litter share not only a common genetic background, but also a similar intrauterine and postnatal environment. Hence, the data from litter mates are likely to be correlated. If few litters are studied and all pups are included, and each is treated as an independent unit, a single aberrant litter could seriously bias the outcome of a study.

The undue influence of a particular litter can be minimized either by randomly selecting one pup of each sex per litter for testing, or by taking the entire litter as the unit of analysis and using the average value for all pups of the same sex in the litter, or by performing a "nested" statistical analysis. The latter will determine whether indeed a litter effect is "nested" within treatment. If the litter effect is not significant, the individual pup can then appropriately be used as the unit of analysis. In some circumstances, the individual pup may be an appropriate unit (Teicher *et al.*, 1981). The overriding consideration is a cause-and-effect relationship free from artifact, statistical or otherwise (Riley & Meyers, 1984).

Litter Size

Litters should be culled to a specific number (e.g., eight pups per dam) to avoid possible confounding of litter size and teratogen exposure, especially because some teratogens may affect the number of animals born in a litter. The way in which such culling is done, however, may affect the outcome of an experiment. This is because *in utero* exposure to teratogens may significantly increase intralitter variability. Pups from control litters are much more homogeneous, and therefore random culling is likely to result in the choice of similar pups. If litter variability is increased among teratogen-exposed pups, random culling may not result in the selection of the most affected pups—animals that are of greatest interest.

One possibility is to select pups on the basis of birthweight, the assumption being that those most likely to exhibit later effects are those below average weight at birth (Lochry *et al.*, 1985). However, very small pups may not survive (cf. Leichter & Lee, 1979), so that the intention of examining offspring at a later time may be frustrated. Furthermore, subsequent behavioral testing may be confounded because the groups will most probably differ significantly in body weight, and differences might be attributed to this variable rather than to the teratogen. One alternative is to keep the median pups and discard the heaviest and lightest.

As a corollary, the increased postnatal mortality in offspring prenatally exposed to teratogens should also be taken into account. Litters may be culled to eight pups at birth, but at weaning there may only be four pups. Prenatal teratogen treatment may then be confounded with litter size. If control animals have much larger litters than experimental animals, these litters might be omitted from further analysis. Alternatively, litters may be culled throughout the postnatal period to minimize litter size differences between groups. Currently, the culling issue has been largely ignored.

Negative and Positive Controls

Negative controls are animals treated like the experimental group except for exposure to the teratogen. Negative controls can be animals whose parents were the same age, weight, genetic background, and the like as experimental animals and that differ only in not receiving the same amount of teratogen. Unfortunately, controls are fickle and it is necessary to include them each time an agent is examined.

Positive controls are rarely used in behavioral teratology but represent a way of comparing a possible teratogen with a known teratogen. A positive control is a standard for comparing how alike or unalike two teratogens are qualitatively and quantitatively. Positive controls also allow for verifying the reliability and sensitivity of test procedures and the apparatus used in testing. For instance, an agent may be a behavioral teratogen but may not be so identified simply because the apparatus used to assess behavior is insensitive. If exposure to a positive teratogen produces an effect whereas exposure to the test agent does not, the possibility that the test is not sensitive is less likely.

Residual Effects

A possible source of confounding in studies of prenatal drug exposure is residual effects on the mother. If there are residual effects of maternal teratogen exposure on postnatal maternal behavior or lactational performance, these maternal effects could affect offspring behavior independent of or in combination with prenatal teratogen exposure. Previous studies have shown that dams given alcohol by intubation may exhibit residual effects such as increased cannibalism (see Figure 7.2).

If offspring are not removed from their biological mothers soon after birth, there is a good possibility that there may not be any offspring to study later on. Furthermore, offspring that are not cannibalized may be "abused" and subsequent behavior may be affected accordingly.

Lactational performance may also be adversely affected because alcohol is known to affect the milk-ejection reflex. Leaving pups with their biological mothers thus increases the potential for confounding between pre- and postnatal factors, although this possibility may have more to do with a particular substance than all test substances. For instance, when the liquid diet method of alcohol administration is employed, there is little evidence of residual effects compared with what is seen when alcohol is administered by intubation (Abel, 1982).

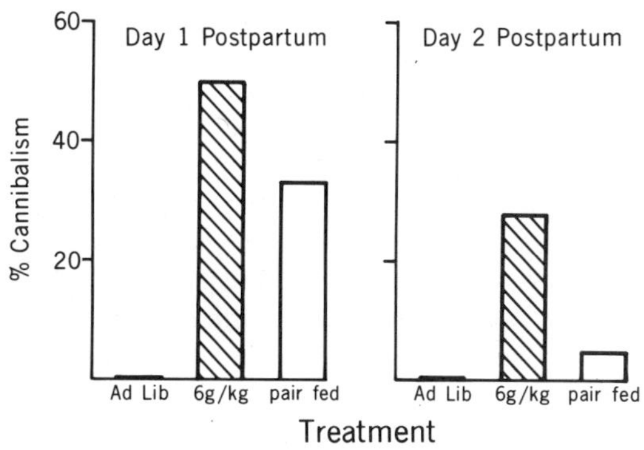

Figure 7.2. Postnatal cannibalism in rats intubated with alcohol only during pregnancy.

One alternative to this problem is to remove offspring as soon as possible from their biological mothers and to place them with nontreated surrogate mothers that have also just given birth. The effects of removal of her own litter and discovery of a new litter may introduce some stress to the surrogate mothers that could affect their maternal behavior. Compared with the neglect and possible cannibalism if pups remain with their own mothers, this altered maternal behavior is a much less serious consideration.

Depending on the aim of the study, the whole litter can be fostered or a postnatal cross-fostering procedure can be used in which half of a litter remains with each dam and half from another dam is given to her to raise (e.g., Martin *et al.,* 1978). This permits some assessment of postnatal variables but introduces the possibility of differential treatment of pups. If prenatal factors are of interest, then the "split litter" technique is not the best method; instead the whole litter should be fostered.

Husbandry Conditions

Animals live in relatively unstimulating environments. Conceivably, exposing animals to more postnatal stimulation via handling or by placing objects to explore in the cage might affect postnatal development and impact on the effects of prenatal exposure to teratogens.

This possibility was examined in conjunction with performance of maze-bright and male-dull rats in behavioral genetics studies (see Chapter 5). Gallo and Weinberg (1982) used a similar procedure in their study of rats prenatally exposed to alcohol. Postnatally, animals were assigned to one of two subgroups in which they were handled or not handled. At 35 days of age, all animals were tested in a step-down passive avoidance task. Nonhandled alcohol-exposed pups did not inhibit responding as well as controls. Alcohol-exposed pups that had been handled did not differ significantly from controls. This suggests that postnatal conditions may indeed affect the "impact" of prenatal exposure to teratogens.

Handling animals prior to weaning has been found to accelerate development (e.g., Levine, 1960) and, conceivably, postnatal handling may be a viable procedure for studying "intervention" techniques for attentuating other effects of prenatal alcohol exposure.

Test Methods

Almost any behavior test can be used for assessing functional deficits in animals exposed to teratogens. The choice of method is governed by the interests and ingenuity of the experimenter. If a previously unstudied agent is being evaluated, the investigator may wish to rely on one of several behavioral test batteries (see Adams, 1986b for review). Most of these are apical test procedures that use behavior as an overall reflection of sensory, motor, learning/memory, and motivational factors. The behavioral tasks used in these batteries are ideal for gaining some insight into the integration of several subsystems (Butcher, 1976). However, if some systems are able to compensate for deficits in other systems, behavioral impairment may not be noted (Buelke-Sam & Kimmel, 1979).

In assessing behavioral consequences of *in utero* exposure, attention must be given to the possibility that behavioral differences between experimentally treated and control animals are due to impaired sensory ability, motor coordination, and the like, rather than direct impairment of a particular function being studied.

Body weight itself is a particularly important pretest variable. Tests that rely on speed of responding or complex motor skills may reflect general physical status rather than learning/memory function. If food is used as a reinforcement, the amount of food reward may be proportionately greater for lighter animals. The greater the difference in body weight, the greater the differences in the motivation for that food reward. If food deprivation is used to motivate responding, a specific period of food deprivation or deprivation that reduces body weight by a given percentage may have a greater impact on lighter animals. For tasks motivated by food reinforcement, it is possible that food reward will be greater for one group of animals than another if one group is smaller in size to begin with. A loss of 10 g in a 300-g animal, for instance, constitutes less of a relative decrease in body weight than the same decrease in a 250 g animal. If the proportionate weight loss is not equal, the food reward will have greater reinforcing value for one group than the other. Lighter animals will thus be more deprived than heavier animals.

Tasks that use aversion stimulation to motivate animals may also be confounded by pretest differences in body weight. Smaller rats have lower shock thresholds than larger rats and this may make them more emotionally responsive. Depending on the task, this could facilitate or interfere with performance (see below).

Several years ago, Buelka-Sam and Kimmel (1979) surveyed 110 behavioral teratologists concerning the behavioral methods they used in their assessments. These methods were then categorized into 13 basic methods (those that could not be readily categorized were labeled "other"). In order of frequency these were as follows: activity, motor coordination, reflex development, maze/instrumental behavior, sensory function, avoidance, observation, operant, sexual behavior, pharmacologic/environmental challenge, aggression, social/group behavior, and classical conditioning. Within each of these main categories several different kinds of methods have been used. A discussion of all of these would require its own text. I have therefore restricted the following review to those methods I am most familiar with either through my own experience or from the frequent references to them in the behavioral teratology literature.

Developmental Landmarks

Many studies have used the appearance of certain developmental landmarks as dependent variables using scales such as those described by Fox (1965). However, specific criteria have not been provided for many of these tests and therefore there is often poor interlaboratory reliability associated with scoring items. To introduce better reliability, Burkhalter and Balster (1979) extracted several tests from Fox's (1965) battery and provided criteria for their evaluation. These criteria are described in Table 7.4.

Table 7.4. Criteria for Recording Appearance of Various Reflexes[a]

Righting reflex

0. Animal remains on the side it was placed and does not try to right itself within 60 sec. Attempts to right are squirming, rolling over, or limb movements which alter body position.
1. Animal attempts to right itself (sqirm, roll, limb movement) but is unable to do so completely within 60 sec.
2. Complete righting in 31–60 sec.
3. Complete righting in 11–30 sec.
4. Complete righting in 2–10 sec.
5. Complete righting in 1 sec or less.

Forelimb placing

0. Forelimb does not move or place.
1. One limb moves but no placing.
2. Both limbs move but no placing.
3. One limb places.
4. Both limbs place.

Forelimb grasp

0. No movement.
1. One or both limbs move but no grasping.
2. Weak grasp—flexion of digits in one paw but no complete grasp.
3. Weak grasp with both paws.
4. Complete grasp with one paw.
5. Complete grasp with both paws.

Rooting reflex

0. No respone in 5 sec.
2. Rooting from 1–2 cm.
3. Rooting from 3–5 cm.
4. Rooting from 6–10 cm.
5. Rooting from 11 cm or more.

Cliff drop

0. Animal does not try to move away from cliff edge in 60 sec. Movement occurs but is not directed at cliff avoidance.
1. Animal tries to move from edge but fails, e.g., falls over edge.
2. Animal moves away in 31–60 sec.
3. Animal moves away in 11–30 sec.
4. Animal moves away in 4–10 sec.
5. Animal moves away in 3 sec or less.

Startle response

0. No response.
3. Mild response, e.g., brief jump or freezing which causes brief discontinuity in ongoing behavior.
5. Strong response, e.g., jump, running, freezing, which disrupts ongoing behavior.

Bar holding

0. No grasping or holding.
1. Latency to fall 1–5 sec.
2. Latency to fall 6–10 sec.
3. Latency to fall 11–20 sec.
4. Latency to fall 21–50 sec.
5. Latency to fall 51 sec or longer.

[a]From Burkhalter and Balster, 1979.

The tests studied by Burkhalter and Balster (1979) were righting reflex, forepaw grasp, rooting reflex, cliff-drop aversion, auditory startle, and bar holding.

Righting reflex was examined by placing the animal on its side and noting the latency to right itself. The criterion for complete righting was turning so that all four paws were in direct contact with the test surface and limbs positioned so that walking was possible.

To test forelimb placing, the animal was held by the nape of the neck and the dorsal surface of each forepaw was stimulated by a light stroke. After the stroke, the bar was kept in contact with the paw for a maximum of 3 seconds and the next paw was then stroked. The criterion for complete forelimb placing was placing the ventral surface of the paw on the instrument within 3 seconds.

To test forepaw grasp the animal was held by the nape of the neck and the ventral surface of each forepaw was prodded and kept in contact for a maximum of 3 seconds. The criterion for complete response was full flexion of the digits around the prod.

To test for rooting reflex, the animal was placed with all four paws on the test surface and its facial area or snout was stimulated by pressing a thumb and index finger around them lightly for a maximum of 5 seconds. The criterion for complete response was rhythmic up-and-down pushing movements of the head as it is stimulated. If the hand was then removed in a linear path and the snout still stimulated, the animal followed the hand while continuing to root. Distance was indicated by a piece of tape marked off in centimeters and rooting distance was measured as an index of reflex intensity.

The cliff-drop response was measured by placing the animal on the edge of a platform so that most of its head and forelimbs extended over the edge and the latency to move its entire body from the edge was determined.

Bar holding was measured by grasping the animal by the nape of the neck and placing its forelimbs in contact with a horizontal bar until it grasped it. The animal was then released and the latency to fall was recorded.

Startle behavior can be evaluated in very young or older animals with primitive or sophisticated equipment. Primitive methods assess reaction to a sudden sound. The sound stimulus, such as a clicker or buzzer, is activated a short distance (e.g., 10 cm) in front of the animal for about 1 second. The criterion for complete response is disruption of ongoing behavior. More sophisticated test procedures for startle are discussed below.

Learning/Memory Tasks

A basic problem is assessing the effects of teratogens on learning/memory processes is that these processes are not directly observable but must be inferred from behavior. In making such inferences, therefore, it is especially important to take into account other effects that could indirectly affect the study problem. For example, before inferring that prenatal drug exposure produces deficits in learning/memory function, effects on general motor activity should be precluded, especially when the required response involves speed. In avoidance learning, it is also worthwhile to determine the latency for escape responding because of the possibility of pain sensitivity affecting behavior. If a drug affects the sensitivity to shock any effects of that drug on learning/memory could be secondary to its effects on performance as a result of differences in the level of motivation.

In addition to simultaneous measurement of the number of conditioned and unconditioned responses (avoidance and escapes) and the latency of response to both the conditioned and unconditioned stimuli, it is also of value to record number of intertrial responses as a measure of overall activity because a more active animal is likely to make more responses than a less active animal.

When dealing with aversely motivated tasks, some attention should also be given to the nature of the task itself. During conditioned avoidance, for instance, the pairing of the conditioned and unconditioned stimuli is presumed to result in a conditioned emotional response (''fear'') that motivates the animal to respond in the presence of the conditioned stimulus alone. If animals are already different in terms of fearfulness prior to such conditioning because of perinatal drug treatment, behavior could be facilitated, due to increased motivation, or impaired if the increase is so great that behavior becomes disorganized.

Active Avoidance

In avoidance learning, there are two basic paradigms: active and passive avoidance. In the active avoidance paradigm an animal does not experience an aversive event like a mild electric shock if it makes some specified response in a limited time. In shuttle box avoidance learning, a rat has to move from one compartment of a two-compartment chamber to the other side after hearing a warning signal. If the animal crosses into the other compartment during the specified time, it ''avoids'' the shock. If it waits too long, it receives the shock and continues to receive it unless it ''escapes'' into the second compartment or the shock is turned off automatically.

Most explanations of avoidance learning are based on Mowrer's theoretical formulations of the avoidance paradigm (e.g., Mowrer & Lamoreaux, 1946). Mowrer postulated that avoidance learning is a combination of classical and instrumental learning. Classical or Pavlovian learning involves the association of fear with the shock stimulus called the *unconditioned stimulus* (abbreviated UCS). The more the animal experiences shock, the more fearful it becomes and this association results in a learned emotional response.

The second component of the theory involves escape from the UCS. Escape is a reflex response called the *unconditioned response* (UCR). Escape into the safe compartment turns off the UCS and reduces fear because the animal removes itself from the shock area and the UCS. Moving to the safe area also reinforces the likelihood of making future escape responses because they reduce fear. Because the signal also acquires the capacity to elicit fear as a result of its association with the UCS, it is called a *conditioned stimulus* (CS). Moving into the safe compartment likewise allows the animal to reduce its fear and avoid being shocked. The avoidance response is called the *conditioned response* (CR).

Literally hundreds of experiments have investigated various aspects of Mowrer's formulation. These studies have shown that the association of a CS (e.g., tone, light) with shock results in the CS taking on aversive properties where none existed prior to the pairing of such stimuli with shock. For example, pairing a CS with shock results in an increase in heart rate and release of stress hormones to the CS. Rats will also bury the metal rods that deliver shock (e.g., Pinel & Triet, 1978) and the light used as the CS (e.g., Anderson *et al.*, 1983). Less clear is the distinction between whether the animal is escaping from the CS or is actually learning to anticipate the future.

There are two subtypes of active avoidance. The easiest to learn is the one-way avoidance task in which the animal is required to cross from the compartment in which the shock is delivered to another in which shock is never encountered. The animal is removed from the safe compartment manually and replaced into the shock compartment. In this situation, either the replacement of the animal or a specific warning signal is the CS. A typical avoidance apparatus is shown in Figure 7.3.

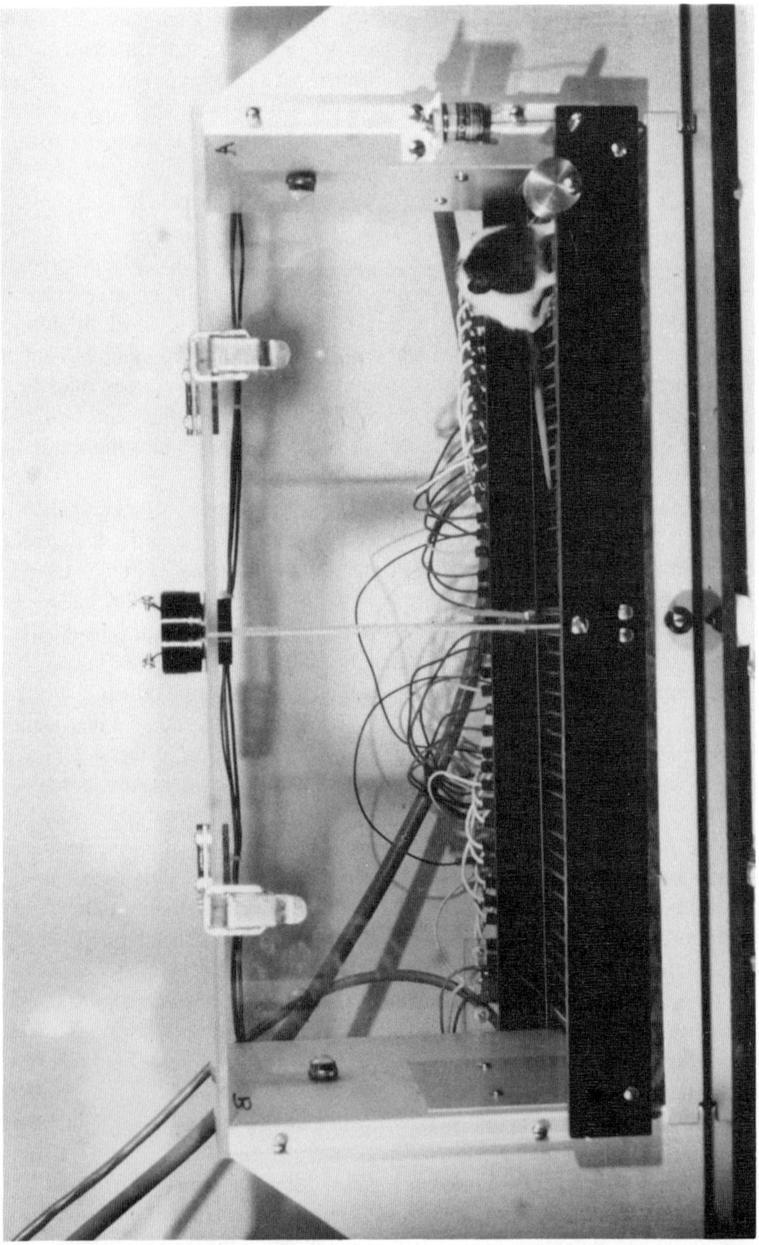

Figure 7.3. Shock avoidance apparatus. In one-way avoidance, the animal is manually removed from the safe compartment and put back into the shock area. In two-way avoidance, the previous safe area now becomes the shock area.

One-way avoidance learning is easily learned and because of this ease the test usually does not distinguish between animals prenatally exposed to teratogens and controls (e.g., Abel & Dintcheff, 1978).

In two-way shock avoidance, shock can be presented to either chamber. After a 5- to 10-second presentation of the CS, the animal is shocked. If the animal crosses over to the other compartment before shock is delivered, it avoids being shocked. If it waits too long, the shock comes on causing it to scurry about the cage until it eventually escapes to the safe side. After a few seconds, the signal comes on again and the previous shock side now becomes the safe compartment. The animal must therefore shuttle back and forth.

The obvious difference between one- and two-way shock avoidance is that in one-way avoidance, the safe compartment is always safe. In two-way avoidance, the safe compartment is safe for only a short period. This uncertainty makes the task of learning to avoid the CS a lot harder and animals differ in their speed of learning. As a result, the two-way avoidance paradigm is more sensitive in detecting differences between animals prenatally exposed to teratogens and controls.

Two-way active avoidance can also be subdivided into discriminated and nondiscriminated avoidance. In discriminated avoidance, a specific CS (light or tone) precedes the presentation of shock by a fixed delay. In nondiscriminated avoidance, there is no stimulus signaling the onset of the noxious stimulus. Instead, presentation of the noxious stimulus occurs on a fixed interval (shock-shock interval). By jumping over the hurdle, the animal postpones the shock. The amount of time of the postponement is called the *response–stimulus interval*.

An example of discriminated two-way avoidance is Abel's (1979) evaluation of the effects of prenatal alcohol exposure. Rats prenatally exposed to alcohol were placed individually into the shuttle box shown in Figure 7.3. They could avoid or escape being shocked through the grid floor by jumping over the barrier dividing the two compartments of the apparatus. A 10-second light and an auditory stimulus was the compound conditioned stimulus (CS), and a 2.0 mA constant current electric shock served as the unconditioned stimulus (US). If an avoidance response was not made during the CS period, the shock was presented and remained on until the animal jumped over the barrier. Each presentation of the CS and the performance of a jumping response constituted a single trial. The intertrial interval was 20 seconds. Fifty such trials were given over a 5-day period. In addition to determining the number of avoidance responses made, records were also kept of the time required to jump following shock presentation (escape latency) and the number of between-trial jumps not associated with the CS as a measure of general motor activity.

The *ad lib* and two pair-fed control groups (4P, 6P) made slightly more avoidances on the first days of testing than the two groups whose mothers received 4 to 6 g/kg/day of alcohol and continued to improve as testing continued. Although group 4 animals did not perform as well as pair-fed animals on the first three days of testing, they continued to improve on days 4 and 5. Animals exposed to 6g/kg, however, did not continue to improve avoidance after day 3 and, in fact, performance declined as testing continued (see Figure 7.4). Avoidance latency scores of the alcohol-treated females were greater than those for control groups, escape latencies were slightly but not significantly greater for the alcohol-treated animals relative to controls.

Passive Avoidance

In contrast with active avoidance learning, in passive avoidance learning, an animal must learn not to enter a second compartment. In this situation, the animal is again placed

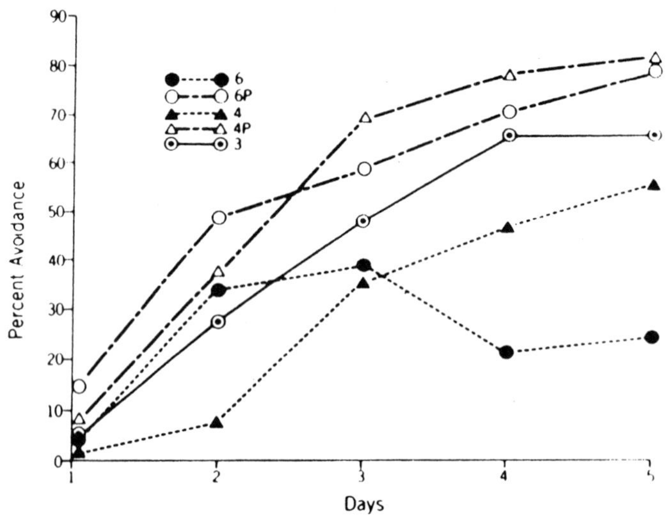

Figure 7.4. Two-way shock avoidance learning in animals prenatally exposed to 4 or 6 g/kg alcohol compared with controls.

in a two-compartment box like that shown in Figure 7.5. In many cases, the side it is placed into is lighted and the other compartment is dark. The two compartments are separated by a gate that is left open until the animal moves into the dark area. After entry, the gate closes and the animal receives a brief shock (UCS) from which there is no escape.

In this situation, the CS is the darkness, and the CR is fear. To avoid or escape this fear, the animal must inhibit its natural tendency to explore novel environments. Because this is a passive response, the paradigm is called *passive avoidance*. Such learning is usually rapid. An example is the evaluation by Riley *et al.* (1979a) of the prenatal effects of alcohol.

A Plexiglas rectangular chamber (18.75 × 7.5 × 7.5 cm) divided into two equal sections by a guillotine door was used as the test apparatus. One side of the chamber was painted black and had a grid floor through which shock could be delivered. The other side was transparent and had a solid floor. A shock generator delivered a 0.5 mA shock for 0.5 seconds when the animal entered the dark chamber. A 28 v light located at the far end of the "safe" chamber signaled the start of a trial.

Offspring were tested at 18 days of age. Animals were removed from their home cage and placed in a holding cage for 2 minutes. They were then put into the transparent side of the chamber facing away from the shock chamber. The light came on signaling the start of the trial. When animals stepped into the black chamber with all four feet, shock was delivered and the light went off. Animals were removed from the chamber and returned to the holding cage for 30 seconds after shock. This procedure was continued until the animal remained on the solid floor for 180 seconds. The results are shown in Figure 7.6.

Maze Learning

Performance in mazes is one of the earliest tests for assessing learning ability in rats (Munn, 1950). Rats use all their senses to find their way through mazes. If visual cues are

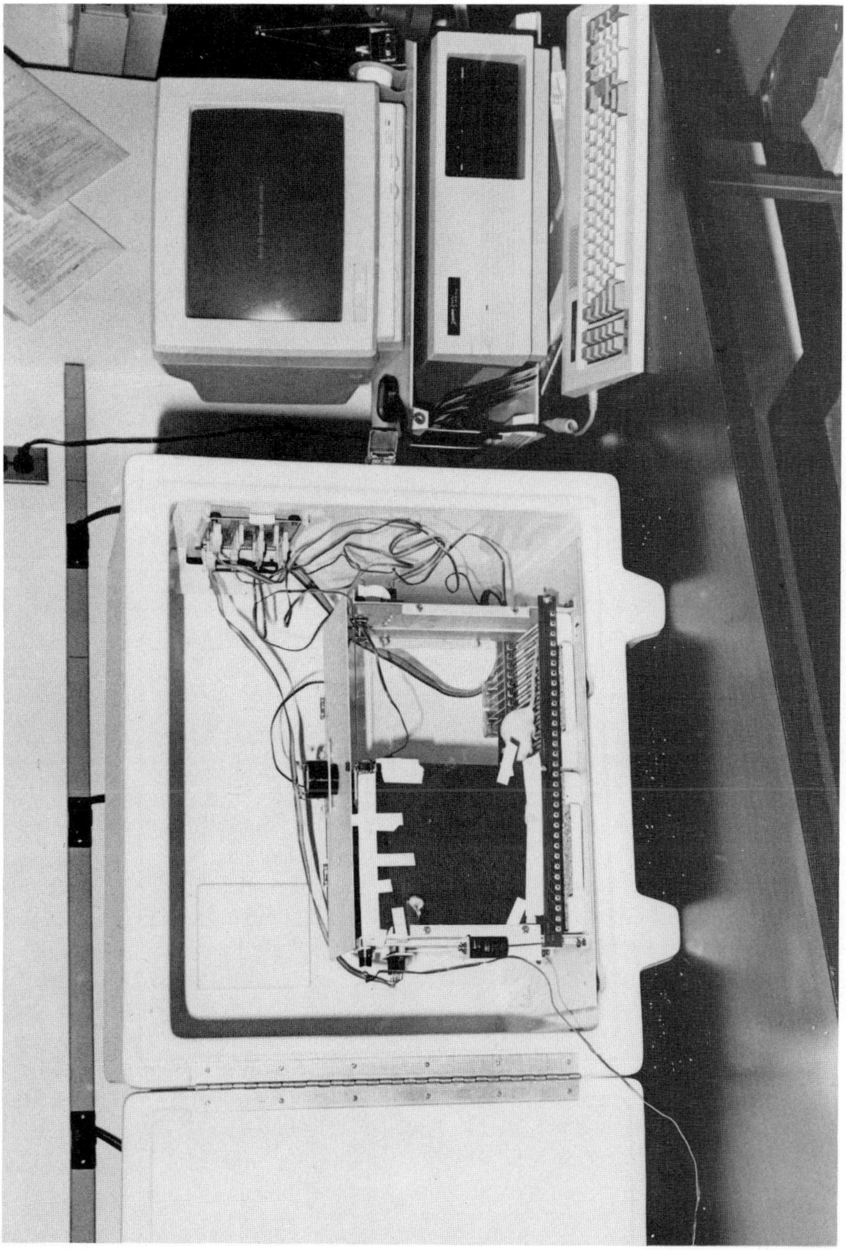

Figure 7.5. Apparatus for studying passive avoidance in rats and mice.

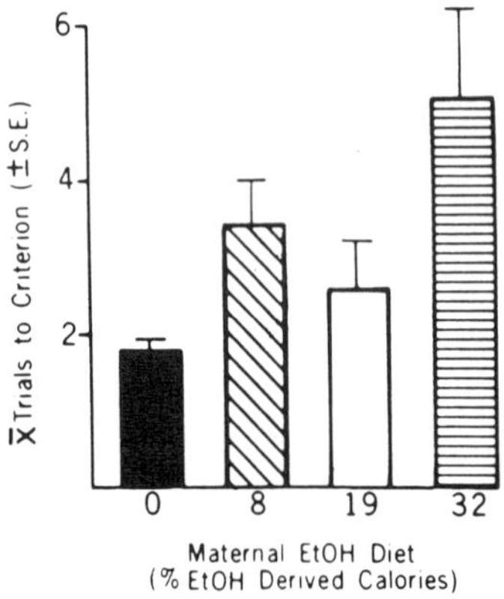

Figure 7.6. Passive avoidance scores in rats prenatally exposed to alcohol.

missing or rats are blinded, they will rely on olfactory and tactile cues. After reviewing hundreds of such studies, Munn (1950) concluded that rats will learn the shortest path in a maze, and blind alleys that point away from the goal are easier to learn to avoid than those that point in the same direction.

The most simple maze is the T maze. In T maze learning, the animal is rewarded for learning to go to one side. The reward can be food or escape from shock or water.

Many researchers prefer to use escape from aversive stimulation as a reinforcer in maze learning rather than food reward because consummatory rewards may not be equal for experimental and control animals. For example, Cravens (1974) found that protein malnutrition during pregnancy and lactation resulted in significantly inferior performance when food was used as a reward but not when escape from water was used to motivate learning. This indicates that inferior performance of one group compared with another may not be due to intelligence but may be a reflection of the motivation to learn the task.

The T maze can also be used to examine an animal's ability to learn a brightness discrimination. In one such procedure, one arm of the maze is painted white and the other black. The arms are switched from trial to trial to make sure the animal is paying attention to the brightness cue and not to a spatial cue (turning left or right).

A variation of the T maze is the Biel maze. This is a multiple T maze containing water; the reinforcement is escape. Vorhees (1983) used this maze to examine the effects of prenatal exposure to anticonvulsant drugs. Pregnant dams were intubated with one of three different anticonvulsant agents: diphenylhydantoin (5, 50, or 200mg/kg), phenobarbital (5, 50, or 80 mg/kg), or trimethadione (5, 50, or 250 mg/kg) on gestation days 7 to 18 or with vehicle. Offspring were examined at 53 days of age.

Treatment did not affect swimming time indicating that motor ability was not impaired. However, animals exposed to the high dose of diphenylhydantoin made about three times as many errors and took about 1.5 times longer to learn the maze than did controls. Animals exposed to the other drugs did not differ significantly from controls.

Spontaneous Alteration

When rats are placed into a T maze, they explore the stem into which they are placed and then enter one of the arms. If, after they explore that arm, they are placed back into the stem, they will explore it again and then usually enter the previously nonentered arm. This change from one arm to the other is called *spontaneous alternation* and is a common characteristic of rats.

Spontaneous alternation has been subjected to intensive behavioral analysis (Douglas, 1967). It is believed to reflect habituation and the integrity of central cholinergic mechanisms thought to be required for habituation.

Abel (1982) used this test to evaluate the prenatal effects of alcohol. Offspring were tested at 16, 63, and 112 or 113 days of age. Animals were removed from their home cage and placed in a holding cage for 30 seconds. They were then placed in the start alley and permitted to explore the T maze. After entry into one of the arms of the maze, a gate was lowered to prevent exit from the alley and the animal was confined for 30 seconds. The animal was then placed back into the holding cage and testing continued as before. Animals were tested until they entered an arm of the maze opposite to that previously entered or until a maximum of five trials.

At 16 days of age, both groups of animals exposed to alcohol *in utero* engaged in more perseverative behavior before alternating than did pair-fed controls. At 63 days of age, group 17% animals continued to persevere more than controls but group 35% became similar to controls. At 112 days of age, there was a trend for alcohol-exposed animals to exhibit more perseverative behavior compared with controls but the trend was not statistically significant.

Reversal Learning

A T maze can also be used to study reversal learning. In this paradigm an animal is trained to turn in one direction to escape shock or to gain a reward (e.g., food). After it learns this response, the goal is switched to the other side and it must learn to reverse its direction. Reversal learning has been interpreted as involving extinction of an initial discrimination combined with acquisition of a reversal discrimination. Riley *et al.* (1979b) used a T maze with an electrifiable floor to assess the effects of prenatal exposure to alcohol on reversal learning in rats. Animals were placed into the stem and three seconds later the grid was electrified with an 0.25 mA shock. On the first trial, animals had to enter both

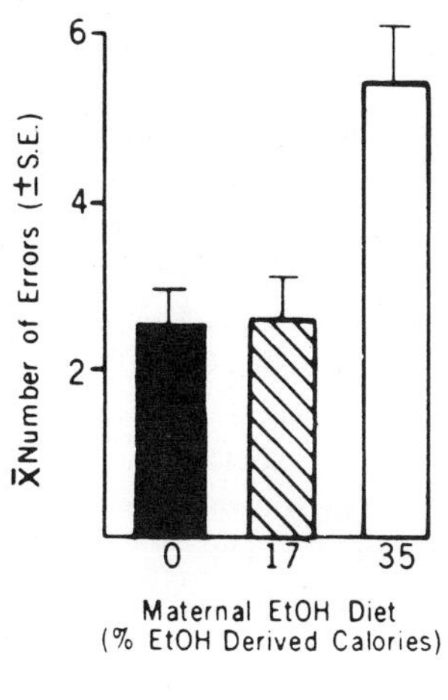

Figure 7.7. Reversal learning in rats prenatally exposed to alcohol.

arms before shock was terminated. The first choice was then made the incorrect choice for further training. After seven correct choices (the criterion for acquisition), the reversal phase was started. Now the formerly incorrect response terminated shock and a response to the previously correct side was ineffective. The criterion for renewal was also seven consecutively correct choices.

The data are shown in Figure 7.7. Animals did not differ in acquisition but during reversal the 35% groups made significantly more mistakes than either the 17% or 0% groups. Riley *et al.* (1979b) interpreted the result as suggesting that prenatal alcohol exposure in some way interfered with the central inhibitory system that controls this behavior.

Activity

By far, the most common procedure in behavioral teratology involves some assessment of spontaneous locomotor activity. Spontaneous activity presumably reflects a state of arousal or "nonspecific excitability" (Ossenkopp and Mazmanian, 1985). The U.S. Environmental Protection Agency (1985) recommends activity testing for any substance suspected of neurotoxicity. Undoubtedly, activity measures will also be required by any regulations covering behavioral teratology (see Chapter 8).

Although activity seems to be a straightforward and unambiguously evaluated behavior, there are many different ways of assessing it and there is little correlation between the results of these methods. An excellent survey of these various methods and the advantages and disadvantages of each can be found in Reiter and Macphail (1979).

The two basic procedures for assessing activity involve observational and automated techniques.

Observational Techniques

Animals can be directly observed in their home cages or in novel environments for the presence or absence of certain motor components of behavior or quantitatively for the frequency, duration, or patterning of behavior. Checklists of behaviors that can be monitored in this way are available for rats and mice (e.g., Irwin, 1968) and can be modified depending on the experimenter's resources and available time.

The disadvantages of observational methods are: (1) They are very labor intensive and require considerable time. If many animals are to be tested, these procedures are not practical. (2) They require subjective evaluation, and interobserver reliability may be poor; e.g., if more than one person is conducting observations, observers may not agree in what they are recording. (3) The presence of the observer may influence the behavior of the animal. For example, when rats are placed in an open field, they tend to remain in the area closest to the observer. This problem can be avoided by use of videotape recording. Using such a recording also will reduce the problem of interobserver reliability.

The most common apparatus for observing activity has been the open field. Basically, this is an enclosed area to prevent the animal from escaping. The field is usually marked off into squares or concentric circles and the number of lines the animal crosses is the measure of activity.

Although the open field is the most commonly used apparatus for measuring activity, it is at the same time the most unreliable as far as interlaboratory comparisons are concerned. This is because it is totally unstandardized as to size, illumination, square versus circular lines, and so on. (For a thorough review of the open field and its shortcomings, see Walsh and Cummins, 1976.)

One reason the open field is being relied on less is the variety of automated devices now available for assessing activity. Among these are photocell cages, mechanical devices, stabilimeters, activity wheels, and field detectors.

Photocell Cages. Photocell cages are the most commonly used of the automated activity monitors. The basic apparatus is a cage in which a beam of light crosses from one side to the other. If the beam is broken, it is recorded as an activity count. The disadvantage associated with some of these monitors is that stereotypic head movements in and out of the beam will be included as activity counts even though the animal is not actually ambulatory. This problem has been overcome by more sophisticated photobeam monitors and stereotypic movements can be recorded separately from ambulatory movements (Sandberg *et al.*, 1985).

One example of an automated activity monitor, the Digiscan Animal Activity Monitor manufactured by Omnitech Electronics (Columbus, Ohio) is shown in Figure 7.8. It is an

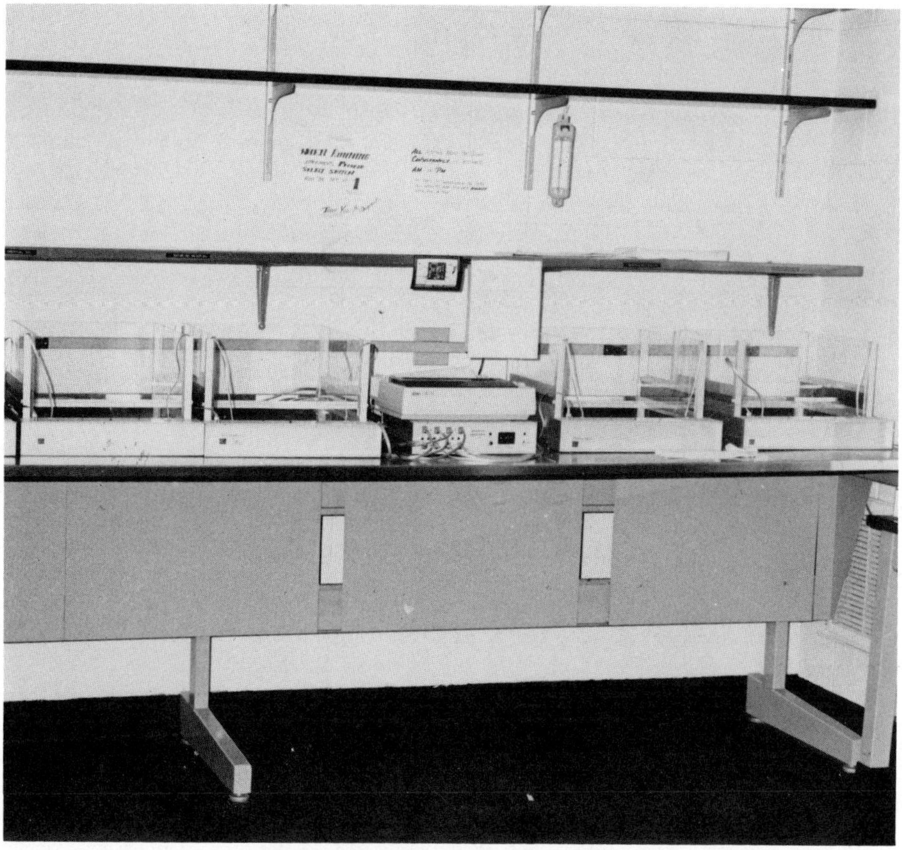

Figure 7.8. Automated activity monitors.

enclosure measuring 40 by 40 by 30.5 cm, surrounded by horizontal and vertical infrared beams spaced 2 inches apart with Plexiglas walls to keep the animal from escaping. When the beams are broken, a signal is sent to an analyser, which collects, sorts, and prints the information.

The activities recorded by the analyser are (1) Horizontal activity: the total number of beam interruptions occurring in the horizontal sensors during the specified test session. (2) Total distance: the distance traveled in inches. This is not the same as the horizontal activity count—even if the animal travels from one corner to another diagonally opposite, it will generate the same horizontal activity count regardless of path. Distance traveled therefore depends on the path the animal takes. (3) Movement time: the apparatus keeps incrementing as long as the animal is moving. If it rests for 1 second or more the incrementing stops. This gives a measure of the amount of time the animal is in constant motion. (4) Rest time: this is the opposite of movement time. (5) Number of movements: each time a break occurs in horizontal movement for more than one second, number of movements is incremented. This indicates separate horizontal movements. (6) Vertical activity: the total number of beam interruptions occurring in the vertical sensor. (7) Vertical time: the apparatus keeps incrementing as soon as the vertical sensor is activated and continues to increment until the animal goes below the sensor. (8) Number of vertical movements: each time the animal rears, this parameter is incremented by one. (9) Stereotypy time: the time spent breaking the same beam repeatedly, for example, by grooming. (10) Number of stereotypies: the number of times the same beam is broken, with a break of one or more seconds between breaks. (11) Clockwise movements: the number of times the animal moves in a clockwise circle. (12) Counterclockwise activity: the number of times the animal moves in a counterclockwise direction.

One way of statistically reducing many of these measures to a few components is by factor analysis. If there are structural similarities between variables, factor analysis offers a way of examining these as a smaller set of variables called *factors*. Such reduction has the advantage of creating a more stable and less biased indicator of a category of responses. If any single component has some bias, the combination of several similar measures will average out these error components, resulting in a better estimate of some parameter (Ossenkopp & Mazmanian, 1985). Basically, this is the same as saying that the larger the sample size, the more representative of the population.

An example of this kind of data reduction is shown in Table 7.5, which categorizes various activity parameters in terms of three factors readily interpreted as ambulation (Factor 1), rearing (Factor 2), and stereotypy (Factor 3). These factors can then used as dependent variables in statistical analyses of group differences.

Mechanical Devices. Mechanical devices respond to horizontal or vertical displacement of a platform in response to an animal's movements. There are several different types of such devices: jiggle cages, tilt cages, and force platforms. Differences between these kinds of mechanical devices have been described in Reiter and Macphail (1979). The advantages of using these devices are their sensitivity to movement that can be adjusted so that some kinds of behavior can be measured in preference to others. The disadvantages are that the apparatus does not distinguish between different movements—it only records cage movement, and boxes have to be constantly readjusted to make them all the same.

Activity Wheels. Activity wheels are among the oldest automated devices for measuring activity. Basically, these are wheels mounted on a horizontal axis. The animal is placed

Table 7.5. Factor Loadings for Various Measures of Activity

	Factors	Factor loadings
Horizontal activity	1	0.919
Total distance	1	0.944
Move time	1	1.940
Number of movements	1	0.745
Rest time	1	−0.942
Vertical activity	2	0.973
Vertical time	2	0.916
Number of vertical movements	2	0.886
Number of stereotypies	3	0.796
Stereotypy time	3	0.743
Number of clockwise movements	1	0.691
Number of anticlockwise movements	1	0.675

inside the wheel or in a cage attached to the wheel area so that it can enter on its own. The number of revolutions are then recorded giving a measure of how much the animal has run. The advantage of this apparatus is that the animal lives in the cage and can eat and drink or rest whenever it chooses. The main disadvantage is that the animal remains in one place and moves in one direction only.

Activity cages were used by Martin *et al.* (1978) to evaluate the effects of prenatal alcohol exposure in rats. Pregnant rats were intubated twice daily with 4.5 g/kg of alcohol. At birth, offspring were cross-fostered to evaluate the possible role of parental rearing influences. At 60 days of age, animals were tested in activity wheels. Animals prenatally exposed to alcohol were significantly more active than pair-fed controls. Differences between fostered and nonfostered animals were not significant.

Rotorod

In contrast to measures of spontaneous activity, some tests force the animal to move in the apparatus. One such test is the rotorod shown in Figure 7.9. The rotorod is a rotating drum that can accelerate linearly until the animal falls onto a platform. Depression of the platform activates a switch halting the drum and a timer. Animals are given an initial test to familiarize them with the apparatus followed by two or more test trials.

Abel and Dintcheff (1978) used this apparatus to evaluate the coordination of rats born to mothers intubated with 4 or 6 g/kg/day throughout pregnancy. Offspring, tested at 20 days of age, could not remain on the drum as long as pair-fed controls.

Preference Tests

In preference testing, animals are typically presented with two bottles, one containing plain water, the other, a test solution. The amount of the test solution the animal consumes relative to water indicates its preference or aversion for that solution. On the basis of such

Figure 7.9. Rotorod apparatus for studying coordination in rats and mice.

preferences, inferences can be made about changes in brain structures due to teratogens. For example, when a solution of saccharin in water is offered to male and female rats, females typically consume much higher amounts of this nonnutritional solution (Valenstein *et al.*, 1967). However, if females are injected with testosterone soon after birth, they resemble males in saccharin preference (Wade & Zucker, 1969). To determine if alcohol would have an effect similar to direct administration of testosterone, McGivern and his colleagues (1984) fed pregnant rats liquid alcohol diets containing either 35% or 0% ethanol-derived calories. Between 90–130 days of age, offspring of these mothers were presented with two bottles containing either plain water or a solution of saccharin. Three different concentrations of saccharin were presented for 2 days each. On alternate days, the position of the bottles was switched to avoid position bias because some animals prefer to drink from bottles placed on the left or right.

Female control animals consumed more saccharin than male controls at each concentration, in keeping with the greater preference of females for saccharin. However, alcohol-exposed males and females did not differ in their saccharin preference, indicating an absence of normal sexual dimorphism. McGivern *et al.* (1984) then repeated this study exposing animals to alcohol only during the last week of gestation, which corresponds most closely to the critical period of neural sexual differentiation. As in the previous study, female pair-fed controls consumed considerably more saccharin compared with males, whereas this pattern was not observed in alcohol-exposed offspring. These results were interpreted as evidence of a feminizing effect of prenatal alcohol exposure. A similar effect has been noted for prenatal nicotine exposure (e.g., Lichtensteiger & Schlumpf, 1985).

Sexual Behavior in the Rat

Male sexual behavior in rats and mice is a relatively fixed action pattern. When a male encounters a female, he usually starts nuzzling her and then tries to sniff and lick her genitals. If she is receptive, he will then mount her several times from the rear and engage in pelvic thrusting (a rhythmic movement of the hindquarters). In most of these mounts, he will achieve intromission (entry of his penis into the female's vagina). After about 8 to 14 intromissions, he will ejaculate. Following ejaculation, the male becomes lethargic and shows little interest in the female, but after about 5 to 8 minutes he will begin mounting the female again (postejaculatory refractory period).

Mating is often studied in a darkened room using only dim illumination from a red light (rats do not see red). However, this is not essential and animals will readily mate in daylight. The male is adapted to the test apparatus for about 5 to 10 minutes, and a female in heat (primed with estrogen and progesterone before testing) is then introduced into the test chamber. The following indices are often recorded over the test session (10–20 minutes):

1. Mount latency. Time to mount the female from the time of her placement into the test chamber. Mounting is defined as clasping and pelvic thrusting of the flanks.
2. Intromission latency. Time from introduction of the female until first intromission. Intromission is defined as the insertion of the penis into the vagina and is typically followed by licking of the genitals.
3. Mount frequency. Number of mounts before ejaculation.
4. Intromission frequency. The number of intromissions before ejaculation.
5. Ejaculatory latency. Time from first intromission to ejaculation.
6. Postejaculation refractory period. Time from ejaculation to next intromission.

Dalterio (1980) used these measures of male sexual behavior to evaluate the effects of perinatal exposure to delta-9-tetrahydrocannabinol (THC), the psychoactive ingredient in marihuana. Around 24 hours prior to parturition, 50 mg/kg of THC or vehicle were administered to pregnant mice. After birth, mothers continued to be injected with the same dose up to lactation day 6. When males were later tested for copulatory behavior, less than half the THC males tried to mount with females, whereas all control males mounted. Mount latency was also significantly increased.

Female sexual behavior in the female rat and mouse is also stereotypic. If the female is in heat, she will exhibit lordosis in response to the male's mounting her. In lordosis the female extends her body, arches her back thereby exposing her genitals, and moves her tail to the side. If the female is not in heat, she will not permit herself to be easily mounted and will not engage in lordosis.

Normally, the female comes into heat as a result of the cyclic secretion of estrogen and progesterone every 4 to 5 days. Females can be brought into heat by injecting them with these hormones. Sexual behavior in females is not studied as frequently as male sexual behavior because there are fewer behavioral parameters to measure.

A test for feminization involves placing a test male with a sexually vigorous male and observing if the male exhibits lordosis. Hard and his co-workers (1984) reported that 45% of 80-day-old male rats prenatally exposed to alcohol were sexually unresponsive compared with only 8% of control males.

Acoustic Startle

The startle reaction to a brief auditory stimulus can be examined in terms of arousal or as a way of assessing auditory function. The response is affected by several procedural variables such as stimulus intensity, stimulus interval, and background noise at the time startle is elicited (see Davis, 1980, for a review).

Acoustic startle is a valuable behavioral test method because the neuroanatomy of the primary acoustic startle circuit is known. Briefly, it consists of auditory nerve input to the cochlear nucleus (first synapse), proceeding to the lateral lemniscus, on to the nucleus reticularis pontis caudalis, which projects down the reticulo-spinal tract to lower motor neurons in the spinal cord, and finally to muscles. From input to response, there are as few as three and no more than five central synapses involved, in addition to the neuromuscular function (Davis, 1980). Because the neuroanatomical circuitry is well known, changes in acoustical responsiveness due to prenatal exposure to teratogens are likely to be amenable to correlative evoked potential and direct neuroanatomical examination.

The auditory reflex has been evaluated in a number of behavioral teratological studies and was a component of the behavioral test battery in the Collaborative Behavioral Teratology Study (see Chapter 8). It was also one of the tests that produced consistent effects across laboratories (Buelke-Sam *et al.*, 1985).

Seizure Susceptibility

Seizure susceptibility is sometimes evaluated as a way of making inferences about the excitability state of the brain. The rationale for such studies is that if prenatally treated animals are hyperexcitable, less of a seizure-inducing stimulus will be needed to elicit seizures.

Pizzi and his co-workers (1979) injected neonatal mice subcutaneously for 10 consecutive days beginning on day 2 after birth with monosodium glutamate. Control animals received saline. At 150 days of age, the animals were injected (i.p.) with pentylenetetrazol (PTZ), a drug that induces convulsions.

Following injection with PTZ, animals were observed for 7 minutes. During each 30 second interval of this 7-minute period, animals were scored using the 6-point rating scale shown in Table 7.6.

Animals exposed to MSG were more likely to have seizures following injection of the test compound than controls and their convulsions were of greater severity.

Golub and Kornetsky (1974) conducted a similar evaluation using the inhalant flurothyl. Chlorpromazine (20 mg/kg) was administered subcutaneously twice daily to pregnant rats. At birth, offspring were cross-fostered and were tested at 30 days of age for seizure susceptibility induced by inhalation of flurothyl. The animals were placed in a

Table 7.6. Six-Point Rating Scale for Evaluating Seizures in Mice[a]

0 = No observable convulsion
1 = Mild intermittent myoclonic spasm
2 = Strong myoclonic spasm (all four feet leave the floor but righting reflex is not lost)
3 = Strong clonic convulsion with loss of righting reflex
4 = Severe tonic/clonic convulsion with prolonged extension of limbs
5 = Convulsions resulting in death

[a]From Pizza *et al.*, 1979.

plexiglas chamber and the volatile convulsant was infused into the chamber at a constant rate by means of a polyethylene tube inserted onto the center of the chamber lid and dropped onto a filter paper below. The dependent variable was the time from exposure to loss of posture accompanied by seizures. Chlorpromazine-exposed animals convulsed much earlier than controls.

Many studies examining seizure susceptibility elicit seizures in mice by exposing them to loud noise. In a typical test situation, a bell is mounted on the inside of the chamber lid or a tone generator is placed into the floor and the sound intensity within the chamber is adjusted to about 100 dB, which is loud enough to produce audiogenic seizures in susceptible animals.

Sound stimulation is initiated around 30 seconds after animals are placed within the apparatus and is continued until the onset of a convulsion or until a maximum of 60 seconds has elapsed. Seizure intensity is rated using scales like that in Table 7.6 or by those specifically devised for sound-induced seizures such as that of Jobe *et al.* (1973). This rating scale monitors the number of running phases and convulsions and ranges from generalized clonus of forelimbs (score of 3) to complete tonic extension of hindlimbs (score of 9). Running (score of 1) is not an integral part of the convulsion but is an essential part of the seizure pattern. Audiogenic seizures typically begin with a period of running, followed by tonic and then clonic seizures. Dependent variables will be latency to convulse (i.e., time from onset of sound stimulus to occurrence of convulsion) and ratings of the general seizure pattern: such as running (score of 1), clonus of forelimbs (2 or 3), complete tonic extension of hindlimbs (score of 8 or 9).

The focus of audiogenic seizures appears to be the brainstem (Browning *et al.*, 1985). The audiogenic stimulus that evokes seizures is channeled through the inferior colliculus (Willott & Lu, 1980) because lesions of the inferior colliculus block audiogenic seizures (Kesner, 1980) whereas medial geniculate and cortical lesions are ineffective (Servit, 1960). The seizure pathway is thought to follow a colliculofugal route to mesencephalic or diencephalic structures (Kesner, 1966).

Operant Behavior

Operant behavior represents a class of behaviors that depend on what happens after the behavior occurs. This is because reinforcement is made contingent on performances. As a result, changes in performance lead to certain conditions that subsequently affect behavior. This is a circular relationship in which behavior is relatively easy to elicit and is recorded by means of automated electronic or electomechanical devices. Data from operant studies usually take the form of rate of occurrence of some act. In most cases, this act is bar pressing in a specially designed apparatus. The reinforcement for such behavior is usually pellets of food for hungry animals or water for animals that have been water deprived.

Although the process of data collection is automated, the experimenter must often devote considerable time, especially at the beginning of the test phase, to train animals to perform. Once the animals have been trained, behavior is relatively stable and data can be collected relatively easily. Slight deviations in level or pattern of performance are readily observed and may be indicative of neurobehavioral deficits. Basic principles and procedures can be found in Dinsmoor (1966).

Once the animal associates lever pressing with delivery of reinforcement, it is given sufficient practice so that whenever it operates the lever, it receives reinforcement. This is termed *continuous reinforcement* (CRF). After this behavior is stabilized, many higher order contingencies can be introduced. The two main reinforcement contingencies are ratio and interval reinforcement.

For ratio reinforcement, a number of lever responses must occur for reinforcement to be delivered. A fixed ratio 2 (FR2), for example, means that two lever responses must occur for reinforcement. An FR 10 means 10 lever presses must occur, an FR 20 means

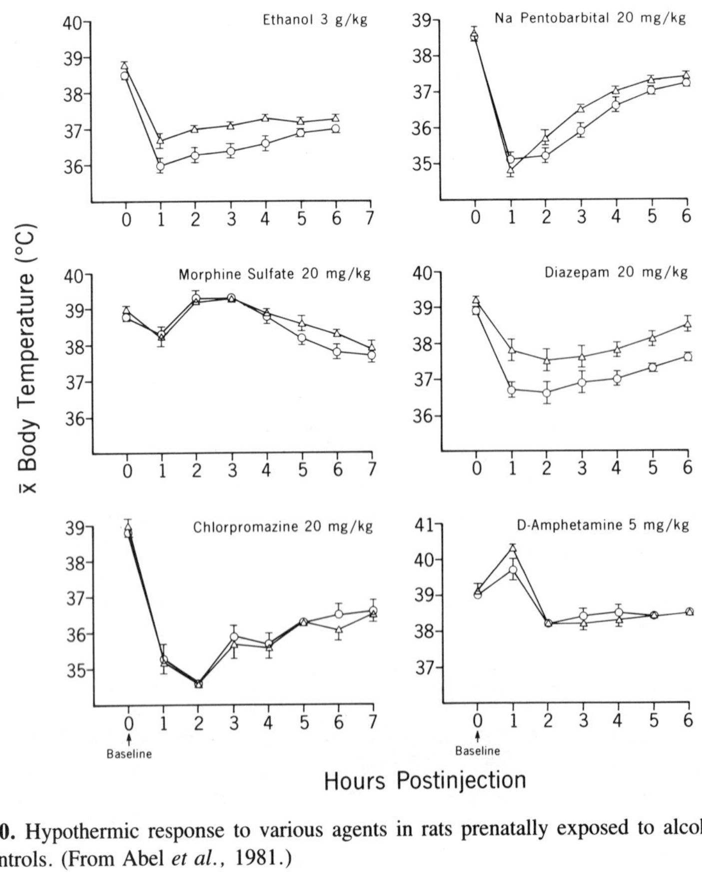

Figure 7.10. Hypothermic response to various agents in rats prenatally exposed to alcohol and their pair-fed controls. (From Abel *et al.*, 1981.)

20 presses, etc. In general, the animal is gradually reinforced to work up to a higher ratio so that performance is steady, with no pauses for more than 5 seconds during lever pressing and no pauses for more than 20 seconds after reinforcement. This may take 2 to 3 days of training.

Middaugh and his co-workers (1975) used an FR operant task to assess the effects of prenatal exposure to phenobarbital (40 mg/kg) administered during the last third of pregnancy. Offspring were placed on a food deprivation schedule to reduce their body weight to 75% to 80% of *ad lib* term levels. After 1 week on the deprivation schedule, they were self-trained to press a metal disc in the operant cage for a food reward. After initial training, they were given two days of practice on a continuous reinforcement (CRF) schedule and then were given five-day sessions each on an FR5, FR20, and FR40. Middaugh found that animals prenatally exposed to phenobarbital responded less than controls on the CRF and the response decrement became more pronounced under the FR40 schedule.

Sensitivity to Drugs

A number of studies have examined subsequent responsiveness to drugs as a consequence of *in utero* exposure to that same drug or other drugs. For example, O'Callaghan and Holtzman (1977) examined postnatal responsiveness to opiates and Harris and Case (1979) evaluated responsiveness to barbiturates after prenatal exposure to these drugs. Abel *et al.* (1981) found that rats prenatally exposed to alcohol experienced a smaller decrease in body temperature in response to alcohol during adulthood than did control animals. Responsiveness to other drugs that exhibit cross-tolerance to alcohol was also reduced, whereas animals did not differ in response to drugs that were not cross-tolerant to alcohol (see Figure 7.10).

Studies showing altered responsiveness to drugs as a result of prenatal exposure are indicative of rather specific changes in those areas of the brain or cell structures that are affected by these agents *in utero*.

References

Abbey, H., and Howard, E. Statistical procedure in developmental studies on a species with multiple offspring. *Developmental Psychobiology,* 1973, *6,* 329–335.

Abel, E. L. Effects of alcohol withdrawal and undernutrition in cannibalism of rat pups. *Behavior and Neural Biology,* 1979a, *25,* 411–413.

Abel, E. L. Prenatal effects of alcohol on adult learning in rats. *Pharmacology, Biochemistry and Behavior,* 1979b, *10,* 239–243.

Abel, E. L. Behavioral teratology of alcohol. *Psychological Bulletin,* 1981, *90,* 564–581.

Abel, E. L. In utero alcohol exposure and developmental delay of response inhibition. *Alcoholism: Clinical and Experimental Research,* 1982, *6,* 369–376.

Abel, E. L. *Fetal Alcohol Syndrome and Fetal Alcohol Effects.* New York: Plenum Press, 1984.

Abel, E. L., Bush, R., and Dintcheff, B. A. Exposure of rats to alcohol alters drug sensitivity during adulthood. *Science,* 1981, *212,* 1531–1533.

Abel, E. L., and Dintcheff, B. A. Effects of prenatal alcohol exposure on growth and development in rats. *Journal of Pharmacology Experimental Therapeutics,* 1978, *217,* 916–921.

Abel, E. L., and Dintcheff, B. A. Effects of prenatal alcohol exposure on behavior of aged rats. *Drug and Alcohol Dependence,* 1986, *16,* 321–330.

Abel, E. L., and Greizerstein, H. B. Ethanol-induced prenatal growth deficiency: Changes in fetal body composition. *Journal of Pharmacology and Experimental Therapeutics,* 1979, *211,* 668–671.

Adams, J. Clinical relevance of experimental behavioral teratology. *Neurotoxicology,* 1986a, *7,* 19–34.

Adams, J. Methods in behavioral teratology. In E. P. Riley and C. V. Vorhees (Eds.), *Handbook of Behavioral Teratology.* New York: Plenum Press, 1986b, pp.67–97.

Anderson, B. J., Nash, S. M., Weaver, M. S., and Davis, S. F. Defensive burying: The effects of multiple stimulus presentation and extinction. *Psychological Record,* 1983, *33,* 185–190.

Barone, E. R., Pirola, C., and Lieber, C. S. Small intestinal damage and changes in cell population produced by ethanol ingestion in the rat. *Gastroenterology,* 1974, *66,* 226–234.

Browning, R. A., Nelson, D. K., Mogharreban, N., Jobe, P. C., and Laird, H. E. Effect of midbrain and pontine ligmental lesions on audiogenic seizures in genetically epilepsy prone rats. *Epilepsia,* 1985, *126,* 1175–1183.

Buelke-Sam, J., and Kimmel, C. A. Development and standardization of screening methods for behavioral teratology. *Teratology,* 1979, *20,* 17–29.

Buelke-Sam, J., Kimmel, C. A., Adams, J., Nelson, C. J., Vorhees, C. V., Wright, D. C., St. Omer, V., Korol, B. A., Butcher, R. E., and Wayner, M. J. Collaborative behavioral teratology study: Results. *Neurobehavioral Toxicology and Teratology,* 1985, *7,* 591–624.

Burkhalter, J. E., and Balter, R. L. Behavioral teratology evaluation of trichloromethane in mice. *Neurobehavioral Toxicology,* 1979, *1,* 199–205.

Butcher, R. E. Behavioral testing as a method for assessing risk. *Environmental Health Perspectives,* 1976, *18,* 75–78.

Cravens, R. W. Effects of maternal undernutrition on offspring behavior: Incentive value of a food reward and ability to escape from water. *Developmental Psychobiology,* 1974, *7,* 61–69.

Dalterio, S. L. Perinatal and adult exposure to cannabinoids alters male reproductive functions in mice. *Pharmacology, Biochemistry and Behavior,* 1980, *12,* 143–153.

Davis, M. Habituation and sensitization of a startle-like response elicited by electrical stimulation at different points in the acoustic startle circuit. *Advances in Physiological Science.* 1980, *16,* 67–78.

Dinsmoor, J. A. Operant conditioning. In J. B. Sidowski, (Ed.), *Experimental Methods and Instrumentation in Psychology.* New York: McGrankell, 1966, pp. 421–445.

Douglas, R. J. The hippocampus and behavior. *Psychological Bulletin,* 1967, *67,* 416–442.

Fox, M. W. Reflex-ontogeny and behavioral development of the mouse. *Animal Behavior,* 1965, *13,* 234–241.

Gallo, P. V., and Weinberg, J. Neuromotor development and response inhibition following prenatal ethanol exposure. *Neurobehavioral Toxicology and Teratology,* 1982, *4,* 505–513.

Golub, M., and Kornetsky, C. Seizure susceptibility and avoidance conditioning in adult rats treated prenatally with chlorpromazine. *Developmental Psychobiology,* 1974, *7,* 79–88.

Gottfried, E. B., Korsten, M. A., and Lieber, C. S. Gastritis and duodenitis induced by alcohol: An endoscopic and histologic assessment. *Gastroenterology,* 1976, *70,* 890.

Greizerstein, H. B., and Abel, E. L. Acute effects of ethanol on fetal body composition and electrolyte content in the rat. *Bulletin of the Psychonomic Society,* 1979, *14,* 355–356.

Hanshaw, J. B., Dudgeon, J. A., and Marshall, W. C. *Viral Diseases of the Fetus and Newborn.* Philadelphia: W. B. Saunders, 1985.

Hard, E., Dahlgren, I. L., Engel, J., Larson, K., Liljequest, S., Linch, A., and Musi, B. Development of sexual behavior in prenatally ethanol-exposed rats. *Drug and Alcohol Dependence,* 1984, *14,* 51–61.

Harris, R. P., and Case, J. Effects of maternal consumption of ethanol, barbital, or chlordiazepoxide on the behavior of offspring. *Behavioral and Neural Biology,* 1979, *26,* 234–247.

Irwin, S. Comprehensive observational assessment: Ia. A systematic, quantitative procedure for assessing the behavioral and physiologic state of the mouse. *Psychopharmacology,* 1968, *13,* 222–257.

Jobe, P. C., Picchioni, A. L., and Chin, L. Role of brain norepinephrine in audiogenic seizure in the rat. *Journal of Pharmacology and Experimental Therapeutics,* 1973, *184,* 1–10.

Kesner, R. P. Subcortical mechanisms of audiogenic seizures. *Experimental Neurology,* 1966, *15,* 192–205.

Leichter, J., and Lee, M. Effect of maternal ethanol administration on physical growth of the offspring in rats. *Growth,* 1979, *43,* 288–297.

Levine, S. Stimulation in infancy. *Scientific American,* 1960, *202,* 80–86.

Lichtensteiger, W., and Schlumpf, M. Prenatal nicotine affects fetal testesterone and sexual dimorphism of saccharin preference. *Pharmacology, Biochemistry and Behavior,* 1985, *23,* 439–444.

Lieber, C. S. Alcohol-nutrition interactions. In T. K. Li, S. Schenker, and L. Luming, (Eds.), *Alcohol and Nutrition.* Rockville, Md.; U.S. Department of Health, Education and Welfare, 1979, pp. 47–63.

Lochry, E. A., Hoberman, A. M., and Christian, M. S. Comparative sensitivity of pups' bodyweight and commonly used developmental landmarks. *Teratology,* 1985, *32,* 28A.

Martin, J. C., Martin, D. C., Sigman, G., and Radow, B. Maternal ethanol consumption and hyperactivity in crossfostered offspring. *Physiological Psychology,* 1978, *6,* 362–365.

McGivern, R. F., Clancy, A. N., Hill, M. A., and Noble, E. P. Prenatal alcohol exposure alters adult expression of sexually dimorphic behavior in the rat. *Science,* 1984, *224,* 896–898.

Means, L. W., Medlin, C. W., Highes, V. D. and Gray, S. L. Rats exposed in utero to ethanol are hyperresponsive to methylphenidate when tested as neonates. *Neurobehavioral Toxicology and Teratology,* 1984, *6,* 187–192.

Meyer, L. S., and Riley, E. P. In E. P. Riley, and C. V. Vorhees (Eds.), *Handbook of Behavioral Teratology.* New York: Plenum Press, 1986, pp. 101–140.

Middaugh, L. D., Santos, C. A., and Zemp, J. W. Effects of phenobarbital given to pregnant mice on behavior of mature offspring. *Developmental Psychobiology,* 1975, *8,* 305–313.

Miller, B. M. (Ed). *Laboratory Safety: Principles and Practices.* Washington, DC, American Society for Microbiology, 1986.

Mowrer, O. H., and Lamoreaux, R. R. Fear as an intervening variable in avoidance conditioning. *Journal of Comparative Psychology,* 1946, *39,* 29–50.

Munn, W. L. *Handbook of Psychological Research on the Rat.* Boston: Houghton Mifflin, 1950.

O'Callaghan, J. P. and Holtzman, S. G. Prenatal administration of levorphanol or dextromethorphan to the rat: Analgesic effect of morphine in the offspring. *Journal of Pharmacology and Experimental Therapeutics,* 1977, *200,* 255–262.

Ossenkopp, K., and Mazmanian, D. S. The measurement and integration of behavioral variables: Aggregation and complexity as important issues. *Neurobehavioral Toxicology and Teratology,* 1985, *7,* 95–100.

Palmer, A. K. The design of subprimate animal studies. In J. G. Wilson and F. C. Fraser (Eds.), *Handbook of Teratology* (Vol. 4). New York: Plenum Press, 1977, pp. 215–253.

Pinel, J. P. J., and Treit, D. Burying as a defensive response in rats. *Journal of Comparative and Physiological Psychology,* 1978, *92,* 708–712.

Pizzi, W. J., Unnerstall, J. R., and Barnhardt, J. E. Neonatal monosodium glutamate administration increases susceptibility to chemically-induced convulsions in adult mice. *Neurobehavioral Toxicology and Teratology,* 1979, *1,* 169–173.

Reiter, L. W., and Macphail, R. C. Motor activity: A survey of methods with potential use in toxicity testing. *Neurobehavioral Toxicology and Teratology,* 1979, *1* (Suppl.), 53–66.

Riley, E. P., Lochry, E. A., and Shapiro, N. R. Lack of response inhibition in rats prenatally exposed to alcohol. *Psychopharmacology,* 1979, *62,* 47–52.

Riley, E. P., Lochry, E. A., Shapiro, N. R., and Baldwin, J. Response perseveration in rats exposed to alcohol prenatally. *Pharmacology, Biochemistry and Behavior,* 1979, *10,* 255–259.

Riley, E. P., and Meyer, L. S. Considerations for the design, implementation, and interpretation of animal models of fetal alcohol effects. *Neurobehavioral Toxicology and Teratology,* 1984, *6,* 97–101.

Sandberg, P. R., Hagenmeyer, S. H., and Henault, M. A. Automated measurement in multivariate locomotor behavior in rodents. *Neurobehavioral Toxicology and Teratology,* 1985, *7,* 87–94.

Servit, Z. The role of subcortical acoustic centers in seizure susceptibility to an acoustic stimulus in symptomatology of audiogenic seizures in the rat. *Physiologia Bohemoslovia,* 1960, *9,* 42–47.

Smith, J. C. Marginal nutritional states and conditioned deficiencies. In T. K. Li, S. Schenker, and L. Lumeng, (Eds.), *Alcohol and Nutrition.* Rockville, Md.: U.S. Department of Health, Education and Welfare, 1979, pp. 23–46.

Teicher, M. H., Pearson, D. E., Shaywitz, B. A., and Cohen, D. J. Identifying experimental units and calculating experimental error. *Science,* 1981, *213,* 931.

United States Protection Agency. Toxic substances control act test guidelines; final rules, 40 CFR Part 798. 6200: Health effects testing guidelines, neurotoxicity, motor activity. *Federal Register,* 1985, *50,* 39460.

Valenstein, E. S., Kakolewski, J. W., and Cox, C. V. Sex differences in taste preference for glucose and saccharin solutions. *Science,* 1967, *156,* 942–943.

Vorhees, C. V. Fetal anticonvulsant syndrome in rats: Dose-and period-response relationships of prenatal diphenylhydantoin, trimethadione and phenobarbital exposure on the structural and functional development of the offspring. *Journal of Pharmacology and Experimental Therapeutics,* 1983, *227,* 274–287.

Wade, G. N., and Zucker, I. Taste preferences of female rats. *Physiology and Behavior,* 1969, *4,* 935–943.

Walsh, R. N., and Cummins, R. A. The open-field test: A critical review. *Psychology Bulletin,* 1976, *83,* 482–504.

CHAPTER 8

Regulatory Protocols for Identifying Teratogens

In 1966, partly in reaction to the thalidomide disaster of the 1950s and early 1960s, the U.S. Food and Drug Administration (FDA) adopted new guidelines for preclinically testing drugs for reproductive and structural defects prior to allowing them to be marketed (U.S. F.D.A., 1966).

The FDA does not assume that these guidelines will result in detection of effects identical to those occurring qualitatively or quantitatively in humans. "What is assumed is the ability of well-designed animal studies to provide an indication of potential risk to humans" (Frankos, 1985, p. 616).

The FDA's guidelines require that every new product to be put through a three-segment screening protocol. Although details of testing vary, the FDA's three-segment approach is currently the cornerstone of all reproductive/teratological screening protocols in all countries that require such screening. The details of this protocol are examined in this chapter along with various modifications.

FDA Guidelines

Segment I (The Fertility Study)

Segment I testing examines fertility and general reproductive effects with respect to gonadal function, estrus cycles, mating behavior, conception rate, and early gestation. Procedural details and differences between U.S. guidelines and those of other countries for segment I testing are shown in Table 8.1.

Male animals (usually rats) are to be treated with a test compound for at least two months before mating; females for 14 days before mating. Females are examined every day for vaginal plugs. Discovery of a copulatory plug is considered day 0 of pregnancy. Pregnant females continue to be treated during gestation. Half the females are sacrificed at midgestation (around gestation day 13) and examined for live and dead embryos. The other half continue to be treated throughout gestation and postnatally up to time of weaning. At birth, newborn pups are counted, sexed, weighed, and inspected for anomalies. Offspring are weighed and counted on days 4 and 21.

Table 8.1. Comparison of Segment I Guidelines for the United States and Other Countries

	U.S.	Canada	Scandinavia	U.K.	Japan
No. of species	1	—	—	2	1
No. of doses	2	—	—	3	3
No. of animals per group					
M	10	15	—	24	20
F	20	30	—		
Treatment period prior to mating					
M	60–80	80	60	—	60
F	14	14	14	—	14

In the United States, only one species of animal has to be tested. This is usually the mouse or rat, although rabbits are also used. These are also the species most often used in other countries. The U.S. guidelines recommend that two species be tested but one should not be a rodent.

The United States requires that only two doses of a compound be tested. The strategy is to use the maximum tolerated dose as a high dose and some fraction of this dose as a low dose. The United Kingdom and Japan require three doses. The high dose is to be toxic but not lethal, the low dose is to be similar to that causing a clinically relevant effect, and the intermediate dose is to be somewhere in between.

Segment II (The Teratology Study)

Segment II examines possible embryotoxic and teratological effects. Comparison of guidelines between countries is shown in Table 8.2.

Drugs are administered during organogenesis (gestation days 6–15 in rats and mice). All countries require that two species be tested. Canada specifically requires that one of these species not be a rodent. Animals are sacrificed on the day before delivery and fetuses are examined for resorptions and gross skeletal and visceral abnormalities. All countries except the United States require that at least three doses be tested; the United States requires only two doses. The United States also requires that fewer animals be treated.

Table 8.2. Comparison of Segment II Guidelines for the United States and Other Countries

	U.S.	Canada	Scandinavia	U.K.	Japan
No. of species	2	2	2	2	2
No. of doses	2	3	3	3	3
No. of animals:					
Mouse	20	30	—	—	30
Rat	20	30	—	—	30
Rabbit	10	15	—	—	12

Table 8.3. Comparison of Segment III Guidelines for the United States and Other Countries

	U.S.	Canada	Scandinavia	U.K.	Japan
No. of species	1	1	2	1	1
No. of doses	2	3	3	3	3
No. of animals	20	20	—	12	20
Behavioral observations required	No	Yes	No	No	Yes

Segment III (The Perinatal Study)

Segment III examines effects on labor and delivery, newborn viability, lactation, and growth. Pregnant females are treated during the last third of pregnancy and during lactation up to weaning. Offspring are monitored for growth and development after birth as in Segment I. Comparison of guidelines is shown in Table 8.3.

The main and most important difference between countries in segment III guidelines is the requirement in Japan and Canada for behavioral evaluation of offspring. In the United States, United Kingdom, and Scandinavian countries, behavioral testing is not mandatory (see below).

In addition to the three-segment protocol for drug testing, the USFDA requires a minimum two-generation study of compounds for testing food and color additives (USFDA, 1970).

In the two-generation reproduction study, some of the weanlings are sacrificed and examined for internal anomalies. Others continue to receive exposure and are mated for a second generation and their offspring are examined at birth or at weaning.

Behavioral Teratology Testing in Japan

In the 1970s, a U.S. government panel considered including behavior in preclinical screening protocols but voted against it. The panel felt that behavioral tests lacked enough standardization, reliability, sensitivity, validity, and applicability to use them in any specific test battery (Collins, 1978). The view of the FDA is still that behavioral teratology test methods are still not "sufficiently mature" for routine screening because of high variability, lack of standardization, skepticism that behavioral defects are real, and difficulty of interpretation in terms of potential human harm (Nolen, 1985; Vorhees, 1986). However, in the mid-1980s, the Federation of the American Societies for Experimental Biology (FASEB) urged the FDA to adopt guidelines for inclusion of behavioral testing in Segment III (FASEB, 1985) and these suggestions are currently under review.

In 1983, another regulatory agency in the United States, the Environmental Protection Agency (EPA), proposed its own guidelines for toxicological and safety assessment standards. These included tests of "motor activity, functional observational battery, and schedule-controlled operant behavior." A year later, these guidelines were revised under the name of the Developmental Toxicity Study (EPA, 1984) and were finally adopted into the *U.S. Federal Register* in 1985.

Although the FDA was unwilling to adopt behavioral testing as an essential part of preclinical evaluation, the Japanese did so in 1974 and the United Kingdom did so the following year.

The reason Japan was much more receptive to inclusion of behavioral testing was its experience with methylmercury contamination of fish in Minimata Bay in the 1950s. The dramatic aspects of Minimata disease, as this poisoning came to be known, were not structural but neurological and psychological—every confirmed case out of the hundreds of newborns whose mothers had eaten fish contaminated with mercury was characterized by mental dysfunction (usually mental retardation), inability to walk properly, and difficulty in swallowing and chewing (Reuhl & Chang, 1979). Investigation showed that the mercury had come from plastics industries located in the area. To the Japanese the behavioral problems due to mercury were as dramatic as the missing limbs associated with thalidomide. The United States did not experience the epidemic of mercury poisoning so there was no pressure to adopt screening procedures.

The Japanese guidelines requiring behavioral teratology testing were introduced in 1974 (Japanese Ministry of Health and Welfare, 1984). Behavioral testing of offspring is required in both segments II and III but the guidelines do not stipulate what methods are to be used. Measures of motor function, learning, ability, or emotion were initially recommended (Yakuji Nippo, 1982) but were subsequently removed, leaving those conducting behavioral tests to choose whatever procedures they feel are adequate (Omori, cited by Vorhees, 1986).

Behavioral Teratology Testing in the United Kingdom

The United Kingdom suggested behavioral teratology testing for screening of new drugs in 1975. Its general test protocol (U.K., 1984) requires that "late effects on the progeny" be included as an endpoint. For "late effects" the guidelines stipulate only that "auditory, visual and behavioral function" be evaluated but do not specify the kinds of tests to be done. These guidelines for behavioral evaluation were kept deliberately vague because at the time they were issued, behavioral teratology was still considered too unsophisticated in its methodology and interpretation of data (Barlow, 1985). Tests of auditory, visual, and behavioral development were only suggested as examples of the kinds of tests that should be done. Currently, visual testing is usually done via a two-choice brightness discrimination swimming Y maze; hearing is assessed via auditory startle.

Although behavioral teratology testing is suggested for new drugs in the United Kingdom, no such testing is presently required for nondrugs. This is because meaningful tests for demonstrating effects have not yet been developed, there is skepticism about the value of such testing, and there is no evidence that these substances are behaviorally teratogenic in humans; therefore the predictive value of results in animals for effects in humans is questionable (Barlow, 1985).

Behavioral Teratology Testing in France and Italy

France adopted behavioral teratology testing in 1977 (Committee for Proprietory Medical Products, 1977); Italy did so in 1978 (Barlow & Sullivan, 1975). Both countries adopted the U.K. guidelines, that is, "late effects of the drug on the progeny in terms of auditory, visual and behavioral impairment should be assessed." The European economic community as a whole also adopted the U.K. guidelines in 1985 (Vorhees, 1986). Scandinavian countries do not require behavioral teratology testing (Nordic Council, 1983).

Efforts to Create a Behavioral Teratology Test Battery in the United States

Behavioral teratologists have struggled for some time to devise a test battery to evaluate substances. Until there is some degree of scientific agreement about appropriate test methods, government regulatory agencies will not require such testing (Hattan *et al.*, 1983), and industry is likely to oppose testing unless the proposed tests are minimally labor intensive for the numbers of animals required in safety studies (Nolen, 1985). However, the FDA Bureau of Foods "views the development of behavioral teratological or neurotoxicological testing as one of the most important and urgent areas for future improvement. We await with keen interest the creation of testing paradigms that can be recommended for routine measurement of the neurotoxical potential of food additive substances" (Hattan *et al.*, 1983, p. 87).

In 1978, in response to the perceived absence of any consensus about behavioral teratology, the Collaborative Behavioral Teratology Study was conceived to determine (1) the reliability/reproducibility of test methods and the data within and between laboratories, (2) the ability of such methods to detect effects of two alleged positive behavioral teratogens, amphetamine and mercury, at subtoxic doses, and (3) the possibility that early testing of animals influences results on later tests, that is, to determine if animals should be tested once or tested on all procedures (Kimmel & Buelke-Sam, 1985).

Five laboratories in addition to an FDA laboratory, participated in the study, representing industry, academia, and government. The study began in 1981 and was completed in 1983. It was obviously influenced by proposed EPA guidelines (see above) because it specifically included evaluation of motor activity, a functional observational battery, and a test of schedule-controlled operant behavior. As recommended in the EPA guidelines, motor activity was assessed using rodents, testing was done in an automated activity cage, and animals were tested after weaning. The functional observational battery included observations of posture, coordination, lacrimation, and other characteristics and tests of sensory function including vision, audition, and pain. The schedule-controlled operant procedures included an operant conditioning paradigm.

All laboratories used the same strain of rats purchased from the same supplier and the same automated and recording equipment and protocols and conducted the studies at the same time (Adams *et al.*, 1985). Personnel involved in the project were all trained together at a special session to maximize standardization of procedures.

Except for activity measures (a key outcome), there were no significant effects of laboratories (Buelke-Sam *et al.*, 1985). The final result was of dubious value for several reasons. One was the choice of compounds that were evaluated. Amphetamine was chosen as a test substance because of its well-known effects on the activity of adult animals. But there was no evidence that it was a behavioral teratogen when it was chosen and in fact it did not produce many noteworthy effects. If a compound has no effect, it is not particularly surprising to find laboratories agreeing that it has no effect. This is the equivalent of proving the null hypothesis. The fact that a compound may be a potent psychopharmacological agent but a weak behavioral teratological agent speaks to the uniqueness of behavioral teratology.

A second problem regarding the results had to do with the standardization of equipment, methodology, and protocols. Although the intent was to show that given the same equipment, method, training, and procedures, similar results could be obtained—this is a "no win" situation. One expects the same outcome under identical conditions. When dog bites man, it ain't news. Moreover, except for the measures of activity, there seems to have been no appreciation of the sensitivity of the proposed tasks except for their superficial face validity (Vorhees, 1986), and for activity, a key outcome, there were differences between laboratories in results.

Summary and Conclusions

Currently, there is little consensus about appropriate behavioral test measures for screening new compounds. There is also no pressure from government to include such testing in the current segment III paradigm in the United States. The EPA guidelines are based on pharmacological and toxicological procedures and principles rather than on behavioral teratology. It is as if we were back in the early years of building rockets and government turned to experts in bicycle or car design rather than airplane engineers for advice. The Collaborative Behavioral Teratology Study was an earnest and sincere effort to bring standardization to behavioral teratology. It had enough expertise to recognize the weaknesses of the protocols being adopted but unfortunately the organizers seem to have followed rather than led in the design of protocols for establishing standardization in behavioral teratology.

References

Adams, J., Buelke-Sam, J., Kimmel, C. A., Nelson, C. J., Reiter, L. W. Sobutka, T. J., Tilson, H. A., and Nelson, B. K. Collaborative behavioral teratology study: Protocol design and testing procedures. *Neurobehavioral Toxicology and Teratology*, 1985, *7*, 579–586.

Barlow, S. M. United Kingdom: Regulatory attitudes toward behavioural teratology testing. *Neurobehavioral Toxicology and Teratology*, 1985, *7*, 643–644.

Barlow, S. M., and Sullivan, F. M. Behavioural teratology. In C. L. Berry, and D. E. Poswillo, (Eds.), *Teratology: Trends and Applications*. New York: Springer-Verlag, 1975, 103–120.

Buelke-Sam, J., Kimmel, C. A., Adams, J., Nelson, C. J., Vorhees, C. V., Wright, D. C., St. Omer, V., Korol, B. A., Butcher, R. E., and Wayner, M. J. Collaborative behavioral teratology study: Results. *Neurobehavioral Toxicology and Teratology*, 1985, *7*, 591–624.

Canada, Health and Welfare. *Preclinical Toxicological Guidelines*. Ottawa: Health and Welfare, 1981. Pp. 34–49.

Collins, T. F. X. Reproduction and teratology guidelines; review of deliberations by the national toxicology advisory committee's reproduction panel. *Journal of Environmental and Pathological Toxicology*, 1978, *2*, 141–147.

Committee for Proprietary Medicinal Products. *Guidance to Applicants for Marketing Authorization of New Drugs on the Conduct of Reproduction Studies*. Paris: Secretariat, 1977.

European Economic Community, Reproduction studies: Notes for guidance concerning the application of Chapter I (C) and (D) of part 2 of the Annex to Directive 75/318/EEC, with a view to the granting of a marketing authorization for a new drug. *Official Journal of the European Communities*, 1983, *322*, 20–22.

Federation of Societies for Experimental Biology, Life Sciences and Research Office. *Neurotoxicity and Behavioral Dysfunction*. Bethesda, Md: Federation of Societies for Experimental Biology, 1985.

Frankos, V. H. FDA perspectives on the use of teratology data for human risk assessment. *Fundamental and Applied Toxicology,* 1985, *5,* 615–626.

Hattan, D. G., Henry, S. H., Montgomery, S. B., Bleiberg, M. J., Rulis, A. M., and Bolger, P. M. Role of the Food and Drug Administration in regulation of neuroeffective food additives. In R. J. Wurtman, and J. J. Wurtman, (Eds.), *Nutrition and the Brain* (vol. 6). New York: Raven Press, 1983, *6,* pp. 31–99.

Japanese Ministry of Health and Welfare. *Information on the Guidelines of Toxicity Studies Required for Applications for Approval to Manufacture (Import) Drugs* (Notification No. 118.) Pharmaceutical Affairs Bureau, Ministry of Health and Welfare, February 15, 1984, Tokyo.

Nolen, G. A. An industrial developmental toxicologist's view of behavioral teratology and possible guidelines. *Neurobehavioral Toxicology and Teratology,* 1985, *7,* 653–657.

Nordic Council on Medicine. *Drug Applications, Nordic Guidelines* (NLN Publication No. 12) 1983. Quoted by Tanimura, T. Guidelines for developmental toxicity testing of chemicals in Japan. *Neurobehavioral Toxicology and Teratology,* 1985, *7,* 647–652.

Reubl, K. R., and Chang, L. W. Effects of methylmercury on the development of the nervous system: A review. *Neurotoxicology,* 1979, *1,* 21–55.

Tanimura, T. Guidelines for developmental toxicity testing of chemicals in Japan. *Neurobehavioral Toxicology and Teratology,* 1985, *7,* 647–652.

United Kingdom, Committee on Safety of Medicine. *Notes for Guidance on Reproduction Studies.* London: Department of Health and Social Security, London. 1973.

United Kingdom, Department of Health and Social Security. *Medicines Act 1968: Guidance Notes on Applications for Product Licenses.* London: HMSO, 1984.

U.S. Environmental Protection Agency. *New and Revised Health Effects Test Guidelines* (PB No. 83–257691). Washington, DC: Office of Pesticides and Toxic Substances, 1983.

U.S. Environmental Protection Agency. *New and Revised Health Effects Test Guidelines.* (PB No. 84–233295). Washington, DC: Office of Pesticides and Toxic Substances, 1984.

U.S. Environmental Protection Agency. Toxic substances control act test guidelines. Final rules. *Federal Register,* 1985, *50,* 39,252–39,516.

U.S. Food and Drug Administration. *Guidelines for Reproduction Studies for Safety Evaluation of Drugs for Human Use.* Washington, D.C.: Food and Drug Administration, 1966.

U.S. Food and Drug Administration. Advisory Committee on Protocols for Safety Evaluation: Panel on reproduction studies in safety. *Toxicology and Applied Pharmacology,* 1970, *16,* 264–296.

Vorhees, C. V. Comparison and critique of government regulations for behavioral teratology. In E. P. Riley and C. V. Vorhees, (Eds.), *Handbook of Behavioral Teratology.* New York: Plenum Press, 1986, pp. 49–66.

Yakuji Nippo, Inc. *Requirements for the Registration of Drugs in Japan.* Tokyo: Yakuji Nippo, 1982.

Index